Handbook of Moth-Flame Optimization Algorithm

Moth-Flame Optimization algorithm is an emerging metaheuristic and has been widely used in both science and industry. Solving optimization problem using this algorithm requires addressing a number of challenges, including multiple objectives, constraints, binary decision variables, large-scale search space, dynamic objective function, and noisy parameters.

Handbook of Moth-Flame Optimization Algorithm: Variants, Hybrids, Improvements, and Applications provides an in-depth analysis of this algorithm and the existing methods in the literature to cope with such challenges.

Key Features:

- Reviews the literature of the Moth-Flame Optimization algorithm
- Provides an in-depth analysis of equations, mathematical models, and mechanisms of the Moth-Flame Optimization algorithm
- Proposes different variants of the Moth-Flame Optimization algorithm to solve binary, multi-objective, noisy, dynamic, and combinatorial optimization problems
- Demonstrates how to design, develop, and test different hybrids of Moth-Flame Optimization algorithm
- Introduces several applications areas of the Moth-Flame Optimization algorithm

This handbook will interest researchers in evolutionary computation and metaheuristics and those who are interested in applying Moth-Flame Optimization algorithm and swarm intelligence methods overall to different application areas.

Advances in Metaheuristics

Series Editor:

Patrick Siarry, Universite Paris-Est Creteil, France
Anand J. Kulkarni, Symbiosis Center for Research and Innovation, Pune, India

Handbook of AI-based Metaheuristics
Edited by Patrick Siarry and Anand J. Kulkarni

Metaheuristic Algorithms in Industry 4.0
Edited by Pritesh Shah, Ravi Sekhar, Anand J. Kulkarni, Patrick Siarry

Constraint Handling in Cohort Intelligence Algorithm
Ishaan R. Kale, Anand J. Kulkarni

Hybrid Genetic Optimization for IC Chip Thermal Control: with MATLAB® applications
Mathew V K, Tapano Kumar Hotta

Handbook of Moth-Flame Optimization Algorithm: Variants, Hybrids, Improvements, and Applications
Edited by Seyedali Mirjalili

For more information about this series please visit:
https://www.routledge.com/Advances-in-Metaheuristics/book-series/AIM

Handbook of Moth-Flame Optimization Algorithm

Variants, Hybrids, Improvements, and Applications

Edited by
Seyedali Mirjalili

CRC Press is an imprint of the
Taylor & Francis Group, an **informa** business

MATLAB® is a trademark of The MathWorks, Inc. and is used with permission. The MathWorks does not warrant the accuracy of the text or exercises in this book. This book's use or discussion of MATLAB® software or related products does not constitute endorsement or sponsorship by The MathWorks of a particular pedagogical approach or particular use of the MATLAB® software.

First edition published 2023
by CRC Press
6000 Broken Sound Parkway NW, Suite 300, Boca Raton, FL 33487-2742

and by CRC Press
4 Park Square, Milton Park, Abingdon, Oxon, OX14 4RN

CRC Press is an imprint of Taylor & Francis Group, LLC

ISBN: 9781032070919 (hbk)
ISBN: 9781032070926 (pbk)
ISBN: 9781003205326 (ebk)

DOI: 10.1201/9781003205326

Typeset in Minion Pro
by codeMantra

Contents

Editor, ix

Contributors, xi

SECTION I **Moth-Flame Optimization Algorithm for Different Optimization Problems**

CHAPTER 1 ▪ Optimization and Metaheuristics 3

SEYEDALI MIRJALILI

CHAPTER 2 ▪ Moth-Flame Optimization Algorithm for Feature Selection: A Review and Future Trends 11

QASEM AL-TASHI, SEYEDALI MIRJALILI, JIA WU, SAID JADID ABDULKADIR, TAREQ M. SHAMI, NIMA KHODADADI, AND ALAWI ALQUSHAIBI

CHAPTER 3 ▪ An Efficient Binary Moth-Flame Optimization Algorithm with Cauchy Mutation for Solving the Graph Coloring Problem 35

YASSINE MERAIHI, ASMA BENMESSAOUD GABIS, AND SEYEDALI MIRJALILI

CHAPTER 4 ▪ Evolving Deep Neural Network by Customized Moth-Flame Optimization Algorithm for Underwater Targets Recognition 53

MOHAMMAD KHISHE, MOKHTAR MOHAMMADI, TARIK A. RASHID, HOGER MAHMUD, AND SEYEDALI MIRJALILI

SECTION II **Variants of Moth-Flame Optimization Algorithm**

CHAPTER 5 ▪ Multi-objective Moth-Flame Optimization Algorithm for Engineering Problems 79

NIMA KHODADADI, SEYED MOHAMMAD MIRJALILI, AND SEYEDALI MIRJALILI

CHAPTER 6 ▪ Accelerating Optimization Using Vectorized Moth-Flame Optimizer (vMFO) 97

AMIRPOUYA HEMMASIAN, KAZEM MEIDANI, SEYEDALI MIRJALILI, AND AMIR BARATI FARIMANI

CHAPTER 7 ▪ A Modified Moth-Flame Optimization Algorithm for Image Segmentation 111

SANJOY CHAKRABORTY, SUKANTA NAMA, APU KUMAR SAHA, AND SEYEDALI MIRJALILI

CHAPTER 8 ▪ Moth-Flame Optimization-Based Deep Feature Selection for Cardiovascular Disease Detection Using ECG Signal 129

ARINDAM MAJEE, SHREYA BISWAS, SOMNATH CHATTERJEE, SHIBAPRASAD SEN, SEYEDALI MIRJALILI, AND RAM SARKAR

SECTION III **Hybrids and Improvements of Moth-Flame Optimization Algorithm**

CHAPTER 9 ▪ Hybrid Moth-Flame Optimization Algorithm with Slime Mold Algorithm for Global Optimization 155

SUKANTA NAMA, SANJOY CHAKRABORTY, APU KUMAR SAHA, AND SEYEDALI MIRJALILI

CHAPTER 10 ▪ Hybrid Aquila Optimizer with Moth-Flame Optimization Algorithm for Global Optimization 177

LAITH ABUALIGAH, SEYEDALI MIRJALILI, MOHAMED ABD ELAZIZ, HEMING JIA, CANAN BATUR ŞAHIN, ALA' KHALIFEH, AND AMIR H. GANDOMI

CHAPTER 11 ■ Boosting Moth-Flame Optimization Algorithm by Arithmetic Optimization Algorithm for Data Clustering 209

LAITH ABUALIGAH, SEYEDALI MIRJALILI, MOHAMMED OTAIR, PUTRA SUMARI, MOHAMED ABD ELAZIZ, HEMING JIA, AND AMIR H. GANDOMI

SECTION IV **Applications of Moth-Flame Optimization Algorithm**

CHAPTER 12 ■ Moth-Flame Optimization Algorithm, Arithmetic Optimization Algorithm, Aquila Optimizer, Gray Wolf Optimizer, and Sine Cosine Algorithm: A Comparative Analysis Using Multilevel Thresholding Image Segmentation Problems 241

LAITH ABUALIGAH, NADA KHALIL AL-OKBI, SEYEDALI MIRJALILI, MOHAMMAD ALSHINWAN, HUSAM AL HAMAD, AHMAD M. KHASAWNEH, WAHEEB ABU-ULBEH, MOHAMED ABD ELAZIZ, HEMING JIA, AND AMIR H. GANDOMI

CHAPTER 13 ■ Optimal Design of Truss Structures with Continuous Variable Using Moth-Flame Optimization 265

NIMA KHODADADI, SEYED MOHAMMAD MIRJALILI, AND SEYEDALI MIRJALILI

CHAPTER 14 ■ Deep Feature Selection Using Moth-Flame Optimization for Facial Expression Recognition from Thermal Images 281

ANKAN BHATTACHARYYA, SOUMYAJIT SAHA, SHIBAPRASAD SEN, SEYEDALI MIRJALILI, AND RAM SARKAR

CHAPTER 15 ■ Design Optimization of Photonic Crystal Filter Using Moth-Flame Optimization Algorithm 313

SEYED MOHAMMAD MIRJALILI, SOMAYEH DAVAR, NIMA KHODADADI, AND SEYEDALI MIRJALILI

INDEX, 323

Editor

Seyedali Mirjalili is a Professor at Torrens University Center for Artificial Intelligence Research and Optimization and internationally recognized for his advances in nature-inspired Artificial Intelligence (AI) techniques. He is the author of more than 300 publications including five books, 250 journal articles, 20 conference papers, and 30 book chapters. With more than 50,000 citations and H-index of 75, he is one of the most influential AI researchers in the world. From Google Scholar metrics, he is globally the most cited researcher in Optimization using AI techniques, which is his main area of expertise. Since 2019, he has been in the list of 1% highly-cited researchers and named as one of the most influential researchers in the world by Web of Science. In 2021, *The Australian* newspaper named him as the top researcher in Australia in three fields of Artificial Intelligence, Evolutionary Computation, and Fuzzy Systems. He is a senior member of IEEE and is serving as an editor of leading AI journals including Neurocomputing, Applied Soft Computing, Advances in Engineering Software, Computers in Biology and Medicine, Healthcare Analytics, and Applied Intelligence.

Contributors

Said Jadid Abdulkadir
Centre for Research in Data Science (CERDAS)
Universiti Teknologi Petronas
Seri Iskandar, Malaysia

Laith Abualigah
Faculty of Computer Sciences and Informatics
Amman Arab University
Amman, Jordan
and
School of Computer Sciences
Universiti Sains Malaysia
Pulau Pinang, Malaysia

Waheeb Abu-Ulbeh
Faculty of IT
Al-Istiqlal University
Jericho, Palestine

Husam Al Hamad
Faculty of Computer Sciences and Informatics
Amman Arab University
Amman, Jordan

Nada Khalil Al-Okbi
Department of Computer Science
College of Science for Women
University of Baghdad
Baghdad, Iraq

Alawi Alqushaibi
Centre for research in data science (CERDAS)
Universiti Teknologi Petronas
Seri Iskandar, Malaysia

Mohammad Alshinwan
Faculty of Computer Sciences and Informatics
Amman Arab University
Amman, Jordan

Qasem Al-Tashi
Department of Imaging Physics
The University of Texas MD Anderson Cancer Center
Houston, Texas

Ankan Bhattacharyya
Department of Computer Science
University of Kentucky
Lexington, Kentucky

Shreya Biswas
Department of Electronics and Telecommunication Engineering
Jadavpur University
Kolkata, India

Sanjoy Chakraborty
Department of Computer Science and Engineering
Iswar Chandra Vidyasagar College
Belonia, India
and
Department of Computer Science and Engineering
National Institute of Technology Agartala
Jirania, India

Somnath Chatterjee
Department of Computer Science and Engineering
Future Institute of Engineering and Management
Kolkata, India

Somayeh Davar
Department of Engineering
Concordia University
Montreal, Canada

Mohamed Abd Elaziz
Department of Mathematics
Faculty of Science
Zagazig University
Zagazig, Egypt

Amir Barati Farimani
Department of Mechanical Engineering
Carnegie Mellon University
Pittsburgh, Pennsylvania
and
Machine Learning Department
Carnegie Mellon University
Pittsburgh, Pennsylvania
and
Department of Biomedical Engineering
Carnegie Mellon University
Pittsburgh, Pennsylvania

Asma Benmessaoud Gabis
Systems Design Methods Laboratory
National School of Computer Science
Algiers, Algeria

Amir H. Gandomi
Faculty of Engineering and Information Technology
University of Technology Sydney
Ultimo, Australia

Amir Pouya Hemmasian
Department of Mechanical Engineering
Carnegie Mellon University
Pittsburgh, Pennsylvania

Heming Jia
School of Information Engineering
Sanming University
Sanming, China

Ala' Khalifeh
Department of Electrical and Communication Engineering
German Jordanian University
Amman, Jordan

Ahmad M. Khasawneh
Faculty of Computer Sciences and Informatics
Amman Arab University
Amman, Jordan

Mohammad Khishe
Department of Marine Electronics and Communication Engineering
Imam Khomeini Marine Science University
Nowshahr, Iran

Nima Khodadadi
Department of Engineering
Florida International University
Miami, Florida

Hoger Mahmud
Department of Computer Science
College of Science and Technology
University of Human Development
Sulaymaniyah, Iraq

Arindam Majee
Department of Electronics and Telecommunication Engineering
Jadavpur University
Kolkata, India

Kazem Meidani
Department of Mechanical Engineering
Carnegie Mellon University
Pittsburgh, Pennsylvania

Yassine Meraihi
LIST Laboratory
University of M'Hamed Bougara Boumerdes
Boumerdes, Algeria

Seyed Mohammad Mirjalili
Department of Engineering
Concordia University
Montreal, Canada

Seyedali Mirjalili
Department of Engineering
Torrens University
Adelaide, Australia

Mokhtar Mohammadi
Department of Information Technology
College of Engineering and Computer Science
Lebanese French University
Erbil, Iraq

Sukanta Nama
Department of Applied Mathematics
Maharaja Bir Bikram University
Agartala, India

Mohammed Otair
Faculty of Computer Sciences and Informatics
Amman Arab University
Amman, Jordan

Tarik A. Rashid
Computer Science and Engineering Department
Science and Engineering Science
University of Kurdistan Hewler
Erbil, Iraq

Apu Kumar Saha
Department of Mathematics
National Institute of Technology Agartala
Jirania, India

Soumyajit Saha
Department of Computer Science
University of Utah
Salt Lake City, Utah

Canan Batur Şahin
Faculty of Engineering and Natural Sciences
Malatya Turgut Ozal University
Malatya, Turkey

Ram Sarkar
Department of Computer Science and Engineering
Jadavpur University
Kolkata, India

Shibaprasad Sen
Department of Computer Science and Technology
University of Engineering and Management
Kolkata, India

Tareq M. Shami
Department of Electronic Engineering
University of York
York, United Kingdom

Putra Sumari
School of Computer Sciences
Universiti Sains Malaysia
Pulau Pinang, Malaysia

Jia Wu
Department of Imaging Physics
The University of Texas MD Anderson Cancer Center
Houston, Texas

I

Moth-Flame Optimization Algorithm for Different Optimization Problems

CHAPTER 1

Optimization and Metaheuristics

Seyedali Mirjalili

Torrens University Australia
Yonsei University

CONTENTS

1.1 Introduction to Optimization 3
1.1.1 Derivative-Based Optimization Algorithms 5
1.1.2 Non-Derivative-Based Optimization Algorithms 7
1.1.3 Metaheuristics 9
Reference 10

1.1 INTRODUCTION TO OPTIMIZATION

Optimization is the process of finding the best set of solutions from all possible solutions for a given optimization problem considering a set of objective functions and constraints. The decision variables (often called variables, parameters, or inputs) are the unknowns of the optimization problem. This is the first fundamental element of any optimization problem. Some examples of such elements are the length of an air craft's wing, the diameter of concrete rod, the sequence of classes in a university time table, or the pixel colors of a synthetic image, just to name a few. The set of all possible values for the decision variable is often called "search space."

The second fundamental element is the set of objectives. This is often called cost function, fitness function, merit function, reward function, or utility function. Objective functions play the role of a target in optimization, and the ultimate goal is to maximize or minimize them depending on their nature. The set of all possible objective functions for all the

DOI: 10.1201/9781003205326-2

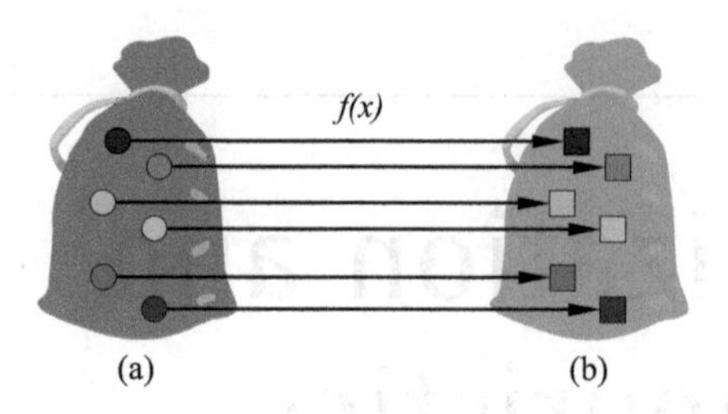

FIGURE 1.1 (a) Decision variables set (search space) and (b) objective values set (objective space).

solutions in the search space is called "objective space." So each solution in the search space is associated with its projection in the objective space. An objective function allows us to evaluate the quality of a solution. It also enables comparison between solutions.

Figure 1.1 shows that for each solution in the search space, there is a corresponding "shadow" in the objective space. In a discrete problem, there is a one-on-one matching as shown in this figure. In a continuous problem, however, there is an infinite number of solutions. Due to the use of a function to calculate objective values for solutions, each element of the left set is set to exactly an element of the right set.

The last fundamental element is the set of constraints, which are applied to the decision variables. The presence of constraints will change the optimization process to constrained optimization, in which the objective function is optimized while satisfying the constraints. Constraints are divided into hard and soft. In the former case, the constraints must be satisfied. Any violation of such constraints at any level will make the solution infeasible. In the latter case, the level of violation of constraints is considered to penalize the decision variable values. Some examples of constraints are the lower/upper bound for decision variables, thresholds of variables, decision variables being equal, etc.

The formulation of optimization problem with discrete variables is as follows. This formulation is done for a minimization problem without the loss of generality:

$$\begin{aligned}
&\text{minimize} \quad f(x) \\
&\text{subject to} \, x \in R, \qquad g_j(x) \leq 0, \quad j = 1,2,\ldots,m \\
&\qquad\qquad\qquad\qquad h_j(x) = 0, \quad j = 1,2,\ldots,p
\end{aligned}$$

where R is a **finite** set

So far, this section introduced problems with discrete search space. In this case, the search space is finite and the optimization process is called discrete optimization or combinatorial optimization. Other type of optimization is called continuous optimization, in which the decision variables can take real-valued numeric values. This will cause the objective values in the objective space to be real-valued numeric too. Such optimization problems are formulated as follows:

$$\begin{aligned} &\text{minimize } f(x) \\ &\text{subject to } x \in R^n, \quad g_j(x) \leq 0, \quad j = 1,2,\ldots,m \\ &\qquad\qquad\qquad\qquad h_j(x) = 0, \quad j = 1,2,\ldots,p \end{aligned}$$

where R^n is an **infinite** set

The search space of a continuous objective function is visualized in Figure 1.2. Note that this figure also shows the objective value for two variables, which is often called search landscape. This book covers both types of optimization.

Optimization algorithms are designed to solve optimization problems. As discussed above, the ultimate goal is to find the best solution to optimize (maximize or minimize) an objective function. Such algorithms are divided into two classes: derivative-based versus non-derivative-based algorithms (see Figure 1.3). In the first class, the optimization problem should be differentiable to be solved by an optimization algorithm. The latter class, however, is often called black-box optimization, in which we do not require to calculate the derivative of the objective function.

1.1.1 Derivative-Based Optimization Algorithms

As discussed above, any algorithm in this class can be applied to differentiable objective functions. By calculating the derivative of a function, an algorithm is able to calculate the amount and range of changes in the objective function for any solution in the search space. This allows the algorithm to move from one solution to the next better solution in the search space. If the objective function is simple, it can be solved analytically using calculus. The majority of real-world problems are not simple and often non-differentiable. Some of the most popular derivative-based optimization algorithms are linear programming, line search, gradient-descent,

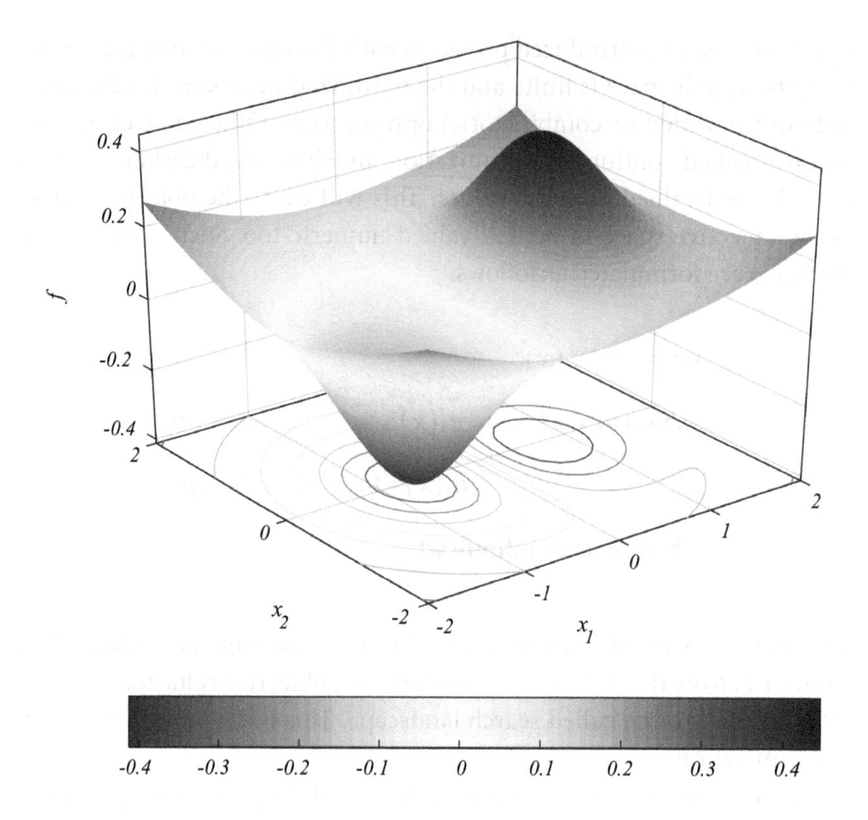

FIGURE 1.2 The search landscape of an optimization problem with two decision variables, one objective, and no constraint.

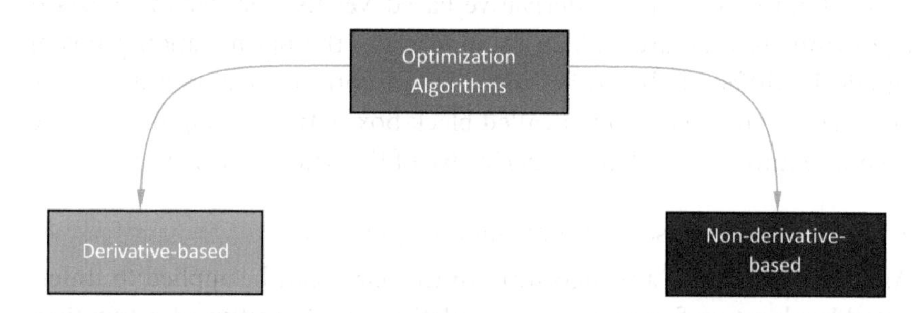

FIGURE 1.3 Classification of optimization problems based on differentiability of search space.

and Adam. An example of how a gradient descent algorithm solves an optimization problem is given in Figure 1.4. It can be seen that the algorithm finds the global minimum when starting from [2,5]. However, when the algorithm starts from [3,3], it gets trapped in a local minimum.

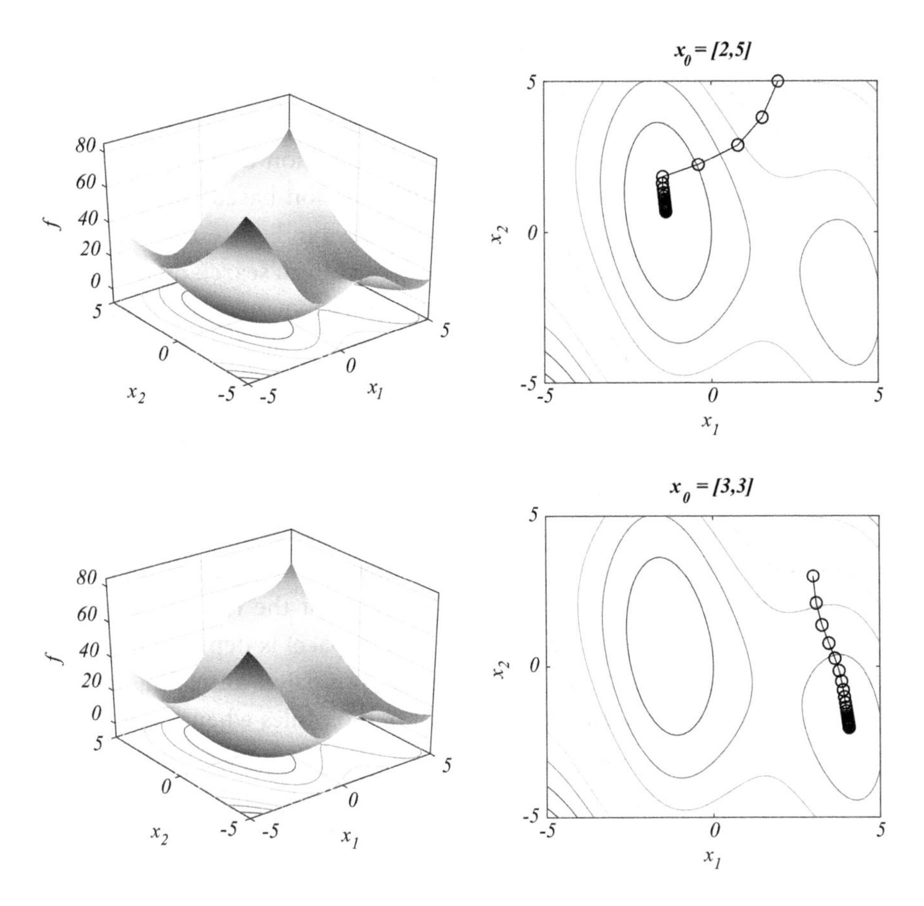

FIGURE 1.4 Gradient descent for solving $f(x_1, x_2) = x_1^2 + x_1x_2 + x_2^2 + 10\sin(x_1)$ starting from two different points.

Note that the partial derivative of f was calculated to be used by the gradient descent algorithm as follows:

$$\frac{\partial f}{\partial x_1} = \frac{\partial\ x_1^2 + x_1x_2 + x_2^2 + 10\sin(x_1)}{\partial x_1} = 2x_1 + x_2 + 10\cos(x_1) \tag{1.1}$$

$$\frac{\partial f}{\partial x_2} = \frac{\partial\ x_1^2 + x_1x_2 + x_2^2 + 10\sin(x_1)}{\partial x_2} = x_1 + 2x_2 \tag{1.2}$$

1.1.2 Non-Derivative-Based Optimization Algorithms

In non-derivative-based algorithms, we do not need derivative information of an objective function. Such algorithms require little to no knowledge of the mathematical models of the objective function and they are considered black boxes. They are more applicable than derivative-based algorithms due to the difficulty in calculating derivative of real-world problems. Non-derivative-based optimization algorithms can be classified into direct algorithms (pattern search), single-solution-based (individual-based) stochastic algorithms, and multi-solution-based (population-based) stochastic algorithms. These are visualized in Figure 1.5.

In direct algorithms, which are often called pattern search methods, an optimization technique generates solutions in the vicinity of the current solution in the search space using patterns (e.g., rectangular, spiral). Based on the objective values, the best one will be chosen and the same process is repeated until the satisfaction of an end condition. Some popular algorithms in this class are Hooke–Jeeves Method, Nelder–Mead Simplex Search, or Cyclic Coordinate Search.

In single-solution stochastic algorithm, a random solution is first generated in the search space. This solution is then iteratively improved until the satisfaction of an end condition. The way that the random solution changes over time ("move around" the search space) is dependent on the mechanism of the algorithm. For instance, simulated annealing uses a decaying variable to mimic the movement of molecules when a metal gets heated and slowly cools down.

As opposed to single-solution stochastic algorithm, population-based algorithms work with a set of solutions. This means that they start the optimization process with multiple random solutions. These solutions are then combined, merged, or evolved to create better set of solutions.

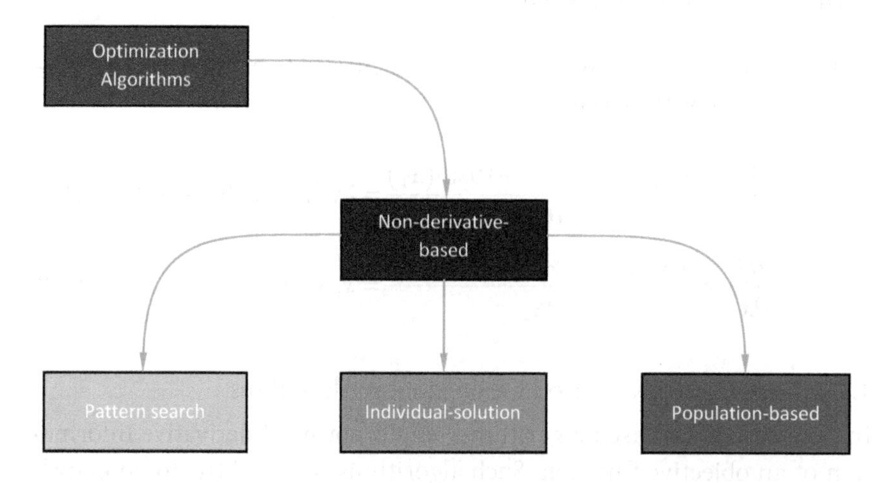

FIGURE 1.5 Classification of non-derivative-based optimization algorithms.

This process is still iterative although the rules to update multiple solutions tend to be more complicated than those in single-solution algorithm.

1.1.3 Metaheuristics

Whether an algorithm works with single or multiple solutions, if it is a metaheuristics it should behave stochastically. This is opposed to derivative-based algorithms that usually tend to make deterministic decision when updating a solution. The stochastic nature of such algorithms allows them to make educated decisions when updating the solutions. This is done using heuristic information that are extracted from the objective function or the position of solutions. For instance, the distance to the best solution obtained so far can be used to choose promising solutions to guide others during the optimization process.

A heuristic algorithm is stochastic but designed to solve a specific problem. This means that the heuristic information is related to the problem that cannot be generalized to a wide range of problems. In robot path planning, for instance, the Euclidean distance of each block in an environment to a target can be calculated and used to optimize the robot's navigator (e.g., A* algorithm). However, this technique will not be applicable to other types of optimization problems. Metaheuristics were proposed in the literature to alleviate this drawback.

A metaheuristic is a general-purpose optimization algorithm that does not require specific heuristic information extracted from the problem. This means that the algorithm uses only the objective values of solutions in the search space to make educated decisions. This makes them highly practical in real-world problems. The literature shows that population-based metaheuristics have been increasingly popular in a wide variety of fields. Some of the reasons for their popularity are simplicity of implementation, high local optima avoidance, and derivative-free mechanism.

There are three classes of metaheuristics: swarm-based, evolution-based, and physics-based. In the former class, the inspiration is the collective intelligence of individual in a swarm. In evolution-based algorithms, the source of inspiration is evolutionary phenomena in nature. Finally, the source of inspiration is physical phenomena in physics-based algorithms.

Moth-Flame Optimization (MFO) algorithm [1] is a recent metaheuristic that mimics the navigation of moths in nature. The main source of inspiration is the attraction and spiral movement of moths around artificial sources of lights. This mechanism has been modeled and implemented in the MFO algorithm to navigate a solution in a search space to find a reasonable estimation of the global optimum for a given optimization problem.

The MFO algorithm is an emerging metaheuristic proposed in 2015. As of mid 2022, with more than 2500 citations, this algorithm has been widely used in both science and industry. This book focuses on the mathematical models, algorithms, variants, improvements, hybrids, and practical applications of this algorithm. Solving optimization problems using MFO algorithm requires addressing a number of challenges including multiple objectives, constraints, binary decision variables, large-scale search space, dynamic objective function, noisy parameters, just to name a few. This handbook provides an in-depth analysis of this algorithm and the existing methods in the literature to cope with such challenges. A tutorial on how to design, adapt, and evaluate this algorithm is also provided, which would be beneficial for the readers interested in developing optimization algorithms for optimization problems.

The structure of this book is as follows:

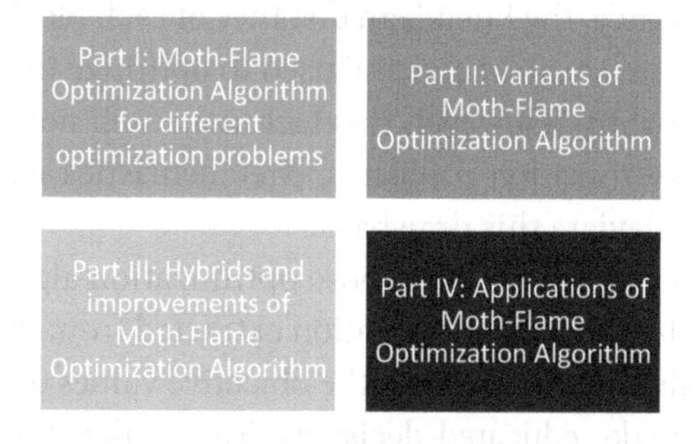

REFERENCE

1. S. Mirjalili, Moth-flame optimization algorithm: A novel nature-inspired heuristic paradigm. *Knowledge-Based Systems*, vol. 89, pp. 228–249, 2015.

CHAPTER 2

Moth-Flame Optimization Algorithm for Feature Selection: A Review and Future Trends

Qasem Al-Tashi

The University of Texas MD Anderson Cancer Center

Seyedali Mirjalili

Torrens University Australia
Yonsei University

Jia Wu

The University of Texas MD Anderson Cancer Center

Said Jadid Abdulkadir

Universiti Teknologi Petronas

Tareq M. Shami

University of York

DOI: 10.1201/9781003205326-3

Nima Khodadadi
Florida International University

Alawi Alqushaibi
Universiti Teknologi Petronas

CONTENTS

2.1 Introduction 12
2.2 Feature Selection 14
2.3 MFO Algorithm 16
2.3.1 MFO Inspiration 16
2.3.2 MFO's Mathematical Model 17
2.4 Feature Selection Using MFO Algorithm 22
2.5 Existing Studies Based on MFO Feature Selection 22
2.6 Discussion and Future Directions 28
2.7 Conclusion 29
References 30

2.1 INTRODUCTION

The development of high-throughput technology has led to an explosive growth in the dimensionality and sample size of the collected data. Managing these data efficiently and effectively has become increasingly challenging. Traditionally, it is impractical to manually manage these datasets [1,2]. Hence, unprecedented obstacles have arisen for researchers, developers, and professionals who are collaborating with data and intend to influence its value. The exponential growth in the amount of data makes the task of processing and analyzing data complicated and requires high computing resources.

A process must be followed in order to attain useful information from the available data. The Knowledge Discovery in the Database (KDD) [3] is a general structure that defines the steps that need to be taken to acquire useful information from a dataset. The core phase in the KDD process is known as data mining [4]. In the context of information technology research, the area of data mining in recent years has witnessed an explosive growth. This rapid growth is due to the massive generation of raw data daily that requires conversion into meaningful information [5].

The technique of identifying patterns and extracting knowledge from large datasets is known as data mining. Anomaly detection, regression, association analysis, clustering, and classification are all components of data mining [6]. Another essential phase in the KDD method is the pre-processing of data, a preparatory yet remarkable phase which, if not performed with care, may make it difficult to acquire useful information from the data. In addition, pre-processing of data is a general phase requiring a variety of strategies that can be utilized to the initial data, and feature selection is considered a main one. Feature selection is a data pre-processing step with the purpose of eliminating irrelevant and redundant attributes from a dataset.

Broadly speaking, data mining utilizes techniques that are derived from different domains of knowledge such as statistics and probability and machine learning especially. Machine learning techniques are grouped into three categories: supervised learning, unsupervised learning, and reinforcement learning. The focus of this chapter is on classification where each instance in the dataset is grouped into different sets according to its characteristics. Redundant or irrelevant feature can significantly degrade model performance [7]. According to Ref. [8], irrelevant features are helpful in the learning process. As a result, feature selection is a technique that acquires the subset from an original feature which eventually ends with the selection of the most appropriate features [9].

The main advantages of feature selection are improving accuracy of results of the produced model and making the training of a model faster or more cost-effective in terms of computational resource consumption [10]. In addition, feature selection improves the resulting models by making them smaller and easier to understand [11]. Generally, feature selection techniques can be grouped into two main methods: wrappers and filters [12,13]. Besides wrappers and filters schemes, some researchers add embedded approaches as a third class of feature selection methods [14,15]. In terms of classifiers, K-Nearest Neighbor (KNN), Artificial Neural Network (ANN) [16–18], Support Vector Machine (SVM), Naïve Bayes, and Decision Tree (DT) are the five most common classifiers that are used with feature selection in order to learn and predict models [19,20]. Inspired by nature, swarm intelligence algorithms have gained considerable attention by the research community due to their effectiveness in solving diverse real-world optimization problems. One of the recent high-performance optimization algorithms is Moth-Flame Optimization (MFO) algorithm developed by Mirjalili in 2015 [21]. Due to its outstanding performance,

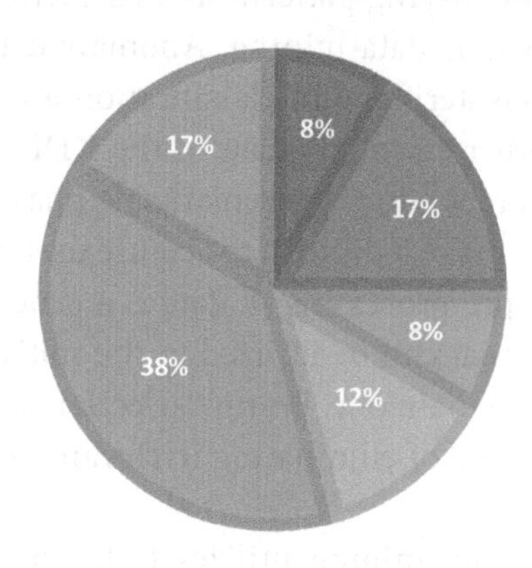

FIGURE 2.1 Studies on feature selection using MFO.

MFO has attracted researchers to utilize it as an optimization tool to solve feature selection problems. This attraction keeps increasing year by year as shown in Figure 2.1, from 2016 to 2021. Figure 2.1 presents the number of research carried out on feature selection utilizing MFO.

The rest of this review is organized as follows. Section 2.2 presents the concept of feature selection and its different methods. Section 2.3 starts by illustrating the MFO inspiration and then it provides the mathematical modeling of MFO. The focus of Section 2.4 is on how MFO can be utilized to solve feature section problems. Section 2.5 critically reviews the performance of MFO when it is applied on feature selection. Finally, Section 2.6 concludes this work and provides potential research directions.

2.2 FEATURE SELECTION

Feature selection classically has been defined as a selection of k features/attributes from the original features/attributes m, ($k<m$), such that the value of a criterion function is optimized over all subsets of size k [22,23]. Moreover, feature selection is known as attribute selection or variable selection or is suggested to minimize or remove unnecessary and redundant features as a pre-processing stage for the data [24].

Feature selection is also known to be a discrete problem by nature, i.e., a binary problem: the representation of the solution is 1 or 0. The most essential step when solving the problem of feature selection is the formation of a solution that can represent the subset of features. Each position in the solution can have binary states: "1" or "0". A value of 0 indicates that a feature is not selected while a value of 1 means a feature is selected. Therefore, the

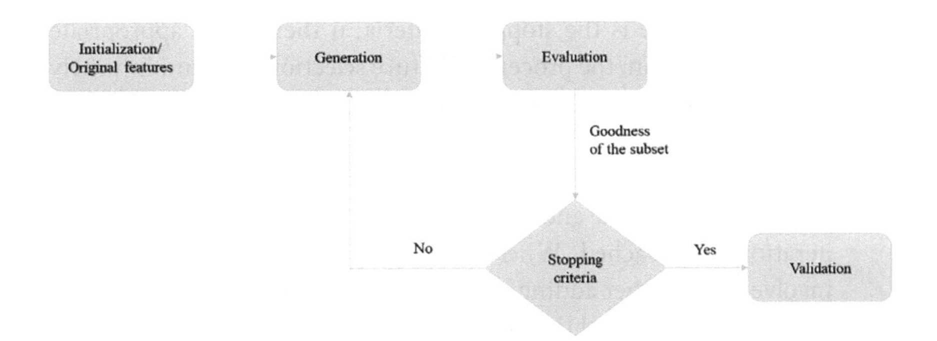

FIGURE 2.2 General process of feature selection.

summation of 1's denotes the subset feature size [25]. Feature selection is a multi-objective optimization problem where the objective is to simultaneously minimize the error rate of classification (maximizing classification performance) and minimize the number of features [26]. This multi-objective formulation helps to obtain a subset of the optimal features to meet different requirements in real-world applications.

Figure 2.2 illustrates the general process of feature selection that consists of five main stages explained as follows:

- The initialization is the first stage of any feature selection method, and it relies on all the original features in the dataset. For instance, in feature selection based MFO algorithm, the search space dimensionality is typically defined as the total number of all features used in this process.
- The second stage is the subset discovery to select candidate subset of feature for evaluation. There are three different ways to do this: starts with no features, starts with all features, or starts with a random subset of features [27–29]. In this stage, several search methods, such as traditional methods, heuristic, and metaheuristic algorithms, can be applied to select the best feature subsets.
- The third stage is the evaluation of the feature subset generated; the subset of features generated by the second stage will be evaluated by an assessment measure to identify their goodness. Such evaluation measures play a significant task in the search algorithm, as it helps direct the algorithm in the quest for obtaining the best subset of features.

- The fourth stage is the stopping criteria; if there is no appropriate stopping criterion, the process of feature selection may run infinitely over the subsets' space. The decision of stopping criteria is influenced by the second or third stage. The influence of the second stage on the stopping criteria involves (a) whether a predefined number of features are selected and (b) whether a predetermined number of iterations are reached. Whereas stopping based on the third stage involves (a) whether adding or removing any feature does not generate a better subset and (b) whether the best subset has been obtained based on a specific evaluation function. Once the stopping criterion is satisfied, the loop will then stop.
- Finally, the validation process is carried out to check the validity of the selected subset. This is not a feature selection stage, but the algorithm of feature selection must be validated. The subset of features that have been selected is validated on the test set. The results are contrasted against the results established previously or with the results of predefined benchmark techniques.

2.3 MFO ALGORITHM

The first part of this section focuses on the inspiration of the MFO algorithm, while the second part provides a detailed explanation of the MFO mathematical model.

2.3.1 MFO Inspiration

A moth is an eye-catching insect that shares a lot of similarities with butterflies. In nature, there exists more than 160,000 various types of species that belong to the moth family. The lifetime of a moth starts with being a larvae and then it spends the rest of its life as an adult. One of the special characteristics of moths is their intelligent navigation behavior at night. Moths rely on the moon light in order to safely fly at night by utilizing a navigation strategy known as transverse orientation. Based on this strategy, a moth travels in the space by keeping its angle with reference to the moon unchanged. This effective approach allows flying in a straight line over long distances [21]. Due to the fact that the distance between a moth and the moon is lengthy, traveling in a straight path is guaranteed. The idea of the transverse orientation mechanism is illustrated in Figure 2.3.

Although transverse orientation is an effective navigation system for moths, it is sometimes observed that moths move spirally around lights.

FIGURE 2.3 Transverse orientation.

This spiral movement by moths occurs because artificial light can easily trick moths. This shows that the transverse orientation is only useful when the source of light is far away from the position of a moth as in the case of a moth and the moon. Although a moth attempts to keep a fixed angle with the light in order to travel in a straight path when it sees an artificial light, it fails due to the extremely short distance. As a consequence, a deadly spiral path is generated. Figure 2.4 presents the spiral behavior of moths.

2.3.2 MFO's Mathematical Model

In MFO, moths are considered candidate solutions while their positions denote the dimensions/variables of the optimization problem. The number of moths and the number of variables are denoted by n and d, respectively. This constitutes an n by d matrix that represents all the moths and it is mathematically written as follows:

$$M = \begin{bmatrix} m_{1,1} & m_{1,2} & \cdots & \cdots & m_{1,d} \\ m_{2,1} & m_{2,2} & \cdots & \cdots & m_{2,d} \\ \vdots & \vdots & \vdots & \vdots & \vdots \\ m_{n,1} & m_{n,2} & \cdots & \cdots & m_{n,d} \end{bmatrix} \tag{2.1}$$

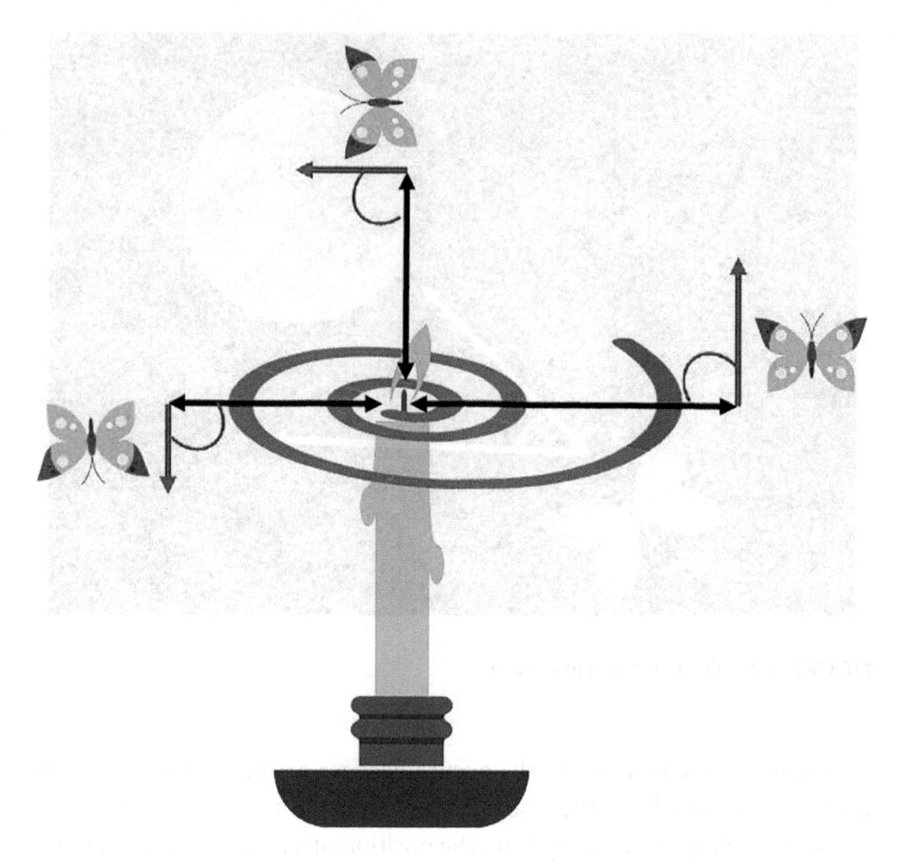

FIGURE 2.4 Spiral flying path around close light sources.

To calculate the fitness of each moth, its position in M is passed to an objective function. The obtained fitness of all moths is stored in a vector as follows:

$$OM = \begin{bmatrix} OM_1 \\ OM_2 \\ \vdots \\ OM_n \end{bmatrix} \tag{2.2}$$

From Equation (2.2), OM_1, OM_2, and OM_3 represent the fitness of the first, second, and third moths, respectively.

Besides moths, flames are another key element of MFO. Similar to the n by d moth matrix shown in Equation (2.1), flames are an n by d matrix that can be written as follows:

$$F = \begin{bmatrix} F_{1,1} & F_{1,2} & \cdots & \cdots & F_{1,d} \\ F_{2,1} & F_{2,2} & \cdots & \cdots & F_{2,d} \\ \vdots & \vdots & \vdots & \vdots & \vdots \\ F_{n,1} & F_{n,2} & \cdots & \cdots & F_{n,d} \end{bmatrix} \tag{2.3}$$

To store the fitness values of flames, a vector is used as follows:

$$\mathrm{OF} = \begin{bmatrix} \mathrm{OF}_1 \\ \mathrm{OF}_2 \\ \vdots \\ \mathrm{OF}_n \end{bmatrix} \tag{2.4}$$

In MFO, both moths and flames are candidate solutions. However, a flame is different from a moth in the way it updates its position. A moth performs the actual search by traveling in the space aiming to find better solutions, while a flame is the best position that has been found so far by the corresponding moth. During the searching process, a moth flies around a flame and it updates its position if it finds a better solution. This strategy avoids losing the best solutions found by moths.

MFO consists of three tuples that are used to search for the global optimum. The MFO three tuples can be represented as follows:

$$\mathrm{MFO} = (I, P, T) \tag{2.5}$$

where I is a function that is responsible to randomly generate an initial population of moths associated with their fitness values, while P represents the MFO main function.

$$I : \varnothing \rightarrow \{M, \mathrm{OM}\} \tag{2.6}$$

The purpose of

P is to control the movement of moths in the search space. The input of P is the current matrix M, while its output is the updated positions of moths:

$$P : M \rightarrow M \tag{2.7}$$

T is a logical function that returns true in case a stopping criterion is met, otherwise it returns false:

$$T: M \rightarrow \{\text{true, false}\} \tag{2.8}$$

As the main components of MFO, the three functions *I*, *P*, and *T* can define the general MFO framework as follows:

```
M=I();
while T(M) is equal to false
M=P(M);
End
```

With the help of *I*, the initial populations of moths and their corresponding fitness values can be generated and evaluated, respectively, as follows:

```
for i = 1: n
for j= 1: d
M(i,j)=(ub(i)-lb(i))* rand()+lb(i);
end
end

OM=FitnessFunction(M);
```

From the initialization step just mentioned above, it is clear that two initialization vectors denoted as *ub* and *lb* are introduced. These two vectors *ub* and *lb* represent the upper bound and lower bound of the problem dimensions or variables, respectively, and their definitions are given as follows:

$$ub = [\, ub_1, ub_2, ub_3 \ldots, ub_{n-1}, ub_n \,] \tag{2.9}$$

where ub_i denotes the upper limit of variable i.

$$lb = [\, lb_1, lb_2, lb_3, \ldots, lb_{n-1}, lb_n \,] \tag{2.10}$$

where lb_i denotes the lower limit of variable i.

The next step after initialization is to iteratively run the *P* function that terminates when the *T* function becomes true. Transverse orientation as the main inspiration of MFO can be mathematically modeled as follows:

$$M_i = S(M_i, F_j) \tag{2.11}$$

where M_i, F_j, and S denote the ith moth, j^{th} flame, and the spiral function, respectively.

Equation (2.11) updates the positions of moths with respect to flames.

The original MFO paper has assumed a logarithmic spiral to update the positions of moths. Nevertheless, it is possible to utilize any kind of spiral as long as it meets the following conditions:

- The initial point of spiral starts from the moth.
- The final point of spiral is the position of the flame.
- It is not allowed for spiral to fluctuate out of the search space.

Meeting the aforementioned conditions, the MFO logarithmic spiral is defined as follows:

$$S\left(M_i,F_j\right)= D_i \cdot e^{bt} \cdot \cos(2nt)+ F_j \tag{2.12}$$

where D_i denotes the distance between the jth flame and the ith moth, and b is a constant value that defines the logarithmic spiral shape while t is a random number that can have a value of [–1,1]. D can be mathematically written as follows:

$$D_i = \left|F_j - M_i\right| \tag{2.13}$$

The spiral motion of a moth around a flame in MFO guarantees a balance between exploitation and exploration. Moreover, in order to successfully escape from being trapped in local optima, the best solutions that have been found are stored in each iteration and moths travel around the flames by utilizing the two matrices: OF and OM.

MFO might experience degraded performance at the exploitation stage because moths update their positions in k different locations. To tackle this issue and enhance the overall performance of MFO, the number of flames is decreased based on the following:

$$\text{Flame no} = \text{Round}\left(N - l * \frac{N-1}{T}\right) \tag{2.14}$$

where N and T denote the maximum number of flames and iterations, respectively, while l indicates the current iteration number.

2.4 FEATURE SELECTION USING MFO ALGORITHM

Feature selection assists with finding an optimal set of features as a pre-processing step before using datasets in Machine Learning algorithms. As it selects features, it is considered as a binary problem [15]. MFO has been designed to solve problems with continuous values, so we need to change it to solve binary problems [2]. Transfer functions map continuous values of algorithm to binaries, which is what we are looking for when using MFO to solve binary problems [30]. Four S-Shaped as well as four V-Shaped transfer functions are presented in this review. The mathematical equations of the eight transfer functions are shown in Table 2.1, and it should be noted that each transfer function provides a different performance. Moreover, only some transfer functions are utilized on MFO which motivated researchers in the literature to apply the rest of the transfer functions on MFO and evaluate their performance.

2.5 EXISTING STUDIES BASED ON MFO FEATURE SELECTION

MFO has attracted several researchers working on feature selection problems and has been utilized as a feature selection approach in various domains including but not limited to medical applications, intrusion detection system, software fault prediction, facial emotion recognition, gene selection, etc. In 2021, a few research have been conducted employing MFO as a feature selection approach. For example, in a research performed by Kumar [31], the MFO was hybridized with hill climbing as a feature selection method to handle the problem of intrusion detection systems. The classifier used in this study was an Optimal Wavelet Kernel Extreme Learning Machine (OWKELM), and the NSL-KDDCup dataset was used for performance evaluation. The finding shows that this

TABLE 2.1 Transfer Function Types of Sigmoid and tanh

S-Shaped (Sigmoid) Families	V-Shaped (tanh) Families
$S_1(x)=\frac{1}{1+e^{x}}$	$V_1(x)=\lvert\tanh(x)\rvert$
$S_2(x)=\frac{1}{1+e^{2x}}$	$V_2(x)=\left\lvert er\ f\left(\frac{\sqrt{\pi}}{2}x\right)\right\rvert$
$S_3(x)=\frac{1}{1+e^{3x}}$	$V_3(x)=\left\lvert\frac{x}{\sqrt{1+x^2}}\right\rvert$
$S_4(x)=\frac{1}{1+e^{x/2}}$	$V_4(x)=\left\lvert\frac{2}{\pi}\arctan\left(\frac{\pi}{2}x\right)\right\rvert$

approach outperforms the other benchmark methods with 99.67% detection accuracy.

In another research by Dabba et al. [32], MFO is hybridized with quantum computation to address the problem of gene selection. SVM classifier is used for evaluating the goodness of the selected genes and 13 microarray datasets are used for validating the classification accuracy performance as well as the best number of genes selected by the proposed method compared with several recent existing algorithms. The findings illustrated that this method had the capability to decrease the number of genes and improve the classification accuracy.

The work in Ref. [33] presented a wrapper feature selection technique based on an enhanced MFO that incorporates the Levy flight to tackle medical applications issues. To convert a normal MFO to a binary MFO, the sigmoid transfer function was utilized. The KNN classifier is used in order to assess the feature subset. Besides, 23 medical datasets were utilized to test their suggested approach. The performance of four alternative feature selection techniques was compared to their method's performance. The findings were superior in terms of decreasing the number of attributes while maximizing classification accuracy. Similarly, the same authors in the same year suggested an improved MFO for wrapper feature selection, with the goal of improving classification tasks in medical applications [34]. They employed eight binary functions from the families of S-shaped and V-shaped transfer functions in their study. They have also included the Levy flight to enhance the MFO exploration capability. Similarly, the suggested techniques are tested on a total of 23 medical datasets, and KNN classifier is used for selected feature evaluations. It has been proven that the suggested technique outperforms previous wrapper approaches significantly; also this study demonstrated that the performance of MFO depends on the choice of the transfer function. The same authors in the previous year 2020 proposed a rank-based feature approach based on the MFO and KNN classifier [35]. Sixteen medical datasets were used for validating the proposed model.

In 2020, a significant number of studies implemented MFO as a method for feature selection task. For instance, the authors in Ref. [36] proposed an enhanced MFO for software fault prediction by integrating an adaptive synthetic sampling. Different transfer functions are used to transfer the algorithm into binary, and the KNN, DTs, and Linear discriminant analysis (LDA) classifiers were used for selected feature evaluations. The finding shows that the proposed method enhanced the performance of all classifiers.

A modified MFO for software usability feature selection problem was proposed in Ref. [37]. The modification was based on transverse orientation method. KNN classifier was used as well as the sigmoid transfer function that enables the algorithm to be converted into a binary version. The finding demonstrated that the modified algorithm could reduce the number of features without degradation in classification accuracy. In Ref. [38], the MFO was fused with a neighborhood search technique with the objective of solving feature selection problems. Eight datasets collected from UCI data repository were used to assess the performance of the proposed method. The SVM classifier was used for evaluating the selected features. The finding demonstrated a competitive performance compared to the original MFO and a set of existing feature selection methods.

In Ref. [39], as a feature selection technique, a hybrid approach between MFO and PSO is proposed to handle the network intrusion detection problem. A weighted KNN was used to perform the classification task. KDD Cup 99 dataset as well as six datasets obtained from UCI were used for validating the proposed approach. The experimental results show that the proposed hybrid method and weighted KNN may enhance both the accuracy as well as the efficiency of network intrusion detection. Another study by Yu et al.[40] presented an enhanced MFO by integrating simulated annealing algorithm to increase the algorithm's lead in the local exploitation phase. The quantum rotation gate is then incorporated to improve the exploration abilities of the algorithm and the moth's variety. Besides feature selection, this study was applied to two engineering problems. They used the tanh transfer function to transfer the real continuous values into binary values, and KNN is used for validating the goodness of the selected features on four medical datasets. This method algorithm achieved particularly notable results.

Khurma et al. [41] used the MFO with time varying flames for the problem of feature selection. The transfer function used is sigmoid in order to form the binary variant of the MFO, and KNN was used for classification of 17 medical datasets. The results were compared against three benchmarking feature selection approaches and demonstrated competitive performance. In Ref. [42], the authors proposed the MFO for hyperspectral band selection problem. SVM classifier was for selected features evaluation and Indian pines experimental dataset was used to validate the performance of the proposed scheme compared to two well-known feature selection methods. The finding demonstrated a competitive accuracy performance. In Ref. [43], the authors proposed a hybrid intelligent approach for evaluating clinical breast cancer data, which combines cluster analysis techniques with metaheuristic algorithms for feature selection. The binary variants of MFO and whale optimization algorithms were proposed. To assess the performance of the proposed scheme, two evaluation criteria were used: cluster-based metrics and statistical-based metrics. KNN is used to verify the quality of the selected features, and the sigmoid transfer function was employed to make the method operate in binary form. The experimental findings positively prove that the proposed method was able to generate meaningful data partitions and critical feature subsets.

Moreover, in 2019, numerous state-of-the-art studies have been performed utilizing MFO to solve various types of feature selection problems. In Ref. [44], the authors proposed an improved MFO for feature selection by hybridizing it with differential evolution and benefiting from opposition-based learning. To enhance the convergence of the MFO, opposition-based learning was utilized to produce an ideal initial population; simultaneously, differential evolution is applied to improve the MFO's exploitation capabilities. A set of experimental findings is utilized as an evaluation step. The presented method was evaluated on numerous CEC2005 benchmark functions. Then 10 UCI datasets are used to validate the classification performance. Threshold values were used to convert the algorithm to binary and KNN classifier used for classification evaluation. The experimental findings demonstrated that the proposed method outperforms benchmarking metaheuristic algorithms. Another study by Ref. [45] integrated the MFO technique with the Spark platform's distributed parallel computing for feature selection. KNN was utilized to verify the quality of the selected features, and the sigmoid transfer function was employed to make the method operate in binary form. Three datasets from UCI were used for validating the proposed method. The finding outperformed the benchmarking methods in terms of feature reduction while maximizing the classification accuracy. In Ref. [46], the authors used MFO for feature selection to identify faulty signals of gear. SVM, KNN, and multilayer perceptron neural network (MLP) were utilized to detect fault patterns. Findings revealed that SVM with MFO performs significantly better than the other compared schemes, where it can achieve a classification accuracy of 99.60.

In 2018, a few number of studies using MFO have been carried out as follows: in Ref. [47], the authors addressed the problem of feature selection by proposing a multi-objective MFO that hybridizes the wrapper and filter approaches. Mutual Information (MI), Random Forests (RFs), and KNN classifiers were used to assess how good the selected features are. The proposed method is validated using 21 datasets collected from UCI data repository. The findings demonstrated that the proposed scheme could obtain optimum feature subset with high classification accuracy. In another study by Darwish et al. [48], the binary variants of MFO and other well-known metaheuristic algorithms were used for feature selection to diagnose the breast cancer disease. Several classifiers were applied such as SVM, KNN, DT, and their variants. The results revealed that MFO took less computing time and it is the most stable method when compared to the other algorithms.

Additionally, in 2017, MFO obtained more attention in feature selection problems. For example, in Ref. [49], the MFO was used as feature selection method for Arabic handwritten letter recognition. Arabic handwritten letter images (CENPARMI) dataset was used as well as three classifiers KNN, RF, and linear discriminant analysis (LDA) were utilized to assess the performance of the proposed scheme in terms of accuracy. The findings demonstrated superior outcomes for the feature selected as well as the accuracy of classification and processing time was reduced. In another interesting study by Hassanien et al. [50], an improved MFO is proposed to help in the detection of tomato diseases. The MFO was hybridized with rough sets dependency and the wrapper SVM was used to evaluate the feature selected. The algorithm performance was tested on seven datasets obtained from UCI data repository and then the algorithm is compared with several existing feature selection approaches. The suggested method was then applied to a real-world problem of identifying tomato illnesses (Powdery mildew and early blight), in which a genuine dataset of tomato disease was manually constructed, and a tomato disease detection method was proposed and assessed utilizing this dataset. Besides its efficiency in reducing feature size and execution time, the findings showed that the suggested method has good accuracy.

In Ref. [51], MFO was utilized as feature selection in breast cancer images for automated mitotic identification. To assess the performance, the KNN classifier was utilized as the fitness function, and a dataset of 50 histopathology pictures was used. When compared to other well-known feature selection methods, the experimental findings demonstrated the effectiveness and resilience of the MFO-based feature selection method. Another research conducted by Ref. [52], the chaotic MFO (CMFO) presented as feature selection method and kernel extreme learning machine (KELM) as classifier. CMFO performed parameter optimization and feature selection

TABLE 2.2 Summary of Feature Selection Methods Based on MFO

No	Study	Technique	Domain	Transfer Function	Feature Selection Method	Classifier
1	[31]	MFO-HC	Intrusion Detection systems	S-shaped	Wrapper	OWKELM
2	[32]	MFO- QC	Microarray Datasets	Threshold value	Wrapper	SVM
3	[33]	MFO-Levy	Medical Applications	S-shaped	Wrapper	KNN
4	[34]	Improved MFO	Medical Applications	S-shaped and V-shaped	Wrapper	KNN
5	[35]	MFO	Medical Applications	S-shaped	Wrapper	KNN
6	[36]	Enhanced MFO	Software Fault Prediction	S-shaped and V-shaped	Wrapper	KNN, DT, LDA
7	[37]	Modified MFO	Software Usability	S-shaped	Wrapper	KNN
8	[38]	Combined MFO	UCI Datasets	Not operator	Wrapper	SVM
9	[39]	MFO-PSO	Intrusion Detection systems	S-shaped	Wrapper	Weighted KNN
10	[40]	MFO-SA	Medical Applications	V-shaped	Wrapper	KNN
11	[41]	MFO	Medical Applications	S-shaped	Wrapper	KNN
12	[42]	MFO	Hyperspectral Band	-	Wrapper	SVM
13	[43]	Hybrid MFO	Clinical Breast Cancer	S-shaped	Wrapper	KNN
14	[44]	MFO-DE	UCI Datasets	S-shaped	Wrapper	KNN
15	[45]	MFO-Spark	UCI Datasets	S-shaped	Wrapper	KNN
16	[46]	MFO	Faulty Signals of Gear	-	Wrapper	SVM, KNN, MLP
17	[47]	MOMFO	UCI Datasets	-	Hybrid	MI, RF, KNN
18	[48]	MFO	Breast Cancer Diagnosis	S-shaped and V-shaped	Wrapper	SVM, KNN, DT
19	[49]	MFO	Arabic Handwriting Recognition	Threshold method	Wrapper	KNN, RF, LDA
20	[50]	MFO-RS	UCI Datasets and Real Data	1 rand () >1, 0 otherwise	Wrapper	SVM
21	[51]	MFO	Breast Cancer Images	Threshold method	Wrapper	KNN
22	[52]	CMFO	Medical Applications	S-shaped	Wrapper	KELM
23	[53]	Modified MFO	Terrorism Prediction	-	Wrapper	KNN and RF
24	[54]	MFO	UCI Datasets	Threshold method	Wrapper	KNN

for medical diagnostic issues of Parkinson's disease and breast cancer. Compared with other approaches, the findings demonstrated that the presented method provides much higher classification accuracy while it also obtained a lesser number of features.

Lastly in 2016 and based on our search, two studies have been performed on feature selection utilizing MFO. In Ref. [53], two modifications were made to the original MFO as feature selection methods for terrorism prediction. KNN and RF classifiers were used for evaluating the selected features. The performance of the proposed approaches is compared to the original MFO as well as with other four well-known feature selection methods. The findings of the proposed methods were competitive to the state-of-the-art methods. The work in Ref. [54] was the first study that used MFO as a feature selection. Eighteen datasets collected from UCI were used to validate the proposed method and compared against two well-regarded feature selection methods. KNN classifier was used to ensure the quality of the selected features. The findings demonstrated the proposed algorithm's efficiency when compared to other methods.

As seen by the literature, a significant number of investigations have been performed using MFO as a feature selection approach. This is owing to the strong features of this algorithm and its capacity to handle such tasks with high performance. Table 2.1 summarizes the existing research work on MFO-based feature selection approaches from 2016 to 2021.

2.6 DISCUSSION AND FUTURE DIRECTIONS

As previously stated, MFO has been widely utilized to tackle numerous feature selection issues in many domains since it was introduced in 2015. As shown in Table 2.1, most of the research studies were based on wrapper-based feature selection and there have been no studies as a filter feature selection method to the wrapper method. The reason is that filters neglect the goodness of the subset of features on the classification algorithm, whereas wrappers evaluate the classification performance-based subsets of features, which usually allow wrappers to achieve better classification performance compared with filters [12,55,56].

The major reasons for the MFO's success are its simplicity, versatility, and ease of implementation. Moreover, the MFO has certain distinct benefits such as it enables fast convergence at an early stage by moving from exploration to exploitation, which increases the efficiency of MFO. On the other hand, MFO has some restrictions and disadvantages. For example, by nature, feature selection is a binary problem, therefore, a transfer function should be integrated to MFO. Furthermore, the No-Free-Lunch theorem asserts that it is impractical to have an optimization algorithm capable of solving all optimization problems which indicates that MFO may need to be improved in order to address various types of optimization issues. Moreover, one of the primary downsides of MFO is that its performance in big datasets is inefficient and may be decreased when compared to other methods, because MFO does not accept new solutions if they are worse than the existing global best. This may limit the number of flames. Also, similar to other optimization algorithms, MFO could be trapped into local optima.

Some of the research recommendations to be further studied by researchers utilizing MFO as feature selection technique can be summarized as follows:

- Investigating the use of MFO in filter-based feature selection approach.
- Applying multi-objective MFO in both wrapper and filter-based feature selection since this has not been fully investigated yet.
- Mechanisms such as Mutations could be integrated to reduce convergence of MFO or to avoid being stuck in local optima.
- Investigating the influence of different transfer functions within MFO such as X-Shaped transfer function.
- Investigating the use of both MFO and MOFO in different domains of feature selection.

2.7 CONCLUSION

This chapter has comprehensively and critically reviewed MFO-based feature selection. It starts with a clear explanation of the feature selection concept and its process. It then provides the inspiration of MFO and its mathematical modeling. The mechanism of utilizing MFO for feature selection problems is discussed in detail. The performance of existing research work that utilized MFO to solve various feature selection problems is critically reviewed. In this chapter, 24 research papers of feature selection based on MFO were collected, examined, and evaluated to highlight the strength, benefits, and drawbacks of MFO for researchers who are interested to work with this algorithm. This study assists researchers working on feature selection by outlining how MFO could be used to solve issues, highlighting its benefits and shortcomings, and demonstrating its efficacy. Lastly, we provided some research directions for future feature selection research using MFO.

REFERENCES

1. J. Tang, S. Alelyani, and H. Liu, "Feature selection for classification: A review," *Data Classification Algorithms and Application*, pp. 37–64, 2014. doi: 10.1.1.409.5195.
2. Q. Al-Tashi, S. J. A. Kadir, H. M. Rais, S. Mirjalili, and H. Alhussian, "Binary optimization using hybrid grey wolf optimization for feature selection," *IEEE Access*, vol. 7, pp. 39496–39508, 2019.
3. U. Fayyad, G. Piatetsky-Shapiro, and P. Smyth, "From data mining to knowledge discovery in databases," *AI Magazine*, vol. 17, no. 3, p. 37, 1996.
4. N. Omar and Q. Al-Tashi, "Arabic nested noun compound extraction based on linguistic features and statistical measures," *GEMA Online Journal Language Stududies*, vol. 18, no. 2, 2018.
5. J. Han, J. Pei, and M. Kamber, *Data Mining: Concepts and Techniques.* Elsevier: Amsterdam, Netherlands, 2011.
6. V. Kotu and B. Deshpande, *Predictive Analytics and Data Mining: Concepts and Practice with Rapidminer.* Morgan Kaufmann: Burlington, MA, 2014.
7. H. Liu and H. Motoda, *Feature Selection for Knowledge Discovery and Data Mining*, vol. 454. Springer Science & Business Media: Berlin/Heidelberg, Germany, 2012.
8. A. L. Blum and P. Langley, "Selection of relevant features and examples in machine learning," *Artificial Intelligence*, vol. 97, no. 1–2, pp. 245–271, 1997.
9. I. Guyon and A. Elisseeff, "An introduction to variable and feature selection," *Journal of Machine Learning Research*, vol. 3, pp. 1157–1182, 2003.
10. Q. Al-Tashi et al., "Binary multi-objective grey wolf optimizer for feature selection in classification," *IEEE Access*, vol. 8, pp. 106247–106263, 2020.
11. R. Al-Wajih, S. J. Abdulkadir, N. Aziz, Q. Al-Tashi, and N. Talpur, "Hybrid binary grey wolf with harris hawks optimizer for feature selection," *IEEE Access*, vol. 9, pp. 31662–31677, 2021, doi: 10.1109/ACCESS.2021.3060096.
12. M. Dash and H. Liu, "Feature selection for classification," *Intelligent Data Analysis*, vol. 1, no. 3, pp. 131–156, 1997.
13. Q. Al-Tashi, H. Rais, and S. Jadid, "Feature selection method based on grey wolf optimization for coronary artery disease classification," in *International Conference of Reliable Information and Communication Technology*, Kuala Lumpur, 2018, pp. 257–266.
14. B. Xue, M. Zhang, W. N. Browne, and X. Yao, "A survey on evolutionary computation approaches to feature selection," *IEEE Transactions on Evolutionary Computation*, vol. 20, no. 4, pp. 606–626, 2016, doi: 10.1109/TEVC.2015.2504420.

15. Q. Al-Tashi, H. M. Rais, S. J. Abdulkadir, S. Mirjalili, and H. Alhussian, "A review of grey wolf optimizer-based feature selection methods for classification," in: S. Mirjalili, H. Faris, and I. Aljarah (eds.), *Evolutionary Machine Learning Techniques*. Springer: Berlin/Heidelberg, Germany, 2020, pp. 273–286.
16. A. Alqushaibi, S. J. Abdulkadir, H. M. Rais, Q. Al-Tashi, M. G. Ragab, and H. Alhussian, "Enhanced weight-optimized recurrent neural networks based on sine cosine algorithm for wave height prediction," *Journal of Marine Science and Engineering*, vol. 9, no. 5, p. 524, 2021.
17. A. Alqushaibi, S. J. Abdulkadir, H. M. Rais, Q. Al-Tashi, and M. G. Ragab, "An optimized recurrent neural network for metocean forecasting," in *2020 International Conference on Computational Intelligence (ICCI)*, India, 2020, pp. 190–195.
18. A. Alqushaibi, S. J. Abdulkadir, H. M. Rais, and Q. Al-Tashi, "A review of weight optimization techniques in recurrent neural networks," in *2020 International Conference on Computational Intelligence (ICCI)*, 2020, pp. 196–201.
19. A. O. Balogun et al., "Impact of feature selection methods on the predictive performance of software defect prediction models: An extensive empirical study," *Symmetry (Basel)*, vol. 12, no. 7, p. 1147, 2020.
20. Q. Al-Tashi, H. Rais, and S. J. Abdulkadir, "Hybrid swarm intelligence algorithms with ensemble machine learning for medical diagnosis," in *2018 4th International Conference on Computer and Information Sciences (ICCOINS)*, 2018, doi: 10.1109/ICCOINS.2018.8510615.
21. S. Mirjalili, "Moth-flame optimization algorithm: A novel nature-inspired heuristic paradigm," *Knowledge-Based Systems*, vol. 89, pp. 228–249, 2015.
22. P. M. Narendra and K. Fukunaga, "A branch and bound algorithm for feature subset selection," *IEEE Transactions on Computers*, vol. 9, pp. 917–922, 1977.
23. B. H. Nguyen, B. Xue, and M. Zhang, "A survey on swarm intelligence approaches to feature selection in data mining," *Swarm and Evolutionary Computation*, vol. 54, p. 100663, 2020.
24. Q. Al-Tashi, H. M. Rais, S. J. Abdulkadir, and S. Mirjalili, "Feature selection based on grey wolf optimizer for oil gas reservoir classification," in *2020 International Conference on Computational Intelligence, ICCI 2020*, 2020, pp. 211–216, doi: 10.1109/ICCI51257.2020.9247827.
25. R. Al-wajih, S. J. Abdulakaddir, N. B. A. Aziz, and Q. Al-tashi, "Binary grey wolf optimizer with K-nearest neighbor classifier for feature selection," in *2020 International Conference on Computational Intelligence (ICCI)*, India, 2020, pp. 130–136.
26. Q. Al-Tashi, S. J. Abdulkadir, H. M. Rais, S. Mirjalili, and H. Alhussian, "Approaches to multi-objective feature selection: A systematic literature review," *IEEE Access*, vol. 8, pp. 125076–125096, 2020.
27. P. Langley et al., "Selection of relevant features in machine learning," in *Proceedings of the AAAI Fall Symposium on Relevance*, California, 1994, vol. 184, pp. 245–271.
28. E. Hancer, B. Xue, and M. Zhang, "A survey on feature selection approaches for clustering," *Artificial Intelligence Review*, vol. 53, pp. 1–27, 2020.

29. A. Mukhopadhyay, U. Maulik, S. Bandyopadhyay, and C.. Coello Coello, "A survey of multiobjective evolutionary algorithms for data mining: Part I," *IEEE Transactions on Evolutionary Computation*, vol. 18, no. 1, pp. 4–19, 2014, doi: 10.1109/TEVC.2013.2290086.
30. K. K. Ghosh, R. Guha, S. K. Bera, N. Kumar, and R. Sarkar, "S-shaped versus V-shaped transfer functions for binary Manta ray foraging optimization in feature selection problem," *Neural Computing and Applications*, vol. 33, pp. 1–15, 2021.
31. B. V. Kumar, "Hybrid metaheuristic optimization based feature subset selection with classification model for intrusion detection in big data environment turkish," *Journal of Computer and Mathematics Education Research Article*, vol. 12, no. 12, pp. 2297–2308, 2021.
32. A. Dabba, A. Tari, and S. Meftali, "Hybridization of Moth flame optimization algorithm and quantum computing for gene selection in microarray data," *Journal of Ambient Intelligence and Humanized Computing*, vol. 12, no. 2, pp. 2731–2750, 2021, doi: 10.1007/s12652-020-02434-9.
33. R. A. Khurma, I. Aljarah, and A. Sharieh, "A simultaneous moth flame optimizer feature selection approach based on levy flight and selection operators for medical diagnosis," *Arabian Journal for Science and Engineering*, 2021, doi: 10.1007/s13369-021-05478-x.
34. R. Abu Khurmaa, I. Aljarah, and A. Sharieh, *An Intelligent Feature Selection Approach Based on Moth Flame Optimization for Medical Diagnosis*, vol. 33, no. 12. Springer: London, 2021.
35. R. A. Khurma, I. Aljarah, and A. Sharieh, "Rank based moth flame optimisation for feature selection in the medical application," *2020 IEEE Congress on Evolutionary Computation (CEC) 2020 Conference Proceedings*, 2020, doi: 10.1109/CEC48606.2020.9185498.
36. I. Tumar, Y. Hassouneh, H. Turabieh, and T. Thaher, "Enhanced binary moth flame optimization as a feature selection algorithm to predict software fault prediction," *IEEE Access*, vol. 8, pp. 8041–8055, 2020, doi: 10.1109/ACCESS.2020.2964321.
37. D. Gupta, A. K. Ahlawat, A. Sharma, and J. J. P. C. Rodrigues, "Feature selection and evaluation for software usability model using modified moth-flame optimization," *Computing*, vol. 102, no. 6, pp. 1503–1520, 2020, doi: 10.1007/s00607-020-00809-6.
38. M. Alzaqebah, N. Alrefai, E. A. E. Ahmed, S. Jawarneh, and M. K. Alsmadi, "Neighborhood search methods with moth optimization algorithm as a wrapper method for feature selection problems," *International Journal of Electrical and Computer Engineering*, vol. 10, no. 4, pp. 3672–3684, 2020, doi: 10.11591/ijece.v10i4.pp3672-3684.
39. H. Xu, K. Przystupa, C. Fang, A. Marciniak, O. Kochan, and M. Beshley, "A combination strategy of feature selection based on an integrated optimization algorithm and weighted k-nearest neighbor to improve the performance of network intrusion detection," *Electronics*, vol. 9, no. 8, pp. 1–22, 2020, doi: 10.3390/electronics9081206.

40. C. Yu, A. A. Heidari, and H. Chen, "A quantum-behaved simulated annealing algorithm-based moth-flame optimization method," *Applied Mathematical Modelling*, vol. 87, pp. 1–19, 2020, doi: 10.1016/j.apm.2020.04.019.
41. R. A. Khurma, P. A. Castillo, A. Sharieh, and I. Aljarah, "Feature selection using binary moth flame optimization with time varying flames strategies," in *IJCCI*, 2020, pp. 17–27.
42. E. Worch, S. Samiappan, M. Zhou, and J. E. Ball, "Hyperspectral band selection using moth-flame metaheuristic optimization," *The International Geoscience and Remote Sensing Symposium (IGARSS)*, pp. 1271–1274, 2020, doi: 10.1109/IGARSS39084.2020.9323754.
43. G. I. Sayed, A. Darwish, and A. E. Hassanien, "Binary whale optimization algorithm and binary moth flame optimization with clustering algorithms for clinical breast cancer diagnoses," *Journal of Classification*, vol. 37, no. 1, pp. 66–96, 2020, doi: 10.1007/s00357-018-9297-3.
44. M. A. Elaziz, A. A. Ewees, R. A. Ibrahim, and S. Lu, "Opposition-based moth-flame optimization improved by differential evolution for feature selection," *Mathematics and Computers in Simulation*, vol. 168, pp. 48–75, 2020, doi: 10.1016/j.matcom.2019.06.017.
45. H. Chen, H. Fu, Q. Cao, L. Han, and L. Yan, "Feature selection of parallel binary moth-flame optimization algorithm based on spark," in *2019 IEEE 3rd Information Technology, Networking, Electronic and Automation Control Conference (ITNEC)*, 2019, pp. 408–412, doi: 10.1109/ITNEC.2019.8729350.
46. P. Ong, T. H. C. Tieh, K. H. Lai, W. K. Lee, and M. Ismon, "Efficient gear fault feature selection based on moth-flame optimisation in discrete wavelet packet analysis domain," *The Journal of the Brazilian Society of Mechanical Sciences and Engineering*, vol. 41, no. 6, pp. 1–14, 2019, doi: 10.1007/s40430-019-1768-x.
47. M. A. S. Ghada, H. M. A.-E.-E. Tarek, E. Emary, and M. H. K. Motaz, "A novel multi-objective moth-flame optimization algorithm for feature selection," *Indian Journal of Science and Technology*, vol. 11, no. 38, pp. 1–13, 2018, doi: 10.17485/ijst/2018/v11i38/128008.
48. A. Darwish, G. Sayed, and A. Hassanien, "Meta-heuristic optimization algorithms based feature selection for clinical breast cancer diagnosis," *Journal of the Egyptian Mathematical Society*, vol. 26, no. 2, pp. 321–336, 2018, doi: 10.21608/jomes.2018.2673.1023.
49. A. A. Ewees, A. T. Sahlol, and M. A. Amasha, "A bio-inspired moth-flame optimization algorithm for arabic handwritten letter recognition," *IEEE 2017 International Conference on Control, Artificial Intelligence, Robotics & Optimization (ICCAIRO)*, vol. 2018-January, pp. 154–159, 2017, doi: 10.1109/ICCAIRO.2017.38.
50. A. E. Hassanien, T. Gaber, U. Mokhtar, and H. Hefny, "An improved moth flame optimization algorithm based on rough sets for tomato diseases detection," *Computers and Electronics in Agriculture*, vol. 136, pp. 86–96, 2017, doi: 10.1016/j.compag.2017.02.026.

51. G. I. Sayed and A. E. Hassanien, "Moth-flame swarm optimization with neutrosophic sets for automatic mitosis detection in breast cancer histology images," *Applied Intelligence*, vol. 47, no. 2, pp. 397–408, 2017, doi: 10.1007/s10489-017-0897-0.
52. M. Wang et al., "Toward an optimal kernel extreme learning machine using a chaotic moth-flame optimization strategy with applications in medical diagnoses," *Neurocomputing*, vol. 267, pp. 69–84, 2017, doi: 10.1016/j.neucom.2017.04.060.
53. A.-E.-E. Soliman, M. Khorshid, and T. Abou-El-Enien, "Modified moth-flame optimization algorithms for terrorism prediction," *International Journal of Application or Innov Engineering is a Management*, vol. 5, no. 7, pp. 47–59, 2016.
54. H. M. Zawbaa, E. Emary, B. Parv, and M. Sharawi, "Feature selection approach based on moth-flame optimization algorithm," in *2016 IEEE Congress on Evolutionary Computation (CEC)*, 2016, pp. 4612–4617, doi: 10.1109/CEC.2016.7744378.
55. H. Liu and Z. Zhao, "Manipulating data and dimension reduction methods: Feature selection," in *Encyclopedia of Complexity and Systems Science*. Springer: New York, 2009, pp. 5348–5359.
56. H. Liu, H. Motoda, R. Setiono, and Z. Zhao, "Feature selection: An ever evolving frontier in data mining," in Z. A. Zhao and H. Liu (eds), *Feature Selection in Data Mining*. CRC Press: Boca Raton, FL, 2010, pp. 4–13.

CHAPTER 3

An Efficient Binary Moth-Flame Optimization Algorithm with Cauchy Mutation for Solving the Graph Coloring Problem

Yassine Meraihi
University of M'Hamed Bougara Boumerdes

Asma Benmessaoud Gabis
National School of Computer Science

Seyedali Mirjalili
Torrens University Australia
Yonsei University

CONTENTS

3.1 Introduction 36
3.2 Graph Coloring Problem Formulation 37
3.3 Moth-Flame Optimization Algorithm and Binary Moth-Flame Optimization Algorithm 37

DOI: 10.1201/9781003205326-4

3.3.1 Moth-Flame Optimization (MFO) Algorithm 37
3.3.2 Binary Moth-Flame Optimization Algorithm 40
3.4 Efficient BMFO Algorithm With Cauchy Mutation for Solving the GCP 41
3.4.1 Efficient BMFO Algorithm With Cauchy Mutation 41
3.4.2 Representation of the Solution 42
3.4.3 Objective Function 43
3.5 Simulation Results 44
3.6 Conclusion 49
References 49

3.1 INTRODUCTION

The Graph Coloring Problem (GCP) is a well-studied and one of the most investigated and challenging optimization problems in operational research especially in graph theory. It has been widely applied to model and solve many real-world problems such as radio frequency assignment [1,2], computer register allocation [3,4], time tabling [5], scheduling [6,7], printed circuit board-testing [8], noise reduction in VLSI circuits [9], channel routing [10], communication networks [11], and many other problems. GCP consists in assigning the vertices of a given graph with colors such that all adjacent vertices have different colors while minimizing the number of used colors. The smallest number of colors that can be used for coloring a graph G is named chromatic number $X(G)$. The problem of determining $X(G)$ of any graph is known to be an NP-complete problem [12] and has attracted the attention of academics and researchers. So, many optimization methods have been introduced to tackle the GCP problem including greedy constructive methods (such as the Largest Saturation degree algorithm (DSATUR) [13] and the Recursive Largest First (RLF) algorithm [14]), local search heuristics (such as Tabu Search (TS) algorithm), metaheuristics (such as Genetic Algorithm (GA) [15], Memetic Algorithm (MA) [16], Dragonfly Algorithm (DA) [17], Selfish Herd Optimizer (SHO) [18], Artificial Bee Colony (ABC) [19], Crow Search Algorithm (CSA) [20], Cuckoo Search (CS) algorithm [21,22], Salp Swarm Algorithm (SSA) [23]), and hybridized optimization algorithms [24–28].

This chapter proposes an efficient binary MFO algorithm, called EC-BMFO, based on Cauchy mutation strategy for solving the graph coloring problem. The binary version of MFO (BMFO) is obtained by using the V-shaped transfer function and the binarization concept. Then, Cauchy mutation strategy is adopted to improve the diversity of solutions when updating the positions of moths. To our knowledge, this is the first work applying MFO to solve the GCP.

The remainder of the chapter is organized as follows. In Section 3.2, we present the graph coloring problem formulation. Section 3.3 describes the moth-flame optimizer and the binary moth-flame optimizer. In Section 3.4, we propose an efficient binary Moth-Flame Optimization algorithm with Cauchy mutation. The simulation results and analysis of different algorithms are given in Section 3.5. This chapter concludes with Section 3.6.

3.2 GRAPH COLORING PROBLEM FORMULATION

The problem formulation of GCP is represented by an undirected graph G consisting of vertices set (V) and arcs set A such as $V(G)=\{v_1,v_2,\ldots,v_n\}$ and $A(G)=\{a_1,a_2,\ldots,a_l\}$, where n and l denote the number of vertices and arcs in the graph G, respectively. $a=(i,j)\in E$ represents the arc that connects the vertex i with the vertex j.

Formally, a legal K-coloring of vertices in the graph G is given by a vector $S=\left[C(v_1),C(v_2),\ldots,C(v_i),C(v_j),\ldots,C(v_n)\right]$ such as $C(v_i)\neq C(v_j)$ for all $a=(i,j)\in E$, where $C(v_i)$ represents the color given to the vertex i. If $C(v_i)=(v_j)$, then vertex i and vertex j are in conflict and are called conflicting vertices. The main objective is to reduce the number of colors necessary to color legally a given graph.

3.3 MOTH-FLAME OPTIMIZATION ALGORITHM AND BINARY MOTH-FLAME OPTIMIZATION ALGORITHM

3.3.1 Moth-Flame Optimization (MFO) Algorithm

Moth-Flame Optimization (MFO) algorithm, developed by Mirjalili in 2015 [29], is one of the interesting population-based optimization algorithms. It is applied to solve many optimization problems, such as image processing [30,31], power dispatch problem [32,33], networks [34,35], electrical engineering [36,37], power flow optimization [38], wind energy [39], PID control [40], and feature selection [41,42]. The main inspiration of MFO algorithm is the navigation strategy (movement) of moths at night around flames [29]. This strategy is called transverse orientation. In the mathematical model of MFO algorithm, moths' positions are the problem variables and flames are the best solutions obtained so far [29].

Moths' positions are represented by the following matrix:

$$\text{Pmo} = \begin{bmatrix} \text{Pmo}_{1,1} & \text{Pmo}_{1,2} & \dots & \dots & \text{Pmo}_{1,d} \\ \text{Pmo}_{2,1} & \text{Pmo}_{2,2} & \dots & \dots & \text{Pmo}_{2,d} \\ \vdots & \vdots & \vdots & \vdots & \vdots \\ \vdots & \vdots & \vdots & \vdots & \vdots \\ \text{Pmo}_{n,1} & \text{Pmo}_{n,2} & \dots & \dots & \text{Pmo}_{n,d} \end{bmatrix}$$

where n represents the number of moths and d refers to the search space dimension. The fitness values of moths are stored in an array as follows:

$$\text{OPmo} = \begin{bmatrix} \text{OPmo}_1 \\ \text{OPmo}_2 \\ \vdots \\ \vdots \\ \text{OPmo}_n \end{bmatrix}$$

On the other hand, flames positions are given by the following matrix:

$$\text{Pfl} = \begin{bmatrix} \text{Pfl}_{1,1} & \text{Pfl}_{1,2} & \dots & \dots & \text{Pfl}_{1,d} \\ \text{Pfl}_{2,1} & \text{Pfl}_{2,2} & \dots & \dots & \text{Pfl}_{2,d} \\ \vdots & \vdots & \vdots & \vdots & \vdots \\ \vdots & \vdots & \vdots & \vdots & \vdots \\ \text{Pfl}_{n,1} & \text{Pfl}_{n,2} & \dots & \dots & \text{Pfl}_{n,d} \end{bmatrix}$$

where n is the number of flames and d represents the search space dimension.

The fitness values of flames are stored in an array as follows:

$$\text{OPfl} = \begin{bmatrix} \text{OPfl}_1 \\ \text{OPfl}_2 \\ \vdots \\ \vdots \\ \text{OPfl}_n \end{bmatrix}$$

The position of the moth is updated around the corresponding flame using the following equation:

$$\text{Pmo}_i = S(\text{Pmo}_i, \text{Pfl}_j) \tag{3.1}$$

where Pmo_i represents the ith moth, Pfl_j refers to the jth flame, and S is the spiral function.

The logarithmic spiral function is chosen for updating the mechanism of MFO algorithm, which is given as follows:

$$S(Pmo_i, Pfl_j) = Dt_i * e^{bt} * Cos(2\Pi t) + Pfl_j \tag{3.2}$$

where b is a constant that defines the shape of the logarithm spiral, t is a random number in the range of $[-1,1]$, and Dt indicates the distance between the ith moth and the jth flame. Dt can be represented as

$$Dt_i = |\text{Pfl}_j - \text{Pmo}_i| \tag{3.3}$$

To obtain a good balance between intensification and diversification, an adaptive scheme is employed to update the number of flames as follows:

$$\text{Numbflames} = \text{round}\left(R - k * \frac{R-1}{it_{\max}}\right) \tag{3.4}$$

where R denotes the maximum number of flames, k is the current iteration number, and $it_{\max}$ is the maximum number of iterations.

The pseudo-code of the MFO is given in Algorithm 3.1 [29].

Algorithm 3.1: The MFO Algorithm

Initialize the MFO search parameters (number of moths n, number of dimensions d, and maximum number of iterations $it_{\max}$)

Initialize moths positions Pmo_i randomly

Calculate the fitness function of each moth Pfl_i

while $(it < it_{\max})$ **do**

Update numbflames using Equation (3.4)

OPmo=Objective function(Pmo)

if $it == 1$ **then**

$\text{Pfl} = \text{sort}(\text{Pmo})$
$\text{OPfl} = \text{sort}(\text{OPmo})$
else
$\text{Pfl} = \text{sort}(\text{Pmo}_{t-1}, \text{Pmo}_t)$
$\text{OPfl} = \text{sort}(\text{OPmo}_{t-1}, \text{OPmo}_t)$
end if
for $i = 1{:}n$ do
for $j = 1{:}d$ do
Update b and t
Calculate Dt using Equation (3.3) with respect to its corresponding moth
Update $Pmo(i, j)$ using Equation (3.2) with respect to its corresponding ~ moth
end for
end for
$it = it + 1$
end while
Return the best solution

3.3.2 Binary Moth-Flame Optimization Algorithm

MFO algorithm was introduced initially for solving continuous optimization problems [43,44]. However, the GCP is considered as a classical combinatorial optimization problem with discrete binary search space in which solutions are also binary consisting of two values "0" and "1". For this reason, some modifications are needed to obtain a binary version of MFO from the original one. So, a transfer function and a binarization concept are employed [43,45–47]. The V-shaped transfer function adopted in this chapter is given by the following equation [47]:

$$V\left(\text{Pmo}_{id}^{(t+1)}\right) = \left|\frac{2}{\pi}\arctan\left(\frac{2}{\pi}\text{Pmo}_{id}^{(t+1)}\right)\right| \tag{3.5}$$

The moth position is updated using the binarization concept as follows [47]:

$$\text{Pmo}_{id}^{(t+1)} = \begin{cases} (\text{pmo}_{id}^{t})^{-1}, \text{if } r \text{ and } () < V\left(\text{Pmo}_{id}^{(t+1)}\right) \\ \text{Pmo}_{id}^{t}, \quad \text{if } r \text{ and } () \geq V\left(\text{Pmo}_{id}^{(t+1)}\right) \end{cases} \tag{3.6}$$

where Pmo_{id}^{t} indicates the position of the ith moth t in the dth dimension and $\left(Pmo_{id}^{t}\right)^{-1}$ is the complement of Pmo_{id}^{t}.

3.4 EFFICIENT BMFO ALGORITHM WITH CAUCHY MUTATION FOR SOLVING THE GCP

3.4.1 Efficient BMFO Algorithm With Cauchy Mutation

In this section, the efficient BMFO algorithm with Cauchy mutation is described. Cauchy mutation derived from Cauchy distribution is integrated into BMFO in order to provide a more suitable balance between the exploitation/exploration and enhance the effectiveness of the BMFO.

Cauchy mutation operation has been derived from Cauchy distribution. It has been integrated favorably into several metaheuristic optimization algorithms to obtain faster convergence, improve the diversity of population, exploit the search space in a much better way, and avoid premature convergence. The Cauchy distribution function is given by the following equation:

$$y = \frac{1}{2} + \frac{1}{\pi} \arctan\left(\frac{\gamma}{g}\right) \tag{3.7}$$

The corresponding probability density function is described as follows:

$$f_{\text{cauchy}(0,g)}(\gamma) = \frac{1}{\pi} \frac{g}{g^2 + \gamma^2} \tag{3.8}$$

where γ is a distributed random number between [0, 1], $g = 1$ represents the proportion (scale) parameter, and $y = \tan\left(\pi\left(y - \frac{1}{2}\right)\right)$.

$$Pmo_{id}^{t+1} = Pmo_{id}^{t}(C(\gamma)+1)$$

The modified position of moth Pmo_{id}^{t} is updated as follows:

$$Pmo_{id}^{t+1} = Pmo_{id}^{t}(C(\gamma)+1) \tag{3.9}$$

In our proposed EC-BMFO algorithm, the position of moths is initialized using the constructive modified RLF model [24]. The pseudo-code of the EC-BMFO algorithm is described in Algorithm 3.2.

Algorithm 3.2: The EC-BMFO Algorithm

Initialize the MFO search parameters (number of moths n, number of dimension d, and maximum number of iteration it_{max})

Initialize the positions of moths Pmo_i using constructive modified RLF model

Calculate the fitness function of each moth f_i

while $(it < it_{max})$ **do**

Update numbflames using Equation (3.4)

OPmo=Objective function(Pmo)

if $it == 1$ **then**

$Pfl = sort(Pmo)$

$OPfl = sort(OPmo)$

else

$Pfl = sort(Pmo_{t-1}, Pmo_t)$

$OPfl = sort(OPmo_{t-1}, OPmo_t)$

end if

for $i = 1{:}n$ **do**

for $j = 1{:}d$ **do**

Update b and t

Calculate D using Equation (3.3) with respect to its corresponding moth

Update $Pmo(i, j)$ using Equation (3.2) with respect to its corresponding moth

Calculate the V-Shaped transfer function value using Equation (3.4)

Update the position of the current moth to the binary position using Equation (3.5)

end for

Update the position of each moth using Cauchy mutation (Equation 3.8)

end for

$it = it + 1$

end while

Return the best solution

3.4.2 Representation of the Solution

The graph coloring solution is given by $V * C$ binary matrix M, where V refers to the number of vertices and C denotes the number of colors by

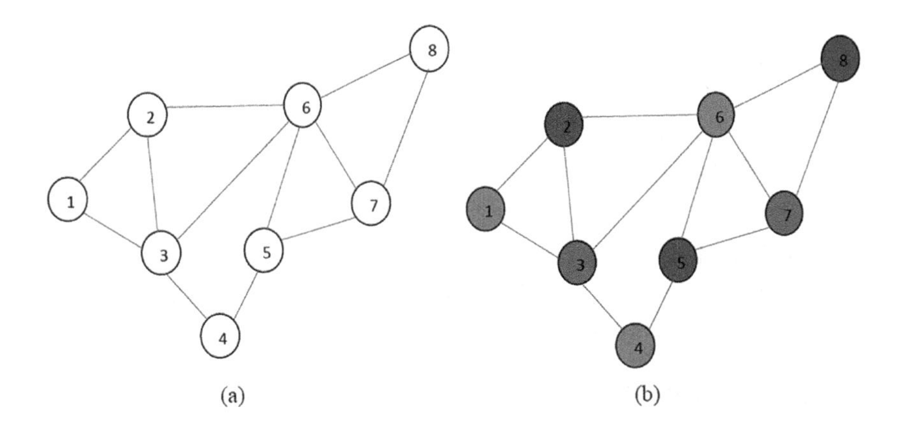

	Green Color	Blue Color	Red Color
Vertex 1	1	0	0
Vertex 2	0	1	0
Vertex 3	0	0	1
Vertex 4	1	0	0
Vertex 5	0	1	0
Vertex 6	1	0	0
Vertex 7	0	0	1
Vertex 8	0	1	0

(c)

FIGURE 3.1 An example of a graph coloring solution representation. (a) Uncolored graph and (b) colored graph.

which the graph is colored. The $V * C$ binary matrix is represented by the following equation:

$$M_{ij} = \begin{cases} 1; & \text{if the color } j \text{ is assigned to the vertex } i \\ 0; & \text{otherwise} \end{cases} \tag{3.10}$$

Figure 3.1 shows an example of a graph coloring solution representation where the vertices 1, 4, and 6 are colored with the green color; the vertices 2, 5, and 8 are colored with the blue color; and the vertices 3 and 7 are colored with the red color.

3.4.3 Objective Function

The objective function $f(S)$ refers to the number of the conflicting vertices produced by S where $S = C(v_1), C(v_2), C(v_i), \ldots, C(v_n)$. It can be given by the following equation:

$$f(s)=\sum_{i=1}^{n}\sum_{j=1}^{n}\text{conflict}_{ij} \tag{3.11}$$

$$\text{conflict}_{ij}=\begin{cases} 0 & \text{if the } C(v_i)\neq C(v_j) \text{ and } a=(i,j)\in A. \\ 1 & \text{otherwise.} \end{cases} \tag{3.12}$$

where conflict_{ij} denotes the conflicting matrix and it is given as follows:

A K-coloring solution S with $f(S)=0$ represents a valid solution. A solution is found if the number of conflicts $f(S)=0$.

3.5 SIMULATION RESULTS

Visual C++ language is used to implement EC-BMFO algorithm and the experiments are performed on a PC with an Intel(R) Core(TM) i5-8350U 1.9 GHz-CPU and 16 GB RAM. In order to testify and evaluate the efficiency and the performance of EC-BMFO, we use 71 challenging instances taken from DIMACS graphs on small $(|V|<100)$, intermediate $(100\leq|V|<500)$, and large $(|V|>500)$ instances [48]. The performance of EC-BMFO is compared with five state-of-the-art coloring algorithms such as MACOL [16], EBDA [17], DBG [25], BEECOL [19], and MCOACOL [21]. The experiment runs 20 times for each instance. The number of individuals is 50. The maximum number of iterations is set to 2,000. Obtained results are shown in Tables 3.1–3.3, where $|V|$ refers to the number of vertices, $|A|$ to the number of arcs, and Kn^* to the best-known result reported in the literature.

Table 3.1 shows the comparison results of EC-BMFO with 5 meta-heuristics on 16 small instances. EC-BMFO obtains similar results with MCOACOL, EBDA, and DBG for, respectively, 12, 15, and 14 instances. Moreover, it has better performance than MCOACOL on 2 instances.

From Table 3.2, it has been shown that EC-BMFO is very competitive and successful on 37 intermediate instances. It has equivalent results with MCOACOL, MACOL, EBDA, DBG, and BEECOL on 29, 9, 32, 24, and 4 instances, respectively, whereas it outperforms MCOACOL, DBG, and BEECOL for, respectively, 5, 1, and 2 instances. Nevertheless, EC-BMFO has worse results with only one instance for both MACOL and DBG algorithms. On the other hand, the comparison of EC-BMFO to the best-known result Kn^* gives equivalent results for 35 instances and worse results for only 2 instances at most.

Experimentation shown in Table 3.3 is conducted on 18 large instances. First, we notice that when compared with the best-known result Kn^*,

TABLE 3.1 Experimental Comparison Results of EC-BMFO with Five Relevant Algorithms on Small Instances ($|V|<100$)

Instance No	Instance Name	$\|V\|$	$\|A\|$	Kn^*	Mahmoudi and Lotfi [21] (MCOACOL)	Faraji and Javadi [19] (BEECOL)	Baiche et al. [17] (EBDA)	Douiri and Elbarnoussi [25] (DBG)	Lu and Hao [16] (MACOL)	Our Algorithm (EC-BMFO)
1	myciel3	11	20	4	4	*	4	4	*	4
2	myciel4	23	71	5	5	*	5	5	*	5
3	myciel5	47	236	6	6	*	6	6	*	6
4	myciel6	95	755	7	7	*	7	7	*	7
5	queen5_5	25	160	5	5	*	5	5	*	5
6	queen6_6	36	290	7	8	*	7	7	*	7
7	queen7_7	49	476	7	7	*	7	7	*	7
8	1-Insertions_4	67	232	4	5	*	5	*	*	5
9	2-Insertions_3	37	72	4	4	*	4	4	*	4
10	3-Insertions_3	56	110	4	4	*	4	4	*	4
11	4-Insertions_3	79	156	3	4	*	4	*	*	4
12	mug88_1	88	146	4	4	*	4	4	*	4
13	mug88_25	88	146	4	4	*	4	4	*	4
14	huck	74	301	11	11	*	11	11	*	11
15	jean	80	254	10	10	*	10	10	*	10
16	david	87	406	11	11	*	11	11	*	11

TABLE 3.2 Experimental Comparison Results of EC-BMFO With Five Relevant Algorithms on Intermediate Instances $(100 \leq |V| < 500)$

Instance No	Instance Name	$\lvert V\rvert$	$\lvert A\rvert$	Kn^*	Mahmoudi and Lotfi [21] (MCOACOL)	Faraji and Javadi [19] (BEECOL)	Baiche et al. [17] (EBDA)	Douiri and Elbarnoussi [25] (DBG)	Lu and Hao [16] (MACOL)	Our Algorithm (EC-BMFO)
17	Anna	138	493	11	11	*	11	11	*	11
18	games120	120	638	9	9	*	9	9	*	9
19	mycile7	191	2,360	8	8	*	8	8	*	8
20	2-Insertions_4	149	541	4	5	*	5	*	*	5
21	mug100_1	100	166	4	4	*	4	4	*	4
22	mug100_25	100	166	4	4	*	4	4	*	4
23	miles250	128	387	8	8	*	8	8	*	8
24	miles500	128	1,170	20	20	*	20	20	*	20
25	miles750	128	2,113	31	31	*	31	*	*	31
26	miles1000	128	3,216	42	42	*	42	*	*	42
27	miles1500	128	5,198	73	73	*	73	*	*	73
28	mulsol.i.1	197	3,925	49	49	*	49	49	*	49
29	mulsol.i.2	188	3,885	31	31	*	31	31	*	31
30	mulsol.i.3	184	3,916	31	31	*	31	31	*	31
31	mulsol.i.4	185	3,946	31	31	*	31	31	*	31
32	mulsol.i.5	186	3,973	31	31	*	31	31	*	31
33	zeroin.i.1	211	4,100	49	49	*	49	49	*	49
34	zeroin.i.2	211	3,541	30	30	*	30	30	*	30
35	zeroin.i.3	206	3,540	30	30	*	30	30	*	30

(*Continued*)

TABLE 3.2 (*Continued*) Experimental Comparison Results of EC-BMFO With Five Relevant Algorithms on Intermediate Instances $(100 \leq |V| < 500)$

Instance No	Instance Name	$\|V\|$	$\|A\|$	Kn^*	Mahmoudi and Lotfi [21] (MCOACOL)	Faraji and Javadi [19] (BEECOL)	Baiche et al. [17] (EBDA)	Douiri and Elbarnoussi [25] (DBG)	Lu and Hao [16] (MACOL)	Our Algorithm (EC-BMFO)
36	DSJC125.1	125	736	5	6	*	5	6	5	5
37	DSJC125.5	125	3,891	17	19	18	*	17	17	17
38	DSJC125.9	125	6,961	44	45	44	*	44	44	44
39	DSJC250.1	250	3,218	8	10	9	8	8	8	8
40	DSJC250.5	250	15,668	28	33	30	30	28	28	30
41	DSJC250.9	250	27,897	72	*	*	72	72	72	72
42	R125.1	125	209	5	5	*	5	*	*	5
43	R125.1c	125	7,501	46	46	*	46	*	*	46
44	R125.5	125	3,838	36	37	*	37	*	*	37
45	R250.1	250	867	8	8	*	8	8	*	8
46	R250.1c	250	30,227	64	64	*	64	64	*	64
47	fpsol2.i.1	496	11,654	65	65	*	*	65	*	65
48	fpsol2.i.2	541	8,691	30	30	*	*	30	*	30
49	fpsol2.i.3	425	8,688	30	30	*	*	30	*	30
50	le450_15a	450	8,168	15	*	*	15	*	15	15
51	le450_15b	450	8,169	15	*	*	15	*	15	15
52	le45_25a	450	8,260	25	25	25	25	*	25	25
53	le450_25b	450	8,263	25	25	25	25	*	25	25

TABLE 3.3 Experimental Comparison Results of EC-BMFO With Five Relevant Algorithms on Large Instances $(|V| > 500)$

Instance No	Instance Name	$\|V\|$	$\|A\|$	Kn^*	Mahmoudi and Lotfi [21] (MCOACOL)	Faraji and Javadi [19] (BEECOL)	Baiche et al. [17] (EBDA)	Douiri and Elbarnoussi [25] (DBG)	Lu and Hao [16] (MACOL)	Our Algorithm (EC-BMFO)
54	DSJC500.1	500	12,458	12	*	14	12	12	12	12
55	DSJC500.5	500	62,624	48	*	53	48	48	48	48
56	DSJC500.9	500	1,12,437	126	*	*	126	126	126	126
57	DSJR500.1	500	3,555	12	12	12	12	*	12	12
58	DSJR500.5	500	58,862	122	*	*	123	124	122	122
59	DSJC1000.1	1000	49,629	20	*	*	20	20	20	20
60	DSJC1000.5	1000	2,49,826	83	*	*	83	83	83	83
61	DSJC1000.9	1000	449,449	222	*	*	223	224	223	223
62	r1000.1	1000	14,378	20	*	*	20	20	20	20
63	r1000.5	1000	2,38,267	234	*	*	244	242	245	240
64	flat1000_5_0	1000	2,45,000	50	*	*	50	50	50	50
65	flat1000_60_0	1000	2,45,830	60	*	*	60	60	60	60
66	flat1000_76_0	1000	2,46,708	76	*	*	*	82	82	82
67	inithx.i.1	864	18,707	54	54	*	54	*	*	54
68	inithx.i.2	645	13,979	31	31	*	31	*	*	31
69	inithx.i.3	621	13,969	31	31	*	31	*	*	31
70	latin_sqr_10	900	3,07,350	98	*	*	99	99	99	99
71	homer	561	1,629	13	13	*	13	*	*	13

EC-BMFO gets similar results in around 78% of cases. In addition, our proposed EC-BMFO algorithm shows its superiority to MACOL, EBDA, DBG, and BEECOL on 1, 2, 3, and 2 instances.

3.6 CONCLUSION

This chapter proposes an efficient binary MFO algorithm (EC-BMFO) based on Cauchy mutation strategy for solving the graph coloring problem. We apply the V-shaped transfer function and the binarization concept to generate the binary version of MFO. Then, Cauchy mutation strategy is adopted to improve the diversity of solutions when updating the positions of moths. The performance of EC-BMFO was investigated based on well-known DIMACS benchmark instances in comparison with five existing algorithms such as DBG, MACOL, MCOACOL, BBCOL, and EBDA. The computational results show that EC-BMFO gives good and competitive performance in solving the graph coloring problem compared to MACOL, DBG, MCOACOL, BEECOL, and EBDA.

REFERENCES

1. Andreas Gamst. Some lower bounds for a class of frequency assignment problems. *IEEE Transactions on Vehicular Technology*, 35(1):8–14, 1986.
2. Derek H. Smith, Steve Hurley, and S.U. Thiel. Improving heuristics for the frequency assignment problem. *European Journal of Operational Research*, 107(1):76–86, 1998.
3. Gregory J. Chaitin, Marc A. Auslander, Ashok K. Chandra, John Cocke, Martin E. Hopkins, and Peter W. Markstein. Register allocation via coloring. *Computer Languages*, 6(1):47–57, 1981.
4. Dominique de Werra, Ch Eisenbeis, Sylvain Lelait, and Bruno Marmol. On a graph-theoretical model for cyclic register allocation. *Discrete Applied Mathematics*, 93(2–3):191–203, 1999.
5. Dominique de Werra. An introduction to timetabling. *European Journal of Operational Research*, 19(2):151–162, 1985.
6. Vahid Lotfi and Sanjiv Sarin. A graph coloring algorithm for large scale scheduling problems. *Computers & Operations Research*, 13(1):27–32, 1986.
7. Kathryn A. Dowsland and Jonathan M. Thompson. Ant colony optimization for the examination scheduling problem. *Journal of the Operational Research Society*, 56(4):426–438, 2005.
8. Michael Garey, David Johnson, and Hing So. An application of graph coloring to printed circuit testing. *IEEE Transactions on Circuits and Systems*, 23(10):591–599, 1976.
9. Timir Maitra, Anindya J. Pal, Debnath Bhattacharyya, and Tai-hoon Kim. Noise reduction in VLSI circuits using modified Ga based graph coloring. *International Journal of Control, Automation and Systems*, 3(2):37–44, 2010.

10. S. Sen Sarma, Ruchira Mandal, and Abhijit Seth. Some sequential graph colouring algorithms for restricted channel routing. *International Journal of Electronics*, 77(1):81–93, 1994.
11. T.-K. Woo, Stanley Y.W. Su, and Richard Newman-Wolfe. Resource allocation in a dynamically partitionable bus network using a graph coloring algorithm. *IEEE Transactions on Communications*, 39(12):1794–1801, 1991.
12. Michael R. Garey and David S. Johnson. *Computers and Intractability*, vol. 174. Freeman, San Francisco, 1979.
13. Daniel Brélaz. New methods to color the vertices of a graph. *Communications of the ACM*, 22(4):251–256, 1979.
14. Frank Thomson Leighton. A graph coloring algorithm for large scheduling problems. *Journal of Research of the National Bureau of Standards*, 84(6):489–506, 1979.
15. Reza Abbasian and Malek Mouhoub. A hierarchical parallel genetic approach for the graph coloring problem. *Applied Intelligence*, 39(3):510–528, 2013.
16. Zhipeng Lü and Jin-Kao Hao. A memetic algorithm for graph coloring. *European Journal of Operational Research*, 203(1):241–250, 2010.
17. Karim Baiche, Yassine Meraihi, Manolo Dulva Hina, Amar Ramdane-Cherif, and Mohammed Mahseur. Solving graph coloring problem using an enhanced binary dragonfly algorithm. *International Journal of Swarm Intelligence Research (IJSIR)*, 10(3):23–45, 2019.
18. Ruxin Zhao, Yongli Wang, Chang Liu, Peng Hu, Hamed Jelodar, Mahdi Rabbani, and Hao Li. Discrete selfish herd optimizer for solving graph coloring problem. *Applied Intelligence*, 50(5): 1633–1656, 2020.
19. Majid Faraji and Haj Seyyed Javadi. Proposing a new algorithm based on bees behavior for solving graph coloring. *International Journal of Contemporary Mathematical Sciences*, 6(1):41–49, 2011.
20. Yassine Meraihi, Mohammed Mahseur, and Dalila Acheli. A modified binary crow search algorithm for solving the graph coloring problem. *International Journal of Applied Evolutionary Computation (IJAEC)*, 11(2):28–46, 2020.
21. Shadi Mahmoudi and Shahriar Lotfi. Modified cuckoo optimization algorithm (MCOA) to solve graph coloring problem. *Applied Soft Computing*, 33:48–64, 2015.
22. Claus Aranha, Keita Toda, and Hitoshi Kanoh. Solving the graph coloring problem using cuckoo search. *In International Conference on Swarm Intelligence*, pp. 552–560. Springer, Japan, 2017.
23. Yassine Meraihi, Amar Ramdane-Cherif, Mohammed Mahseur, and Dalila Achelia. A chaotic binary salp swarm algorithm for solving the graph coloring problem. *In International Symposium on Modelling and Implementation of Complex Systems*, pp. 106–118. Springer, Algeria, 2018.
24. Bchira Ben Mabrouk, Hamadi Hasni, and Zaher Mahjoub. On a parallel genetic–tabu search based algorithm for solving the graph colouring problem. *European Journal of Operational Research*, 197(3):1192–1201, 2009.
25. Sidi Mohamed Douiri and Souad Elbernoussi. Solving the graph coloring problem via hybrid genetic algorithms. *Journal of King Saud University-Engineering Sciences*, 27(1):114–118, 2015.

26. Halima Djelloul, Abdesslem Layeb, and Salim Chikhi. Quantum inspired cuckoo search algorithm for graph colouring problem. *International Journal of Bio-Inspired Computation*, 7(3):183–194, 2015.
27. Mohamed Amine Basmassi, Sidina Boudaakat, Lamia Benameur, Omar Bouattane, Ahmed Rebbani, and Jihane Alami Chentoufi. Hybrid genetic approach for solving fuzzy graph coloring problem. *Sensor Network Methodologies for Smart Applications*, pp. 54–64. IGI Global, 2020.
28. Tansel Dokeroglu and Ender Sevinc. Memetic teaching–learning-based optimization algorithms for large graph coloring problems. *Engineering Applications of Artificial Intelligence*, 102: 104282, 2021.
29. Seyedali Mirjalili. Moth-flame optimization algorithm: A novel nature-inspired heuristic paradigm. *Knowledge-Based Systems*, 89:228–249, 2015.
30. Shereen Said, Abdalla Mostafa, Essam H. Houssein, Aboul Ella Hassanien, and Hesham Hefny. Moth-flame optimization based segmentation for mri liver images. *In International Conference on Advanced Intelligent Systems and Informatics*, pp. 320–330. Springer, Egypt, 2017.
31. Yongquan Zhou, Xiao Yang, Ying Ling, and Jinzhong Zhang. Meta-heuristic moth swarm algorithm for multilevel thresholding image segmentation. *Multimedia Tools and Applications*, 77(18):23699–23727, 2018.
32. Rebecca Ng Shin Mei, Mohd Herwan Sulaiman, Zuriani Mustaffa, and Hamdan Daniyal. Optimal reactive power dispatch solution by loss minimization using moth-flame optimization technique. *Applied Soft Computing*, 59:210–222, 2017.
33. P. Anbarasan and T. Jayabarathi. Optimal reactive power dispatch using moth-flame optimization algorithm. *International Journal of Applied Engineering Research*, 12(13):3690–3701, 2017.
34. Puja Singh and Shashi Prakash. Optical network unit placement in fiber-wireless (fiwi) access network by whale optimization algorithm. *Optical Fiber Technology*, 52:101965, 2019.
35. Hossam Faris, Ibrahim Aljarah, and Seyedali Mirjalili. Evolving radial basis function networks using moth–flame optimizer. In: *Handbook of Neural Computation*, pp. 537–550. Elsevier: Amsterdam, Netherlands, 2017.
36. More Raju, Lalit Chandra Saikia, and Debdeep Saha. Automatic generation control in competitive market conditions with moth-flame optimization based cascade controller. *In 2016 IEEE Region 10 Conference (TENCON)*, pp. 734–738. IEEE, Japan, 2016.
37. D.A. Yousri, Amr M. AbdelAty, Lobna A. Said, Ahmed Abo Bakr, and Ahmed G. Radwan. Biological inspired optimization algorithms for cole-impedance parameters identification. *AEU-International Journal of Electronics and Communications*, 78:79–89, 2017.
38. Indrajit N. Trivedi, Pradeep Jangir, Siddharth A. Parmar, and Narottam Jangir. Optimal power flow with voltage stability improvement and loss reduction in power system using moth-flame optimizer. *Neural Computing and Applications*, 30(6):1889–1904, 2018.
39. L.N. Huang, Bo Yang, X.S. Zhang, L.F. Yin, Tao Yu, and Z.H. Fang. Optimal power tracking of doubly fed induction generator-based wind turbine using swarm moth–flame optimizer. *Transactions of the Institute of Measurement and Control*, 41(6):1491–1503, 2019.

40. B.V.S. Acharyulu, Banaja Mohanty, and P.K. Hota. Comparative performance analysis of PID controller with filter for automatic generation control with moth-flame optimization algorithm. *In Applications of Artificial Intelligence Techniques in Engineering*, pp. 509–518. Springer, 2019.
41. Ahmed A. Ewees, Ahmed T. Sahlol, and Mohamed A. Amasha. A bio-inspired moth-flame optimization algorithm for Arabic handwritten letter recognition. *In 2017 International Conference on Control, Artificial Intelligence, Robotics & Optimization (ICCAIRO)*, pp. 154–159. IEEE, 2017.
42. G.M. Soliman, M.M. Khorshid, and T.H. Abou-El-Enien. Modified moth-flame optimization algorithms for terrorism prediction. *International Journal of Application or Innovation in Engineering and Management*, 5(7):47–58, 2016.
43. Iyad Tumar, Yousef Hassouneh, Hamza Turabieh, and Thaer Thaher. Enhanced binary moth flame optimization as a feature selection algorithm to predict software fault prediction. *IEEE Access*, 8:8041–8055, 2020.
44. Srikanth Reddy, Lokesh Kumar Panwar, Bijaya K. Panigrahi, and Rajesh Kumar. Solution to unit commitment in power system operation planning using binary coded modified moth flame optimization algorithm (BMMFOA): A flame selection based computational technique. *Journal of Computational Science*, 25:298–317, 2018.
45. Seyedali Mirjalili and Andrew Lewis. S-shaped versus v-shaped transfer functions for binary particle swarm optimization. *Swarm and Evolutionary Computation*, 9:1–14, 2013.
46. Shahrzad Saremi, Seyedali Mirjalili, and Andrew Lewis. How important is a transfer function in discrete heuristic algorithms. *Neural Computing and Applications*, 26(3):625–640, 2015.
47. Yassine Meraihi, Dalila Acheli, and Amar Ramdane-Cherif. Qos multicast routing for wireless mesh network based on a modified binary bat algorithm. *Neural Computing and Applications*, 31(7):3057–3073, 2019.
48. David S. Johnson and Michael A. Trick. *Cliques, Coloring, and Satisfiability: Second DIMACS Implementation Challenge*, vol. 26. American Mathematical Society, Providence, Rhode Island, 1996.

CHAPTER 4

Evolving Deep Neural Network by Customized Moth-Flame Optimization Algorithm for Underwater Targets Recognition

Mohammad Khishe

Imam Khomeini Marine Science University

Mokhtar Mohammadi

Lebanese French University

Tarik A. Rashid

University of Kurdistan Hewler

Hoger Mahmud

University of Human Development

Seyedali Mirjalili

Torrens University
Yonsei University

DOI: 10.1201/9781003205326-5

CONTENTS

4.1 Introduction 54
4.2 Methods 57
4.2.1 Deep Neural Network (DNN) Architectures 57
4.2.2 Convolutional Neural Networks 58
4.2.3 Sonar Dataset 59
4.2.4 MFO Algorithm 61
4.3 Methodology 64
4.3.1 Customized MFO 64
4.3.2 Presentation of Searching Agents 64
4.3.3 Loss Function 65
4.4 Simulation Results and Discussion 66
4.4.1 The Analysis of Time Complexity 70
4.4.2 Sensitivity Analysis of Designed Model 71
4.5 Conclusions 72
References 73

4.1 INTRODUCTION

Underwater target classification is a complex and challenging process that has recently been the subject of many studies [1–3]. A practical classification task requires careful consideration for mechanisms that can differentiate between target and non-target object echoes as well as the echoes of objects that lurk in the background; classifications can also be affected by dependencies such as operation mode and engine procedures. Moreover, according to Ref. [3], the phase and amplitude of the acoustic signal received from objects can be altered by signal characteristics such as the temporal variation of propagation channel, dispersion, and ambient noises. To address the challenge, researchers have proposed two main classification methods, which are deterministic and stochastic. The first method utilizes statistical processing, oceanography, and models generated by sonar technologies to perform the classification [4]. The second method makes use of artificial intelligence to extract underwater features and predict and estimate underwater phenomena [5]. Deterministic classification calculates the statistical distribution of several parameters obtained from sources such as environmental conditions, sound propagations, topographic maps, and background clutters [6]. However, the authors of Ref. [7] have stated several disadvantages that may affect the method's applicability, such as the high cost of required equipment and human resource and the lack of guarantee in terms of robustness and generality. The stochastic approach focuses on cost and complexity reduction in sonar target classification using Intelligent Machine Hearing (IMH) to locate and identify acoustic signals coming from underwater objects [8].

Despite the excellent accuracy, the IMH-based methods are tedious because they first divide a problem into several sub-sections and then accumulate the result. Besides, IMH cannot handle big datasets, resulting in low accuracy. Moreover, canonical IMHs may not be able to learn clutters that are backscattered so they rely more on the objects' features than employing the whole background data [9].

These shortcomings can be addressed using Deep Learning (DL) approaches, which are end-to-end models efficiently working with enormous data [10]. However, although DL's outstanding features enable it to resolve different learning issues, training is challenging [11]. By reviewing the literature, it can be realized that eminent algorithms for training DL are Gradient-descendent [12], conjugate gradient algorithm [13], Hessian-free optimization approach [14], and Krylov subspace descent algorithm [15].

Although stochastic gradient descent-based training algorithms have simple structures and fast convergence rates for large training samples, gradient descent-based approaches need numerous manual parameter tuning for optimality. Their design is also sequential, which makes parallelizing with Graphics Processing Units (GPUs) challenging. Simultaneously, though conjugate gradient methods are stable for training, they are almost slow, demanding multiple CPUs and many memory resources [13]. Deep auto-encoders (AEs) utilized Hessian-free optimization for training the weights of typical Convolutional Neural Network (CNN) [14], which is more effective in pre-training and fine-tuning deep AEs than the model introduced by Hinton and Salakhutdinov [16]. On another note, the Krylov subspace descent algorithm is more stable and more straightforward than Hessian-free optimization. Thereby, it presents a more accurate classification rate and optimization speed rather than Hessian-free optimization. On the contrary, the Krylov subspace descent algorithm needs more memory than Hessian-free optimization [16].

Recently, evolutionary and Metaheuristics algorithms have been widely applied to optimize different engineering optimization tasks [17]. Nevertheless, research studies on the Metaheuristics algorithm for optimizing DL models require more studies. The hybrid Genetic Algorithm (GA) and CNN, introduced in Ref. [18], was the first study that initiates this kind of optimization model using Metaheuristics optimization algorithms. This model selects the CNN parameters utilizing the GA's cross-over and mutation operation, in which the structure of DCNN is modeled as a GA's chromosome. Besides, in the cross-over phase, the first convolution layer's biases and weights (C_1) and the third convolution layer (C_3) are utilized as chromosomes. Reference [19] proposed an evolutionary fine-tuning method to optimize the parameters of a DCNN using the Harmony Search (HS) algorithm and some of its improved versions in the field of handwritten digit and fingerprint recognition. To avoid local minima and improve performance, the authors of Ref. [20] combined back-propagation with GSA and Particle Swarm Optimization (PSO) algorithms to develop a new CNN algorithm. To improve canonical PSO's exploration capability, the authors of Ref. [21] proposed cPSO-CNN, a variant of PSO, to enhance architecture-determined CNNs in terms of hyper-parameter configuration. Their reported results show that the cPSO-CNN enhances hyper-parameter configuration and reduces overall computation costs compared to several proposed algorithms in the literature.

The hybrid tabu-genetic algorithm [22], gray wolf optimizer [23], simulated annealing and differential evolution algorithms [24], micro-canonical annealing algorithm [25], improved cuckoo search algorithm [26], whale optimization algorithm [27], etc., are other novel Metaheuristics and nature-inspired algorithms, which have been used to optimize the CNN's performance. However, according to the theory of No Free Lunch (NFL) [28], a comprehensive optimization algorithm capable of addressing all optimization issues is yet to be developed, leading to the utilization of nature-inspired algorithms to develop novel and more capable algorithms by researchers. Following their directions, this chapter proposes the application of Moth-Flame Optimization (MFO) algorithm which is a variant of nature-inspired Metaheuristics algorithm to improve DCNN classification performed on datasets obtained from underwater sonar equipment.

MFO algorithm is developed based on navigation paths that moths carve in nature. Researchers' experimental results show that MFO is highly effective for avoiding local minima [29–31] and MLP and RBF training [29, 30]. The results also show that MFO produces excellent results in the exploration phase; however, its performance is not equally excellent in the exploitation phase of sonar dataset search spaces. This chapter modifies MFO with spiral motions to optimize the algorithm for sonar dataset classification and DCNNs training.

To evaluate the effectiveness of the modified MFO, three underwater acoustic datasets, including synthetic, Sejnowski & Gorman's dataset [31], passive sonar dataset [32], and active sonar dataset [33], are exploited. To ensure a thorough comparison, the efficiency of the designed model is compared with ten recently proposed benchmark models.

This chapter is organized as follows: Background materials, including DCNN, sonar datasets, and MFO, will be reviewed in Section 4.2. Section 4.3 introduces the customized MFO and its fine-tuning scheme. Section 4.4 represents the simulation results and discussion, and conclusions are finally described in Section 4.5.

4.2 METHODS

This section provides background information, such as the DCCN model, underwater sonar datasets, and MFO.

4.2.1 Deep Neural Network (DNN) Architectures

Neurons connect and collaborate to accomplish intelligence functionalities known as Neural Network (NN). DNN is a form of NN which is classified into discriminative and generative categories. The discriminative category utilizes a bottom-up approach to generate labels for hidden classes of data using maximum calculated similarities among them. The second category utilizes a top-down method to create a model for hidden data classes based on the distribution of observable classes.

DNN may be classified further into four general classes based on architectural properties as presented in Figure 4.1: Unsupervised Pre-trained Networks (UPNs) [34], CNNs [35], Recurrent Neural Networks (RNNs) [36], and Recursive Neural Networks. UPNs include AEs [37], Restricted Boltzmann Machines (RBMs) [38], Generative Adversarial Networks

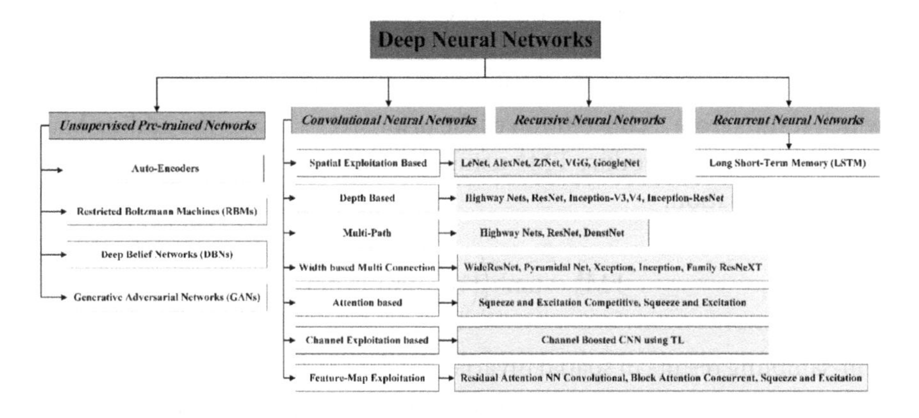

FIGURE 4.1 A general block diagram for DNN architectures.

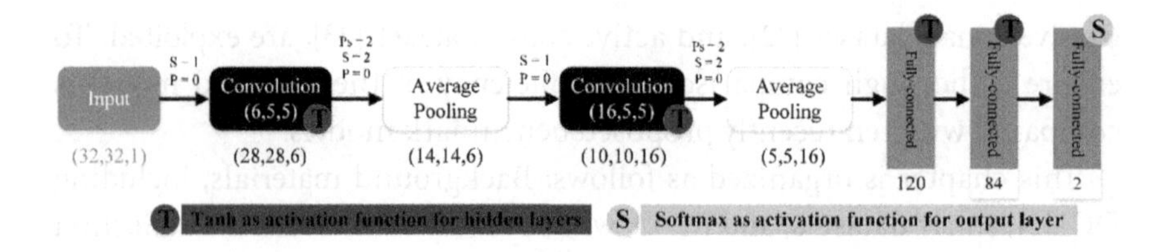

FIGURE 4.2 The LeNet-5 DCNN.

(GANs) [39], and Deep Belief Networks (DBNs) [40]. Long Short-Term Memory (LSTM) [41] is also an implementation of RNNs.

4.2.2 Convolutional Neural Networks

One of the most widely used DNN models in studies related to data structures in grids such as images is CNN. The model consists of several layers: dense, dropout, reshape, flatten, convolution, and pooling. In the model, calculation and transformation tasks are performed independently and passed on to the next layer. The dimension of the hidden layer specifies the model's width, and the depth of the layer determines the length of the model. Figure 4.2 shows the feature map of a CNN model (LeNet-5 DCNN) comprised of two convolution and pooling layers [42]. Each CNN layer performs some specific tasks; for instance, the convolution layer uses probability density function to weight elements in a data matrix singly. Layers are connected through neurons; for instance, convolution layer neurons are connected to the receptive points of the previous layer; any neurons with matching Feature Maps (FMs) can receive data from multiple input neurons until the input is wholly swiped to share identical connection weights. The pooling layers down-sample the FMs by a factor of 2. For example, in layer (C_3), the FM of size 12×12 is sub-sampled to the FM of size 6×6 in the following layer (S_4). The structure is presented in Figure 4.2 in which linear filter and kernels are used to calculate each FM, which results from a convolution of the prior layer's maps. The adding bias *bk* and weights w^k produce FM_{ij}^k using the *tanh* function as shown in the following equation:

$$FM_{ij}^k = \tanh\left(\left(W^k \times x\right)_{ij} + b_k\right) \tag{4.1}$$

Sub-sampling results in spatial invariance when FMs resolution is reduced, and each FM refers to the prior layer. The following equation defines the pooling model:

$$\alpha_j = \tanh\left(\beta \sum_{N \times N} \alpha_i^{n \times n} + b\right) \tag{4.2}$$

where β is a trainable scalar, $\alpha_i^{n \times n}$ are the inputs, and b is a bias. The last layer, which is the fully connected layer, performs the classification and comes after several convolution and sub-sampling layers. It should be noted that for each output class, there is one neuron.

4.2.3 Sonar Dataset

For the aim of having a complete inquiry, the designed classifiers are tested against three underwater sonar datasets as follows:

- Sejnowski & Gorman's Underwater Sonar Dataset [31]: This freely available dataset is produced in Sejnowski & Gorman's experiment [31], which is used in this chapter to evaluate the suggested classifier in comparison to other classifiers presented in the literature. Their experiment collected data regarding two echoes coming from a 5 feet metal cylinder and a rock with the same property. The metal cylinder served the purpose of a real target, and the rock served the purpose of the false target. In order to collect the data, a wideband linear modulated chirp ping is aimed toward the objects, which are positioned on a sandy bottom. During the experiment, Sejnowski & Gorman's collected 1,200 echoes from which 208 were selected, out of which 111 echoes belong to the metal cylinder and 97 are related to the same sized rock. In selecting the 208 echoes, 4–15 dB Signal to Noise Ratio (SNR) is used as the threshold level.
- Passive Sonar Dataset [32]: The authors of Ref. [32] generated a new passive underwater sonar dataset from their experiment in which they used different water velocities and densities to test seven different types of propellers at various Revolutions Per Minute (RPM). The propellers were laboratory samples that were the same as the originals, and the experiment was carried out in a testing tunnel model B&K_8103. Figure 4.3 shows a typical propeller noise sample, Fourier transform, hydrophone noise level, and Power Spectral Density (PSD) estimate. In order to test the proposed classifier in this chapter, the spectrogram of the input signal is required. Figure 4.4 shows a sample of a typical spectrogram prior and post noise removal phase.

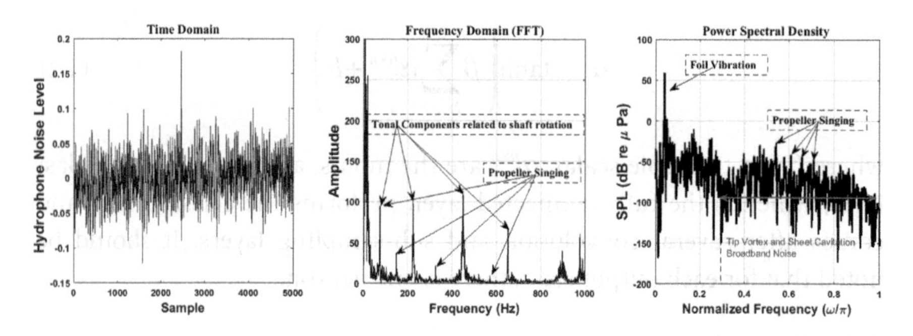

FIGURE 4.3 A typical hydrophone received propeller sound, Fourier transform, and its PSD.

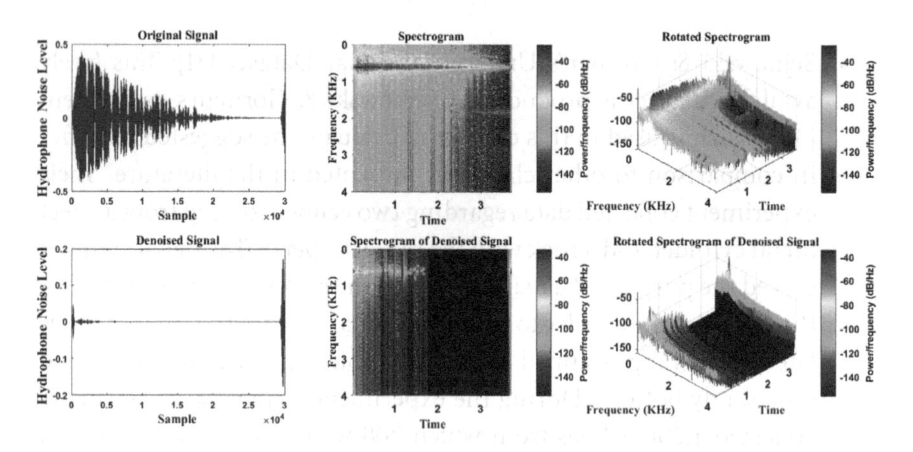

FIGURE 4.4 A typical spectrogram of radiated noise and its denoised spectrogram.

- Active Sonar Dataset [33]: This dataset is generated by Khishe et al. in an experiment they carried out for four targets and two non-target objects placed in a sandy seabed. They used a multi-task sonobuoy owned by the Port and Maritime Organization to collect the data from the objects. In order to collect the data, they rotated the objects 180° with 1° precision using an electric motor. The transmitted signal toward the objects covered a frequency range of 5–110 Hz and used a wideband linear frequency modulated ping signal. The process requires complex calculations as the amount of raw data obtained is large, and to lessen the complexity, the authors have relied on a detection process presented in Ref. [43]. A sample of typical echoes received from the target and non-target objects is shown in Figure 4.5. In Ref. [43], the authors have also described a feature extraction

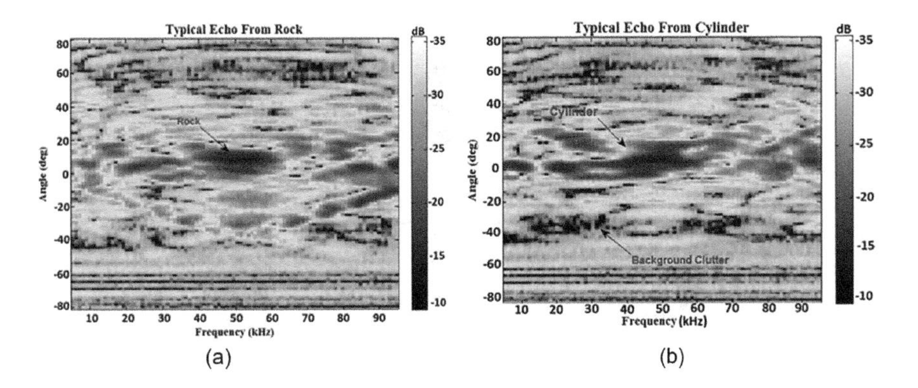

FIGURE 4.5 Typical received echoes belong to a target and a non-target: (a) from rock and (b) from cylinder.

TABLE 4.1 The Description of the Utilized Datasets

Datasets	**Sejnowski & Gorman**	**Active Sonar Dataset**	**Passive Sonar Dataset**
No. training samples (after data augmentation)	150 (1,200)	200 (1,600)	400 (3,200)
No. test samples (after data augmentation)	58 (464)	150 (1,200)	250 (2,000)
No. classes	2	6	7

process to put together the dataset, such as the one presented in Table 4.1, which includes the number of samples, training and test instances, and classes.

It is worth stating that Sejnowski & Gorman sonar dataset is selected to be used as a benchmark to compare the proposed algorithm with others in the literature, while the passive and active datasets presented in [32–34] are selected to be used as a real-world practical experiment.

4.2.4 MFO Algorithm

MFO is a population-based nature-inspired algorithm introduced by Seyedali Mirjalili in Ref. [44]. It is inspired by the way moths move at night in which they keep a constant angle with the moonlight in order to stay on a straight path; this approach is known as traversing orientation for navigation. MFO is mathematically modeled using the movement of moths in nature when subjected to artificial light. In such a case, moths converge with the artificial light and proceed in a closed spiral motion, as pictured in Figure 4.6.

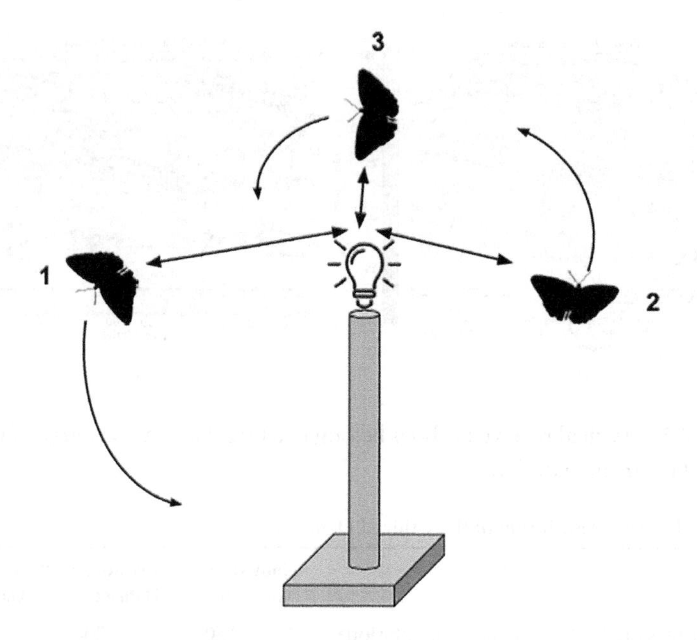

FIGURE 4.6 Diagrammatic illustration of moths' spiral movement when exposed to close artificial light source (see [44]).

It is clear by now that moth and flame are the two main components of MFO; the moth represents the search agent and the position of the moth concerning the light source represents problem variables. Thus the targeted optimization problem's dimension is the length of the position; based on that, the moths' population is expressed in the following equation [44]:

$$\text{Position} = \begin{bmatrix} \text{moth}_{11} & \text{moth}_{12} & \cdots & \text{moth}_{1\text{dim}} \\ \text{moth}_{21} & \text{moth}_{22} & \cdots & \text{moth}_{2\text{dim}} \\ \vdots & \vdots & \vdots & \vdots \\ \text{moth}_{n1} & \text{moth}_{n2} & \cdots & \text{moth}_{n\text{dim}} \end{bmatrix} \tag{4.3}$$

In the equation, the moths' number in a population is represented by n and the number of variables in a problem is represented by *dim*. The algorithm utilizes a fitness function to investigate each moth's fitness in a population and uses a matrix to store the obtained fitness value. For this research, let us call the matrix Outcome, which can be formulated as Equation (4.4) [44], and $Outmoth_n$ represents the fitness value of the moth number n:

$$\text{Outcome} = \begin{bmatrix} \text{Outmoth}_1 \\ \text{Outmoth}_2 \\ \vdots \\ \text{Outmoth}_n \end{bmatrix} \tag{4.4}$$

As flames are the second component, MFO preserves an equal number of flames to that of the number of moths in a population. Flames are also described as a matrix with the same property of Equation (4.3). Flames indicate the best position each moth has obtained, and the fitness value for each flame is stored in the same manner as the population *Position* in Equation (4.4).

MFO creates a random population of moths through its repeated search process, and moths in a population memorize their position in the form of a flame. The algorithm then performs a fitness evaluation for each moth and stores its fitness value. The moths' positions and their flame values are updated in each repeated search process, and the variable lower and upper bound that represents the solution is checked after each update. The updating process ends when the maximum number of iteration is achieved.

Equation (4.5) presented in Ref. [44] is used to calculate each moth's position based on a logarithmic spiral. The distance between a moth (represented as $moth_i$) and its flame (represented as $flame_j$) is calculated and stored as D_{ij} using Equation (4.6) [44]. In the following equation, q represents the logarithmic spiral shape and t denotes a random number in the range of $[r, 1]$:

$$F(\text{moth}_i, \text{flame}_j) = D_{ij} \cdot e^{qt} \cos(2 \times \pi \times t) + \text{flame}_j \tag{4.5}$$

$$D_{ij} = |\,\text{flame}_j - \text{moth}_i\,| \tag{4.6}$$

As explained in Ref. [44], to preserve the exploitation capability of MFO, Equation (4.7) is used to reduce the number of flames (represented as $flames_{No}$) as the number of iterations increases. Thus, in the following equation, the current iteration number is represented as l, and the flames' maximum number and iteration are represented as n and T, respectively:

$$\text{flames}_{No} = \text{round}\left(n - 1 \times \frac{n-1}{T} \right) \tag{4.7}$$

4.3 METHODOLOGY

In the following, the concepts of the customized MFO, the presentation of search agents, and the loss function are described.

4.3.1 Customized MFO

As stated, there is a gap in the literature when it comes to selecting a suitable spiral shape for investigating the aptitude of MFO in the exploitation phase. In this regard, this section investigates several renowned spiral shapes. Generally speaking, a spiral shape is formed when a line is circling outward around a center point, and they come in different types and shapes. Table 4.2 presents six renowned types of spirals and their mathematical models presented by the authors of Refs. [3, 45], and Figure 4.7 shows the two- and three-dimensional view of each of the six spirals.

4.3.2 Presentation of Searching Agents

In order to use an optimization algorithm in the DCNN fine-tuning task, two conditions have to be fulfilled: first, the search agent must represent the parameters of the structure distinctly, and second, the investigated problem must be the base for defining the fitness function. Furthermore, to use MFO in fine-tuning DCNN setup to achieve the maximum detection accuracy, network parameters such as weights and biases must be presented clearly for the fully connected layers.

In such a case, the weight and bias of the last layer are optimized by MFO, and the loss function then calculates the fitness function. Furthermore, MFO uses the bias and connection weight values as moths. In general, binary-based, vector-based, and matrix-based [24] are the schemes used to present a DCNN weight and biases as the MFO possible solution. Since

TABLE 4.2 The Utilized Spirals and Their Mathematical Models

Row	Spirals	Equation
1	Archimedean	$r=a\times\theta$
2	Logarithmic	$\log(r)=a\times\theta$
3	Fermat	$r^2=a^2\times\theta$
4	Lituus	$r^2=a^2/\theta$
5	Equiangular	$\mathrm{Ln}(r)=a\times\theta$
6	Random	$r=\mathrm{rand}()\times\theta$

r represents the radius of spirals, θ denotes the polar angle, and a is any constant [3, 45].

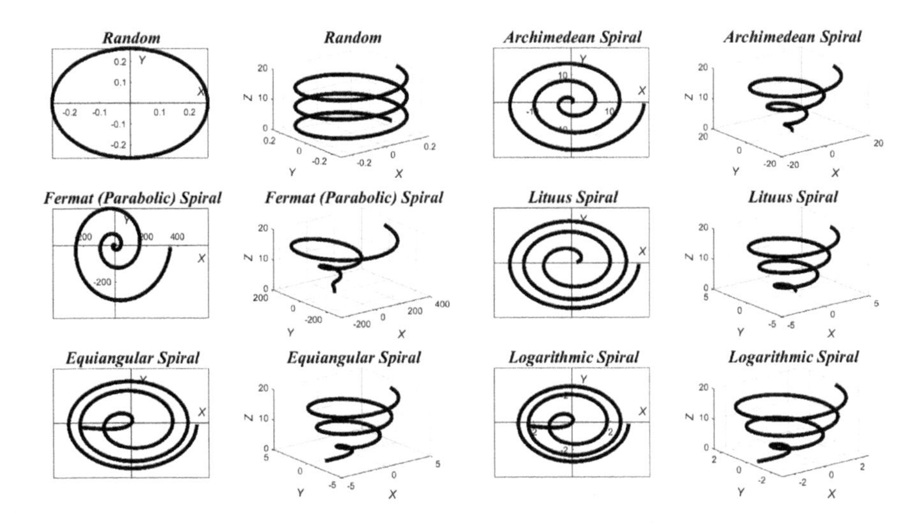

FIGURE 4.7 The illustrations of various spiral motions.

MFO requires the parameters in vector-based form, the following equation is used in this chapter to represent the possible solutions:

$$\text{Moths} = \left[W_{11}, W_{12}, \ldots, W_{nh}, b_1, \ldots, b_h, M_{11}, \ldots, M_{hm}\right] \tag{4.8}$$

In this equation, W_{ij} denotes the connection weight from the i_{th} input neuron to the j_{th} hidden node, n indicates the input neurons' number, b_j shows the j_{th} bias of hidden neuron, and Mjo denotes the connection weight between the j_{th} hidden neuron and the o_{th} output neuron. As previously stated, the proposed architecture is a simple LeNet-5 structure. The convolution layer has a kernel size of 5×5 and the down-sampling is performed by a factor of two.

4.3.3 Loss Function

The main objectives of using MFO to train DCNN are to minimize classification errors and acquire maximum accuracy. The Mean Square Error (MSE) is usually considered as loss function in classification tasks; in this regard, the following Equation shows the loss function as the fitness function for DCNN-MFO:

$$y = \frac{1}{2}\sqrt{\frac{\sum_{i=0}^{N}(o-d)^2}{N}} \tag{4.9}$$

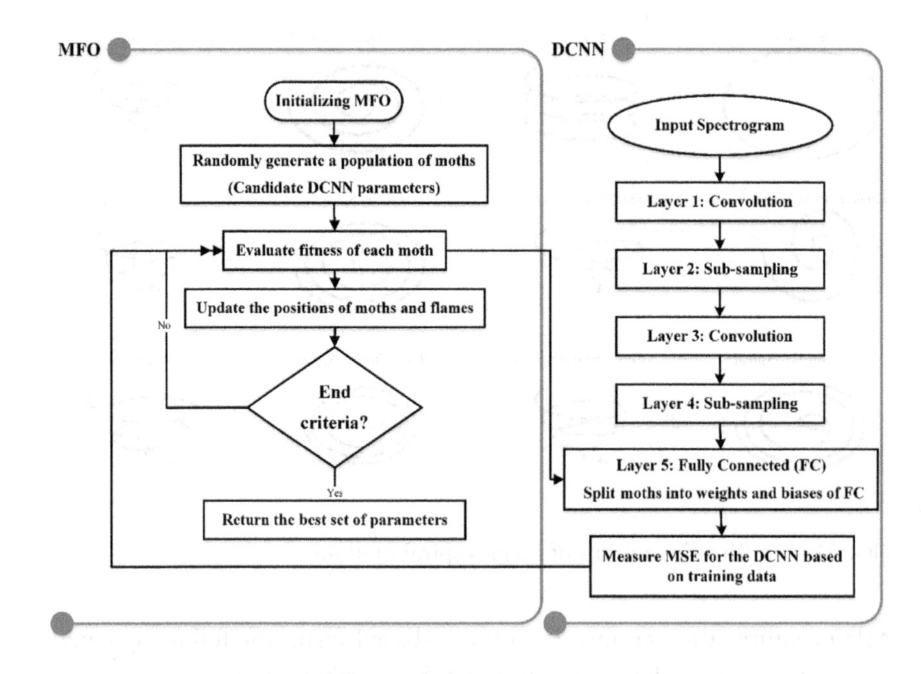

FIGURE 4.8 The DCNN-MFO's flowchart.

In the equation, the calculated output is represented by o, the desired output is represented by d, and the training sample is represented by N. The MFO uses maximum iteration reached or specified loss function as termination criteria. Figure 4.8 shows the flowchart of the designed DCNN-MFO.

4.4 SIMULATION RESULTS AND DISCUSSION

In order to examine the performance of the proposed DCNN-MFO, the datasets presented in Table 4.1 are used and grouped as Sejnowski & Gorman, active, and passive to identify the best spiral shape for enhancing the performance of the classifier. The results of the examination are compared with the results of MFO [44], PSO [46], Ant Lion Optimization (ALO) [47], Stochastic Fractals Search (SFS) [48], Heap-Based Optimizer (HBO) [49], and Chimp Optimization Algorithm (ChOA) [50] for validation purposes. Table 4.3 presents the required primary parameters and values for each of the algorithms. Figures 4.9–4.12 show the convergence curves and MSE boxplot after running the algorithms 20 times for each sonar datasets. MATLAB® on a PC with 16 GB RAM and 3.8 GHz is used to run the algorithms. The results show that no compared algorithms can continue after 500 iterations; Tables 4.4–4.6 present the statistical results of the examination.

TABLE 4.3 The Initialization for Comparison Algorithms

Algorithms	Parameters	Value
PSO	C_1, C_2	2
	Population size	50
ALO	w	[2,6]
	k	500
ChOA	m	Chaotic
	r_1, r_2	Random [0,1]
	Number of chimps	50
HBO	$C; p_1; p_2$	From corresponding equations [49]
	N	50
SFS	Maximum Diffusion Number (MDN)	1
MFO	Number of moths	50

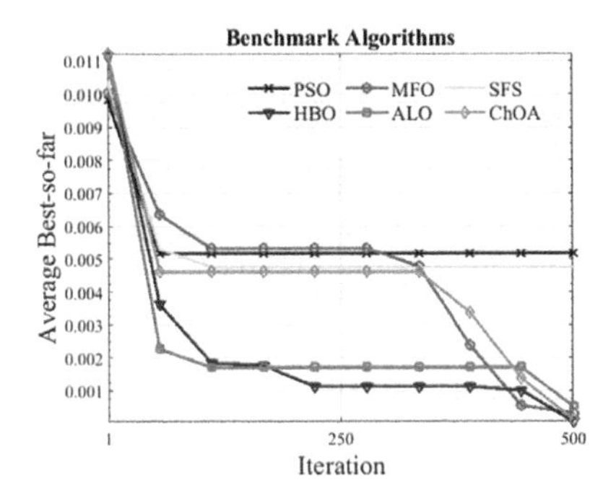

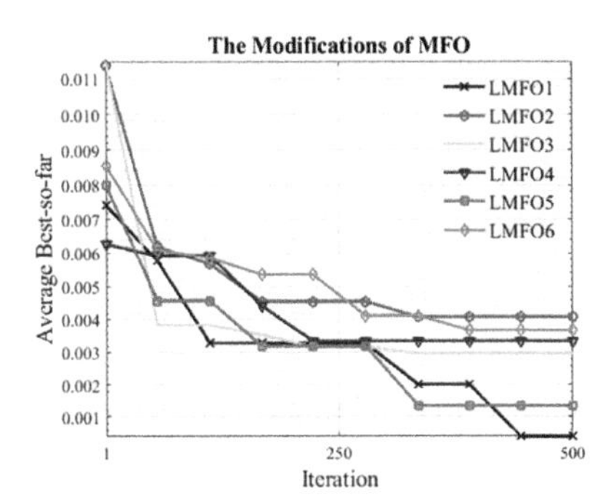

FIGURE 4.9 Convergence curves for Sejnowski & Gorman dataset.

Table 4.4 presents the modified MFO naming for comparison purposes. The Average (AVE) ± Standard Deviation (STD) values must be lower to avoid local minima and achieve a better capability. Tables 4.5–4.10 show each algorithm's calculated AVE and STD; the bolded results indicate the best results achieved. The authors of Ref. [51] state the importance of statistical tests, including AVE±STD, in a balanced performance evaluation of optimization algorithms. The tables also show the significant statistical values ($P \leq 0.05$) computed per Wilcoxon's rank-sum test for each benchmark algorithm and its achieved classification rates. A significant P value provides a statistical indication of whether to accept or reject a null hypothesis; for a hypothesis to be accepted, its computed P value should be ≤ 0.05; otherwise, it is rejected. In the tables, N/A indicates that the determined algorithm cannot be compared with itself (i.e., Not Applicable).

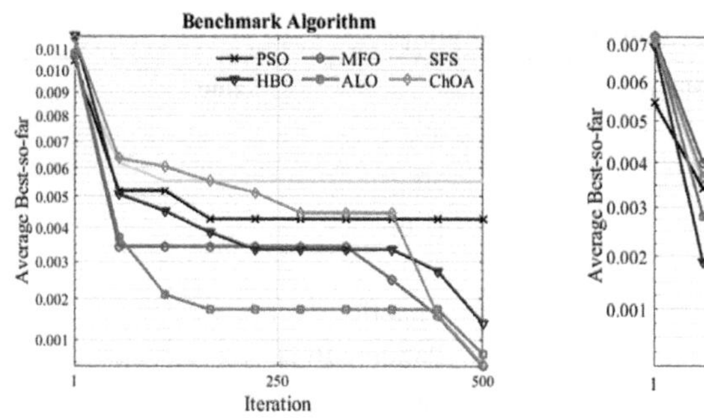

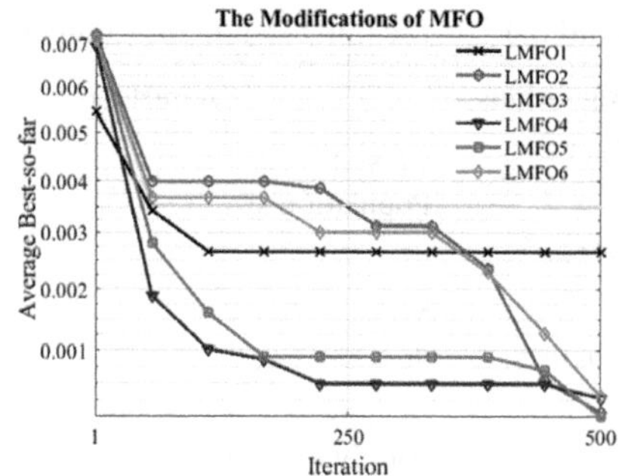

FIGURE 4.10 Convergence curves for passive dataset.

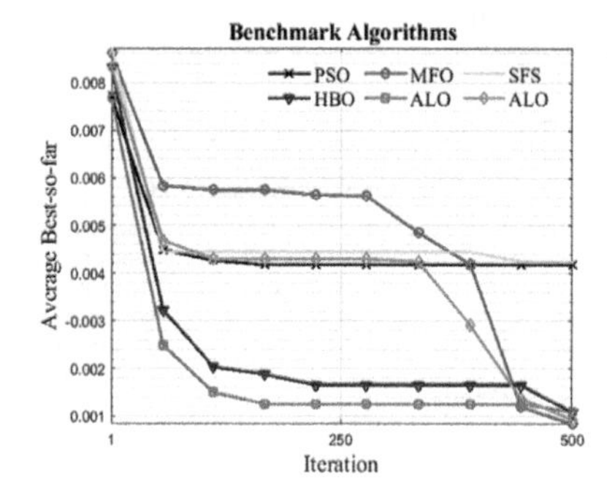

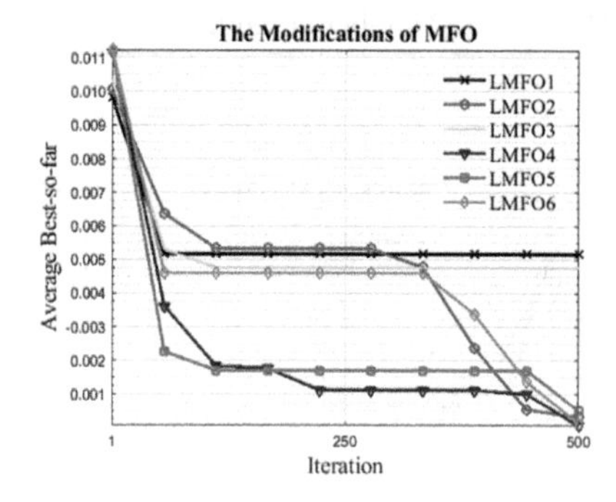

FIGURE 4.11 Comparison of convergence curve for active dataset.

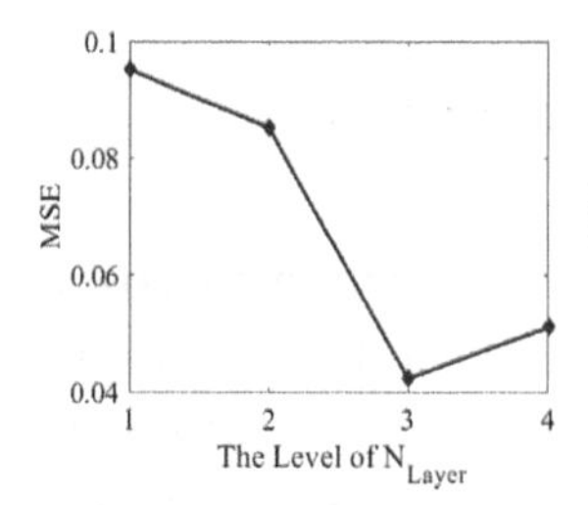

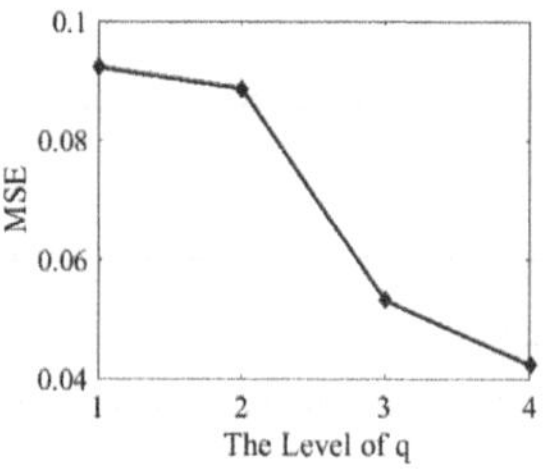

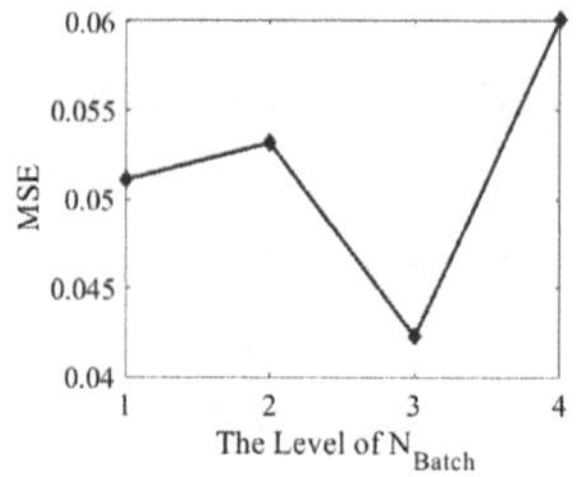

FIGURE 4.12 Level trends of the analyzed parameters.

Results presented in Tables 4.5 and 4.6 and Figure 4.9 show that among the benchmark algorithms, LMFO has achieved the best results among comparison algorithms, and also among the modified algorithms, LMFO1 and LMFO5 have achieved better results for the Sejnowski & Gorman dataset. As for the best spiral shape, the results show that Equiangular and Archimedean spiral can enhance the performance of the MFO-DCNN. The passive dataset results presented in Tables 4.5 and 4.8 and Figure 4.11

TABLE 4.4 The Naming Style for the Modification of MFO

Spiral Types	Archimedean	Logarithmic	Fermat	Lituus	Equiangular	Random
Name	LMFO1	LMFO2	LMFO3	LMFO4	LMFO5	LMFO6

TABLE 4.5 Results for Sejnowski & Gorman Dataset (Benchmark Algorithm)

Algorithms	PSO	ALO	ChOA	HBO	SFS	MFO
MSE (AVE)	0.0051	0.00051	0.00062	0.00096	0.0051	0.00045
MSE (STD)	0.000075	0.000021	0.0000002	0.000052	0.00014	0.000052
P-values	9.28E-18	1.33E-12	N/A	0.00021	0.00052	0.065
Classification rate (%)	93.0214	94.5412	97.2548	96.3254	93.2177	96.9852

TABLE 4.6 Results of the Modifications of MFO for Sejnowski & Gorman Dataset

Algorithms	LMFO1	LMFO2	LMFO3	LMFO4	LMFO5	LMFO6
MSE (AVE)	0.00058	0.0048	0.0031	0.0033	0.0018	0.0044
MSE (STD)	3.73E-11	0.0011	0.000035	0.000045	0.00022	0.0009
P-values	N/A	1.44E-09	0.00052	0.00019	0.066	0.0027
Classification rate (%)	98.8527	91.5018	95.2005	95.0011	97.2177	92.3344

TABLE 4.7 Results for Passive Dataset (Benchmark Algorithm)

Algorithms	PSO	ALO	ChOA	HBO	SFS	MFO
MSE (AVE)	0.0041	0.00079	0.00073	0.0012	0.0052	0.00074
MSE (STD)	1.75E-07	2.01E-08	1.47E-07	0.00033	0.00013	0.000001
P-values	0.000007	0.00031	N/A	0.00072	0.00048	0.122
Classification rate (%)	92.4414	97.8520	98.0014	96.5237	91.9999	98.0009

lead us to similar conclusions as MFO, and Equiangular spiral shapes have achieved the best results. The Equiangular and Archimedean shapes presented in Figure 4.7 can converge faster than the rest and they have a distinctive exploitation ability; these are the reasons why these shapes have a more significant improvement with impact on the performance of MFO. The convergence curves of the algorithms are shown in Figures 4.9–4.11, and the graphs show that Equiangular and Lituus spiral shapes have the best convergence rates among all shapes for the selected datasets in descending order.

Contrary to the two previous datasets, MFO and LMFO4 provide the best results on benchmark algorithms and improved versions of MFO for the active sonar dataset. These results were predictable because the active sonar contains inhomogeneous phenomena such as false alarm rate and

TABLE 4.8 Results of the Modifications of MFO for Passive Dataset

Algorithms	LMFO1	LMFO2	LMFO3	LMFO4	LMFO5	LMFO6
MSE (AVE)	0.0028	0.00046	0.0034	0.000055	0.00045	0.00058
MSE (STD)	0.0004	1.61E-09	0.000023	0.000005	1.51E-09	1.12E-07
P-values	0.00024	4.22E-07	0.00042	0.00033	N/A	3.24E-06
Classification rate (%)	96.3324	98.6302	95.9865	98.0321	98.9998	98.0044

TABLE 4.9 Results for Active Dataset (Benchmark Algorithm).

Algorithms	PSO	ALO	ChOA	HBO	SFS	MFO
MSE (AVE)	0.0042	0.00110	0.00092	0.00112	0.0045	0.0009
MSE (STD)	0.000037	0.00075	0.00071	0.00073	0.00029	0.00033
P-values	0.0054	1.82E-08	0.081	0.00002	0.00085	N/A
Classification rate (%)	92.9004	94.5023	95.9743	94.5284	90.0028	96.0002

TABLE 4.10 Results of the Modifications of MFO for Active Dataset

Algorithms	LMFO1	LMFO2	LMFO3	LMFO4	LMFO5	LMFO6
MSE (AVE)	0.0050	0.0009	0.0048	0.0008	0.0018	0.0009
MSE (STD)	0.00012	0.00002	0.000022	2.66E-12	0.00022	0.000001
P-values	0.022	0.081	0.0023	N/A	0.066	0.085
Classification rate (%)	94.9937	97.4448	95.2117	98.0881	97.2177	98.0004

clutter; therefore, this dataset has a different nature; thereby, different algorithms present better results than other datasets.

Finally, from the results presented, it can be noted that the Equiangular spiral shape can converge better and boost the search performance of the MFO, especially in the exploitation phase. This means the searching capability of the algorithm is not easily overcome by local minima on the one hand and reaches global minima in less time on the other.

4.4.1 The Analysis of Time Complexity

Time complexity measurement is an essential step in analyzing classifiers in real-time in the sonar signal processing domain. In order to measure the time complexity of the MFO-DCNN, it is implemented and used to test a dataset containing 3,364 images. The implementation was carried out in a device that had Tesla K20 as the GPU and an Intel Core i7–7700HQ up to a 3.8 GHz processor as the CPU. The result of the test is presented in Table 4.11. The results show that the modified DCNNs outperform the classic DCNN algorithms concerning training and testing time. More precisely, considering that 90% of the required processing time is spent on feature extraction, the DCNNs take less than a millisecond to process an image.

4.4.2 Sensitivity Analysis of Designed Model

Here, we analyze the control parameters of the proposed classifier model, which are the shape of the logarithmic spiral represented by q, the number of layers represented by N_{layer}, and the number of batches represented by N_{batch}; the last two parameters belong to the network structure.

The sensitivity analysis aims to investigate the sensitivity and robustness of each parameter for various inputs. Following a similar direction as Ref. [52], four-parameter levels were used to conduct several experiments to generate an orthogonal array to characterize various parameter combinations. Table 4.12 shows the four-parameter levels used, and Table 4.13 shows the results of various parameter combinations and calculated MSEs obtained during the experiments. Also, Figure 4.12 shows the parameter level trends. To summarize the results, the parameter combinations $N_{layer}=5$, $q=1$, and $N_{batch}=10$ achieve the best performance.

TABLE 4.11 The Training and Test Time of Comparison Models Implemented on CPU and GPU

Model	CPU vs. GPU	Training Time	Testing Time	P-value
DCNN	GPU	342 s	2225 ms	N/A
	CPU	304 min	3.5 min	N/A
DCNN-MFO	GPU	355 s	2936 ms	1.11E-06
	CPU	325 min	4.23 min	1.22E-04
DCNN-LMFOs	GPU	354 s	2932 ms	1.33E-05
	CPU	325 min	4.22 min	1.71E-06
DCNN-ALO	GPU	401 s	3002 ms	1.08E-07
	CPU	333 min	4.71 min	1.84E-04
DCNN-PSO	GPU	359 s	2999 ms	0.002
	CPU	327 min	4.31 min	0.005
DCNN-HBO	GPU	425 s	3652 ms	2.18E-03
	CPU	364 min	5.02 min	3.56E-09
DCNN-SFS	GPU	411 s	3014 ms	1.04E-07
	CPU	341 min	4.33 min	1.38E-05
DCNN-ChOA	GPU	358 s	2998 ms	1.84E-04
	CPU	327 min	4.21 min	1.73E-06

TABLE 4.12 The Specification of Parameters

Level	N_{layer}	q	N_{batch}
1	3	0.20	6
2	4	0.40	8
3	5	0.80	10
4	6	1	12

TABLE 4.13 Results of Various Parameter Combinations

Experiment Number	Parameters			Result (MSE)
	N_{layer}	q	N_{batch}	
Number 1	1	0.20	1	0.0573
Number 2	1	0.40	2	0.0451
Number 3	1	0.80	3	0.0255
Number 4	1	1	4	0.0142
Number 5	2	0.20	2	0.0421
Number 6	2	0.40	1	0.0352
Number 7	2	0.80	4	0.0123
Number 8	2	1	3	0.0092
Number 9	3	0.20	1	0.0213
Number 10	3	0.40	4	0.0101
Number 11	3	0.80	2	0.0052
Number 12	3	1	3	0.0013
Number 13	4	0.20	4	0.0267
Number 14	4	0.40	3	0.0117
Number 15	4	0.80	2	0.0059
Number 16	4	1	1	0.0014

4.5 CONCLUSIONS

In order to recognize different underwater sonar datasets, this chapter proposed using the MFO algorithm for fine-tuning a DNN. Besides, this chapter investigated the efficiency of six spiral motions with different curvatures and slopes in the performance of the MFO (i.e., LMFO) for underwater sonar dataset classification tasks. In order to evaluate the performance of the customized model, in addition to benchmark Sejnowski & Gorman's dataset, two experimental sonar datasets, i.e., the passive sonar dataset and the active dataset, were exploited. The results of MFO and its modifications were compared to four novel nature-inspired algorithms, including HBO, ChOA, ALO, SFS, and the classic PSO. The results confirmed that the LMFO shows better performance than the other state-of-the-art models so that the classification rates were increased by 1.5979, 0.9985, and 2.0879 for Sejnowski & Gorman, passive, and active datasets, respectively. The results also approved that time complexity is not significantly increased by using different spiral motions.

For future research direction, it would be probable to employ LMFO in solving real-world engineering problems. Furthermore, another future research might use other spiral forms with a wider range of curvatures, slopes, and interception spots. Moreover, it can be attractive to mathematically calculate the best spiral motion for the best DCNN's performance.

REFERENCES

1. Ravakhah S, Khishe M, Aghababaie M, Hashemzadeh E. Sonar false alarm rate suppression using classification methods based on interior search algorithm. *IJCSNS* 2017;17:58.
2. Qiao W, Khishe M, Ravakhah S. Underwater targets classification using local wavelet acoustic pattern and multi-layer perceptron neural network optimized by modified whale optimization algorithm. *Ocean Eng* 2021;219:108415. doi: 10.1016/j.oceaneng.2020.108415.
3. Khishe M, Mosavi MR. Improved whale trainer for sonar datasets classification using neural network. *Appl Acoust* 2019. doi: 10.1016/j.apacoust.2019.05.006.
4. Pearce SK, Bird JS. Sharpening sidescan sonar images for shallow-water target and habitat classification with a vertically stacked array. *IEEE J Ocean Eng* 2013;38:455–69.
5. Mosavi MR, Khishe M, Hatam Khani Y, Shabani M. Training radial basis function neural network using stochastic fractal search algorithm to classify sonar dataset. *Iran J Electr Electron Eng* 2017. doi: 10.22068/IJEEE.13.1.10.
6. Fialkowski JM, Gauss RC. Methods for identifying and controlling sonar clutter. *IEEE J Ocean Eng* 2010;35:330–54.
7. Fandos R, Zoubir AM, Siantidis K. Unified design of a feature-based ADAC system for mine hunting using synthetic aperture sonar. *IEEE Trans Geosci Remote Sens* 2013;52:2413–26.
8. Lyon RF. *Human and Machine Hearing.* 2017. doi: 10.1017/9781139051699.
9. Demertzis K, Iliadis L, Anezakis V.-D. A deep spiking machine-hearing system for the case of invasive fish species. *2017 IEEE International Conference on INnovations in Intelligent SysTems and Applications*, IEEE, Turkey; 2017, pp. 23–8.
10. Guo Y, Liu Y, Oerlemans A, Lao S, Wu S, Lew MS. Deep learning for visual understanding: A review. *Neurocomputing* 2016. doi: 10.1016/j.neucom.2015.09.116.
11. Goodfellow I, Bengio Y, Courville A, Bengio Y. *Deep Learning.* vol. 1. MIT Press Cambridge, MA, 2016.
12. Ruder S. An overview of gradient descent optimization algorithms. ArXiv Prepr ArXiv160904747, 2016.

13. Shewchuk JR. An introduction to the conjugate gradient method without the agonizing pain, Technical report, 1994.
14. Martens J. Deep learning via Hessian-free optimization. *ICML 2010-Proceedings, The 27th International Conference on Machine Learning*, Israel, 2010.
15. Vinyals O, Povey D. Krylov subspace descent for deep learning. *J Mach Learn Res*, 2012.
16. Hinton GE, Salakhutdinov RR. Reducing the dimensionality of data with neural networks. *Science* 2006, 80. doi: 10.1126/science.1127647.
17. Abdel-Basset M, Abdel-Fatah L, Sangaiah AK. Metaheuristic algorithms: A comprehensive review. In: *Computational Intelligence for Multimedia Big Data on the Cloud with Engineering Applications*. Elsevier, Amsterdam, The Netherlands, 2018:185–231.
18. Kozek T, Roska T, Chua LO. Genetic algorithm for CNN template learning. *IEEE Trans Circuits Syst I Fundam Theory Appl* 1993;40:392–402.
19. Rosa G, Papa J, Marana A, Scheirer W, Cox D. Fine-tuning convolutional neural networks using Harmony Search. *CIARP 2015: Progress in Pattern Recognition, Image Analysis, Computer Vision, and Applications*, 2015. doi: 10.1007/978-3-319-25751-8_82.
20. Fedorovici L-O, Precup R-E, Dragan F, Purcaru C. Evolutionary optimization-based training of convolutional neural networks for OCR applications. *2013 17th International Conference on System Theory, Control and Computing (ICSTCC)*, IEEE, Romania; 2013, pp. 207–12.
21. Wang Y, Zhang H, Zhang G. cPSO-CNN: An efficient PSO-based algorithm for fine-tuning hyper-parameters of convolutional neural networks. *Swarm Evol Comput* 2019;49:114–23.
22. Guo B, Hu J, Wu W, Peng Q, Wu F. The tabu_genetic algorithm: A novel method for hyper-parameter optimization of learning algorithms. *Electronics* 2019;8:579.
23. Kumaran N, Vadivel A, Kumar SS. Recognition of human actions using CNN-GWO: A novel modeling of CNN for enhancement of classification performance. *Multimed Tools Appl* 2018;77:23115–47.
24. Rere LM, Fanany MI, Arymurthy AM. Metaheuristic algorithms for convolution neural network. *Comput Intell Neurosci* 2016;2016:1537325.
25. Ayumi V, Rere LMR, Fanany MI, Arymurthy AM. Optimization of convolutional neural network using microcanonical annealing algorithm. *2016 International Conference on Advanced Computer Science and Information Systems (ICACSIS*, IEEE, Indonesia; 2016, pp. 506–11.
26. Mohapatra P, Chakravarty S, Dash PK. An improved cuckoo search based extreme learning machine for medical data classification. *Swarm Evol Comput* 2015;24:25–49.
27. Li L-L, Sun J, Tseng M-L, Li Z-G. Extreme learning machine optimized by whale optimization algorithm using insulated gate bipolar transistor module aging degree evaluation. *Expert Syst Appl* 2019;127:58–67.
28. Webb GI, Keogh E, Miikkulainen R, Miikkulainen R, Sebag M. No-free-lunch theorem. *Encycl Mach Learn*, 2011. doi: 10.1007/978-0-387-30164-8_592.

29. Yamany W, Fawzy M, Tharwat A, Hassanien AE. Moth-flame optimization for training multi-layer perceptrons. *2015 11th International Computer Engineering Conference (ICENCO)*, IEEE, Egypt; 2015, pp. 267–72.
30. Faris H, Aljarah I, Mirjalili S. Evolving radial basis function networks using moth-flame optimizer. *Handb Neural Comput*, 2017. doi: 10.1016/B978-0-2-811318-9.00028-4.
31. Gorman RP, Sejnowski TJ. Analysis of hidden units in a layered network trained to classify sonar targets. *Neural Netw* 1988. doi: 10.1016/0893-6080(88)90023-8.
32. Khishe M, Mohammadi H. Passive sonar target classification using multi-layer perceptron trained by salp swarm algorithm. *Ocean Eng* 2019. doi: 10.1016/j.oceaneng.2019.04.013.
33. Mosavi MR, Kaveh M, Khishe M. Sonar data set classification using MLP neural network trained by non-linear migration rates BBO. *Fourth Iran Conference on Engineering Electromagnetic (ICEEM 2016)*, Iran, 2016, pp. 1–5.
34. Patterson J, Gibson A. *Deep Learning: A Practitioner's Approach*. O'Reilly Media, Inc.: Newton, MA, 2017.
35. Fukushima K, Miyake S. Neocognitron: A self-organizing neural network model for a mechanism of visual pattern recognition. *Compet Coop Neural Nets*, Springer; 1982, pp. 267–85.
36. Rumelhart DE, Hinton GE, Williams RJ. Learning representations by back-propagating errors. *Nature* 1986;323:533–6.
37. Kramer MA. Nonlinear principal component analysis using autoassociative neural networks. *AIChE J* 1991;37:233–43.
38. Smolensky P. Information processing in dynamical systems: Foundations of harmony theory. Colorado University at Boulder Department of Computer Science; 1986.
39. Goodfellow I, Pouget-Abadie J, Mirza M, Xu B, Warde-Farley D, Ozair S, et al. Generative adversarial nets. In: *Advances in Neural Information Processing Systems (NIPS)*, 2014.
40. Hinton GE, Osindero S, Teh Y-W. A fast learning algorithm for deep belief nets. *Neural Comput* 2006;18:1527–54.
41. Hochreiter S, Schmidhuber J. Long short-term memory. *Neural Comput* 1997;9:1735–80.
42. LeCun Y. LeNet-5, convolutional neural networks. URL Http//Yann LecunCom/Exdb/Lenet, 2015;20:14.
43. Mosavi MR, Khishe M, Akbarisani M. Neural network trained by biogeography-based optimizer with chaos for sonar data set classification. *Wirel Pers Commun* 2017. doi: 10.1007/s11277-017-4110-x.
44. Mirjalili S. Moth-flame optimization algorithm: A novel nature-inspired heuristic paradigm. *Knowl. Based Syst* 2015. doi: 10.1016/j.knosys.2015.07.006.
45. Cundy HM, Lockwood EH. A book of curves. *Math Gaz* 1963. doi: 10.2307/3612643.
46. Kennedy J, Eberhart R. Particle swarm optimization. *Proceedings of ICNN'95-International Conference on Neural Networks*, vol. 4, IEEE; 1995, p. 1942–8.

47. Mirjalili S. The ant lion optimizer. *Adv Eng Softw* 2015. doi: 10.1016/j.advengsoft.2015.01.010.
48. Salimi H. Stochastic Fractal Search: A powerful metaheuristic algorithm. *Knowl Based Syst* 2015. doi: 10.1016/j.knosys.2014.07.025.
49. Askari Q, Saeed M, Younas I. Heap-based optimizer inspired by corporate rank hierarchy for global optimization. *Expert Syst Appl* 2020;161:113702.
50. Khishe M, Mosavi MR. Chimp optimization algorithm. *Expert Syst Appl* 2020. doi: 10.1016/j.eswa.2020.113338.
51. Derrac J, García S, Molina D, Herrera F. A practical tutorial on the use of nonparametric statistical tests as a methodology for comparing evolutionary and swarm intelligence algorithms. *Swarm Evol Comput* 2011. doi: 10.1016/j.swevo.2011.02.002.
52. Wu C, Khishe M, Mohammadi M, Karim SHT, Rashid TA. Evolving deep convolutional neutral network by hybrid sine–cosine and extreme learning machine for real-time COVID19 diagnosis from X-ray images. *Soft Comput* 2021:1–20, https://link.springer.com/article/10.1007/s00500-021-05839-6.

II

Variants of Moth-Flame Optimization Algorithm

CHAPTER 5

Multi-objective Moth-Flame Optimization Algorithm for Engineering Problems

Nima Khodadadi

Florida International University

Seyed Mohammad Mirjalili

Concordia University

Seyedali Mirjalili

Torrens University Australia

Yonsei University

CONTENTS

5.1 Introduction 80

5.2 Related Works 83

5.3 MMFO Algorithm 85

5.4 Results and Discussion 87

5.4.1 Performance Metrics 87

5.4.1.1 Generational Distance (GD) 88

5.4.1.2 Spacing (S) 88

DOI: 10.1201/9781003205326-7

5.4.1.3 Maximum Spread (MS) 88
5.4.1.4 Inverted Generational Distance (IGD) 89
5.5 Discussion of the Engineering Design Functions 89
5.6 Conclusion and Future Directions 94
References 94

5.1 INTRODUCTION

Mathematics is the base of analytical approaches. These methods can find the exact solution to the problems and require less computational effort compared with approximate methods. However, these approaches are sensitive to the initial starting point, and just a correct starting point can lead to a high-quality solution. In addition, these methods when faced with complex optimization problems are not capable of providing the global minima and only reached local ones.

Most of the approximate techniques are global search methods, which are called Metaheuristics. These methods are developed to address the weaknesses of classical approaches. The metaheuristic algorithms can find near-global or global solutions by employing intelligence of natural phenomena. In the last decades, various metaheuristic algorithms are proposed based on natural processes, collective behavior, or scientific rules. Genetic Algorithm (GA) [1] and Differential Evolution (DE) [2] are called evolutionary algorithms that follow the process of evolution. Ant Colony Optimization (ACO) [3], Particle Swarm Optimization (PSO) [4], Cuckoo Search (CS) [5] algorithm, Crow Search Algorithm (CSA) [6], Stochastic Paint Optimizer (SPO) [7], and Gray Wolf Optimizer [8] are named swarm-based algorithms, which mimic the process of the natural behavior of creatures. Advanced Charged System Search (ACSS) [9], Chaos Game Optimization (CGO) [10], and Arithmetic Optimization Algorithm (AOA) [11] are called physics- or mathematics-based algorithms, which obey the rules in physics or mathematics.

Many researchers utilized these algorithms for solving different structural optimization problems such as trusses [12,13], frames [14]–[16], and real application of engineering [17, 18]. These studies demonstrate that the metaheuristic algorithms can solve problems with good accuracy in a reasonable time when employed to deal with complex optimization problems. Ease of implementation, simple framework, good accuracy, and reasonable execution time are some advantages of metaheuristic algorithms compared with the analytical techniques.

In the real world, the vast majority of optimization problems are multi-objective. Multi-objective optimization (MOO) is a branch of decision-making that involves balancing various goals that are frequently in dispute [19]. It has been used in various research fields, including physics, economics, and logistics, where optimum choices must be taken in the face of trade-offs between two or more competing goals. A single-objective optimization approach will only represent one part of a multi-objective problem if it is used to solve it. The single-objective optimization algorithm, for example, cannot wholly describe the hydrological mechanism and function in the optimization of hydrological model parameters [20].

The task of solving problems with only one objective function is easy. The optimal solution is the product of such problems and the algorithms are assessed on the quality of the optimal solutions [21]. In multi-objective challenges, there are multiple competing objective functions and there is no one best solution that maximizes all of them at once. In these cases, decision-makers look for the desired option rather than the optimal one. These sorts of solutions are subsets of the Pareto solutions. As a result, Pareto optimality takes the role of optimality in MOO issues. The Pareto optimum (also known as effective, non-dominated, or non-inferior) solutions cannot improve one objective function without impacting at least one of the others.

The majority of recent algorithms have rules in place to solve problems with multiple goals. However, the well-known No-Free-Lunch (NFL) [22] theorem theoretically proves that none of these algorithms can solve all optimization problems. This means that the existing algorithms can be improved or new ones proposed to help solve particular problems. New algorithms are best suited to unconstrained problems and cannot handle various kinds of constraints without the use of particular components. The multi-objective variant of the recently introduced Moth-Flame Optimization (MFO) Algorithm is proposed in this chapter to solve both unconstrained and constrained problems. The Multi-objective Moth-Flame Optimization (MMFO) uses the following contributions:

- An archive in the MFO algorithm has been added to keep undominated solutions.
- A grid system was applied to MFO to improve the undominated solutions in the archive.
- A leader selection mechanism has been recommended based on best flame position in MFO.

Nanda [23] proposed multi-objective version of MFO on six benchmark mathematical functions. The suggested technique differs from the original single-objective MFO due to features such as the introduction of an archive grid, coordinate-based distance for sorting, and non-dominance of solutions. In comparison to some multi-objective methods, the performance of the proposed MOMFO is proven in terms of higher accuracy and reduced computing time.

An improved MMFO based on R-domination (R-IMOMFO) has been presented by Zhang et al. [24]. This method strengthens the ability of the MFO to escape dropping into the local optimum by refining three aspects: the update formula, the inspiration of the moth linear flight path, and the flame population update technique. The performance of this method is evaluated with five algorithms on reservoir operation model.

Soussi et al. [25] added quantum behaved moth to the single version of MFO. Then, multi-objective version of MFO based on quantum behaved moth (MOQMFO) is developed to obtain clustering problems. As objective functions, MOQMFO employed three cluster validity criteria.

Water resource utilization problems were solved using a unique approach known as the Multi-objective Moth-Flame Optimization Algorithm (MOMFA) by Ref. [26]. The suggested approach incorporates the effective qualities of the original MFO algorithm, as well as two efficient processes called opposition-based learning and indicator-based selection, which aid the system in accelerating convergence while maintaining variety. The proposed MOMFA's performance was evaluated using a set of benchmarks.

In this work, multi-objective version of MFO with three different mechanisms such as archive, grid, and leader selection mechanisms is applied to some benchmarks. The proposed MFO proved its ability to solve multiple complex problems compared to other well-known multi-objective methods. The experiments are conducted on eight multi-objective engineering design problems. The proposed MFO proved its ability to solve multiple complex problems compared to other well-known multi-objective methods. The remainder of this chapter is as follows: Section 5.2 presents the MOO theory. Section 5.3 introduces MFO and its multi-objective version. Performance metrics and results of test function problems are discussed in Section 5.4. This chapter concludes with Section 5.5.

5.2 RELATED WORKS

The MOO theory and current metaheuristic techniques are discussed in this section. Due to the unary goal and the existence of just one most satisfactory solution in single-objective problems, the global optimum in single-objective problems is only one solution. When only one target is in consideration, comparing solutions with relational operators is easy. Because of the characteristics of such problems, evaluating potential solutions and selecting the best one is easy for optimization problems. MOO is a branch of multiple-criteria decision-making that deals with mathematical optimization problems that have many objective functions to optimize at the same time. Francis Ysidro [27] proposed comparing two solutions in multi-objective problems in 1881 and Vilfredo Pareto [28] expanded on this in 1964. The following is MOO for minimization problem [28]:

$$\text{Maximization}: \ F(x)=\left\{f_1(\vec{x}),\ f_2(\vec{x}),\ldots,f_z(\vec{x})\right\} \tag{5.1}$$

$$\text{Subject to}: \quad g_i(\vec{x})\geq 0 \ \ i=1,2,\ldots,n \ \ (5.2)$$

$$h_i(\vec{x})=0 \ \ i=1,2,\ldots,p \ (5.3)$$

$$L_i \leq x_i \leq U_i \ \ i=1,2,\ldots,r \ (5.4)$$

where the number of variables, inequality constraints, equality constraints, and objective functions are described by m, p, n, and z, respectively. The boundaries of the ith variable are indicated by L_i and U_i.

Definition 5.1: Pareto Dominance

Let us consider two solutions $\vec{x}$ and $\vec{y}$ with cost function values $\vec{x}=(x_1,\ x_2,\ldots,\ x_k)$ and $\vec{y}=(y_1,\ y_2,\ldots,y_k)$ that fulfill the constraints of a MOO. Providing that MOO is a minimization problem, the solution $\vec{x}$ is said to dominate $\vec{y}$(denoted as $\vec{x}\prec \vec{y}$) if and only if no component of $\vec{y}$ is smaller than the corresponding cost component of $\vec{x}$, and at least one component of $\vec{x}$ must be smaller than that of $\vec{y}$ (see Figure 5.1). This definition can be mathematically presented as follows [29]:

$$\forall i\in\{1,2,\ldots,k\}: \ \ f_i(\vec{x})\leq f_i(\vec{y}) \wedge \exists i\in\{1,2,\ldots,k\}: \ \ f_i(\vec{x})\leq f_i(\vec{y}) \tag{5.5}$$

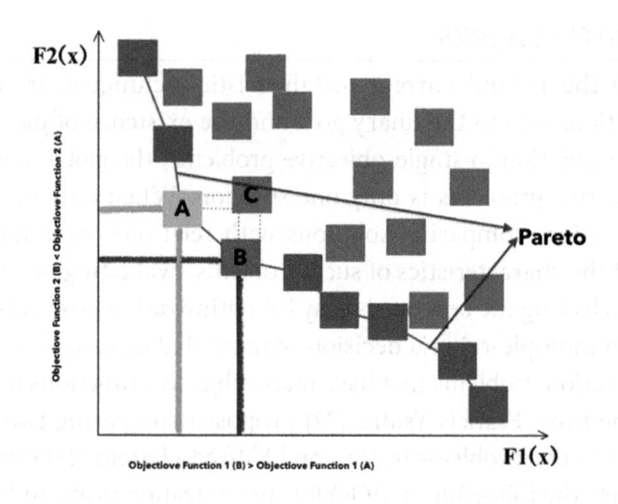

FIGURE 5.1 Pareto dominance.

According to this definition of Pareto dominance, the definition of Pareto optimality is defined as follows [29]:

Definition 5.2: Pareto Optimality

A solution $\vec{x} \in X$ is called Pareto-optimal if and only if

$$\nexists \vec{y} \in X \mid \vec{y} \prec \vec{x} \tag{5.6}$$

The Pareto optimal contains a set that includes all non-dominated solutions to a problem and is defined as follows:

Definition 5.3: Pareto Optimal Set

For a given MOO, the Pareto optimal set (P_s) is defined as Equation (5.7). This set contains all feasible solutions, none of which can be dominated by any other feasible solution. Figure 5.2 depicts the set of Pareto optimal solutions.

$$P_s := \{ \vec{x}, \vec{y} \in X \mid \nexists \, \vec{y} \prec \vec{x} \} \tag{5.7}$$

The following is the definition and description of the Pareto optimal front:

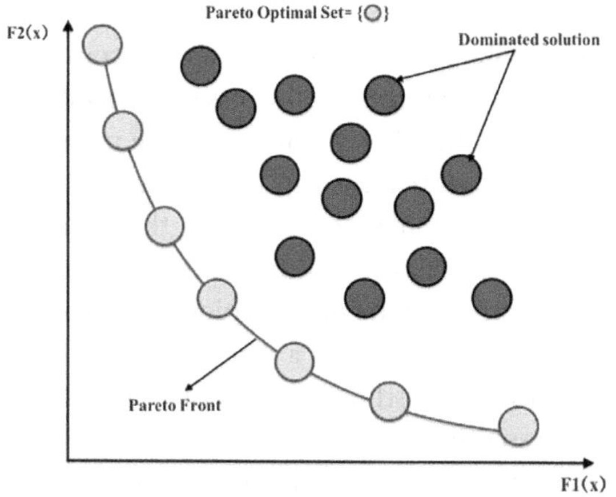

FIGURE 5.2 Pareto optimal solutions.

Definition 5.4: Pareto Optimal Front

A Pareto front (P_f) illustrates the Pareto optimal set in the objective space as shown in Figure 5.2. According to the definitions mentioned above, it can be expressed as follows:

$$P_f = \{F(\vec{x}), \vec{x} \in P_s\} \tag{5.8}$$

One must determine the Pareto optimal set, which is a collection of solutions that represent the best equilibrium between goals, in order to solve a multi-objective problem.

5.3 MMFO ALGORITHM

In the following section, the MMFO is developed and proposed for addressing MOO problems.

To perform MOO, two features (Archive and Leader Selection) have been added to MFO. The archive stores non-dominated Pareto optimum solutions, and leader selection aids in selecting current best place solutions from the archive as the search process leaders. These functions are the same as in MOPSO algorithm [23].

The archive is a basic storage device for non-dominated Pareto optimal answers obtained up to this stage. An archive controller is a critical component of the archive since it controls the archive when a solution is intended to be archived or when the archive is full. Remember that the archive will only be accessible to a restricted number of individuals. Non-dominated solutions obtained up to this step are compared against the existence archive during the training process of an iteration.

The probability of eliminating an answer to the number of answers in the hypercube (section) is increased in direct proportion. If the archive is full, one of the most populated areas is in the beginning picked for eliminating solutions. A solution is additionally arbitrarily removed from among them to make area for the new solution. When a solution is found outside of the hypercubes, it creates a unique scenario. All components have been updated to include the new solutions in this case. As a result, different alternative solutions' components can be altered.

In MMFO, the best answers obtained thus far are used as the best position; according to the second feature, it is the leader selection function. This function leads the other search agents to exciting parts of the search space in order to find a solution close to the global optimum. However, because of the Pareto optimality ideas discussed in the preceding subsection, in a multi-objective search space, it is impossible to compare the results fast. The leader selection function was created to address this problem. As previously mentioned, there is a repository of the best non-dominated answers gathered thus far. The leader selection part selects the least-crowded sections of the search space as well as supplies among its non-dominated answers. For each hypercube, a roulette-wheel method determines the hypercube with the following probability:

$$P_i = \frac{C}{N_i} \tag{5.14}$$

where C is a constant number higher than one as well as N is the variety of acquired Pareto optimal answers in the ith section.

Equation (5.14) presents thus far less populated hypercubes that have a greater chance of suggesting new leaders. As the hypercube's range of obtained solutions is reduced, the chance of picking a hypercube from which leaders are picked increases.

Obviously, the MMFO algorithm acquires its convergence from the MFO algorithm. If we choose one solution from the archive, the MFO algorithm will certainly converge towards it. Finding the Pareto optimal answers developed with a large variety, on the other hand, is challenging. We used the leader function collection as well as archive maintenance to overcome this challenge. Figure 5.3 shows MMFO's pseudo-code.

```
Initialize the MMFO Parameters
Initialize the Solutions' Positions
Evaluate the Fitness Function for Given Solutions
Obtain the Non-dominated Solutions and Create the Archive
While (Iter < MaxIter)
Update Number of  Flames
Calculate the Linearly Decreases Function
     For Each Solution
          Obtain Leader
             for Each Position
                 if r1<= Number of Flames
                         Update the position
                 End if
                 if r1> Number of Flames
                         Update the position
                 End if
             End for
     End for
Evaluate the Fitness Function for New Position of All Objectives
Obtain the Non-dominated Solutions
Update the Archive Based on the Non-dominated Solutions
     If the Archive = Full
         Eliminating the Current Archives According to Grid Mechanism
         Add the New Solution to the Archive
     Endif
End while;
Return Archive
```

FIGURE 5.3 Pseudo-code of MMFO.

The exploration and the exploitation phase in MMFO algorithm are the same as its single-objective version. Searching based on the archive members and saving the non-dominated solution are the main difference between MFO and MMFO.

5.4 RESULTS AND DISCUSSION

In this section, four performance metrics and unconstrained and constrained engineering design problems are used to assess the efficiency of the proposed method. These problems are used to test the capabilities of multi-objective optimizers to tackle problems with a variety of objectives. The method was written in MATLAB® 2021a and includes the following properties: On a computer with Macintosh (macOS BigSur), the CPU is 2.3 GHz (based on an Intel Core i9 computer platform) and the RAM is 16 GB 2,400 MHz DDR4.

5.4.1 Performance Metrics

Four metrics are used for evaluating the algorithms' results as follows [29,30]:

5.4.1.1 Generational Distance (GD)

This index calculates the distance between the true Pareto (PF_{true}) and the obtained Pareto front (PF_{known}). This distance is calculated by the following equation:

$$GD = \left(\frac{1}{n_{PF}} \sum_{i=1}^{n_{PF}} dis_i^2 \right)^{\frac{1}{2}} \tag{5.15}$$

where the number of non-dominated solutions is defined as n_{PF} in PF_{known} and dis_i is the Euclidean distance between the solution i in PF_{known} and its nearest solution in PF_{true}. It should be noted that a lower GD value indicates a more favorable algorithm convergence.

5.4.1.2 Spacing (S)

This measure indicates how evenly the obtained solutions are distributed along PF_{known} This measure is formulated as follows:

$$S = \left(\frac{1}{n_{PF}} \sum_{i=1}^{n_{PF}} \left(d_i - \bar{d}\right)^2 \right)^{\frac{1}{2}} \quad \text{Where} \quad \bar{d} = \frac{1}{n_{PF}} \sum_{i=1}^{n_{PF}} c \tag{5.16}$$

where the number of non-dominated solutions is defined as n_{PF} in PF_{known}, and d_i is the Euclidean distance between the solution i in PF_{known} and the nearest solution in PF_{true}. Convergence is more robust when S is smaller.

5.4.1.3 Maximum Spread (MS)

This metric evaluates how "well" PF_{true} is covered by PF_{known} through hypercubes formed by the extreme cost values observed in PF_{true} and PF_{known}. It is defined as follows:

$$MS = \left[\frac{1}{m} \sum_{i=1}^{m} \left[\frac{\min\left(f_i^{\max}, F_i^{\max}\right) - \max\left(f_i^{\min}, F_i^{\min}\right)}{F_i^{\max} - F_i^{\min}} \right]^2 \right]^{\frac{1}{2}} \tag{5.17}$$

where m is the number of objectives; $f_i^{\max}$ and $f_i^{\min}$ are the maximum and minimum of the ith objective in PF_{known}, respectively; and $F_i^{\max}$ and $F_i^{\min}$ are the maximum and minimum of the ith objective in PF_{true}, respectively.

A more enormous value of Maximum Spread (MS) indicates a better spread of solution as well.

5.4.1.4 Inverted Generational Distance (IGD)

It is a measure for evaluating the quality of MOO algorithms' approximations to the Pareto front. The following is the definition of this metric:

$$\text{IGD} = \frac{\sqrt{\sum_{i=1}^{n} d_i^2}}{n} \tag{5.18}$$

here, n is the number of elements in PF_{true}, and d_i is the Euclidean distance between the solution i in and its nearest solution in PF_{true}. The Zero value shows that all of the generated elements are in the true front of Pareto [31].

5.5 DISCUSSION OF THE ENGINEERING DESIGN FUNCTIONS

In this section, MMFO is compared with the MOPSO, Multi-objective Gray Wolf Optimizer (MOGWO), and Multi-objective Ant Lion Optimizer (MOALO) and the best figures of obtained pareto front compared with true pareto front for all examples are shown in Figures 5.3 and 5.4. The starting parameter settings for all of the algorithms are shown in Table 5.1. Furthermore, all instances are limited to 100 populations and 1,000 iterations. This section assesses the efficacy of the suggested technique by evaluating eight engineering design problems (Figure 5.5).

Engineering design problems contain eight MOO problems; each problem has several variables. The details of these problems can be found in Ref. [32]. The statistical (average and standard deviation) results of engineering design problems are shown in Table 5.2 for the inverted generational distance metric. It is obvious that the proposed MMFO has better results compared to the MOPSO, MOGWO, and MOALO. The proposed MMFO obtained the best results in all test cases except the Welded Beam design problem. Thus, the results illustrated the ability of the mentioned method in solving complex mathematical MOO problems. It acquired the most excellent results in most of the problems tested with low SD values as well.

The convergence and distribution convergence of Pareto optimal solutions for algorithms is calculated by generational distance metric. The GD performance metric of all algorithms is shown in Table 5.3. It is clear that the results of MMFO are better than MOPSO, MOGWO, and MOALO

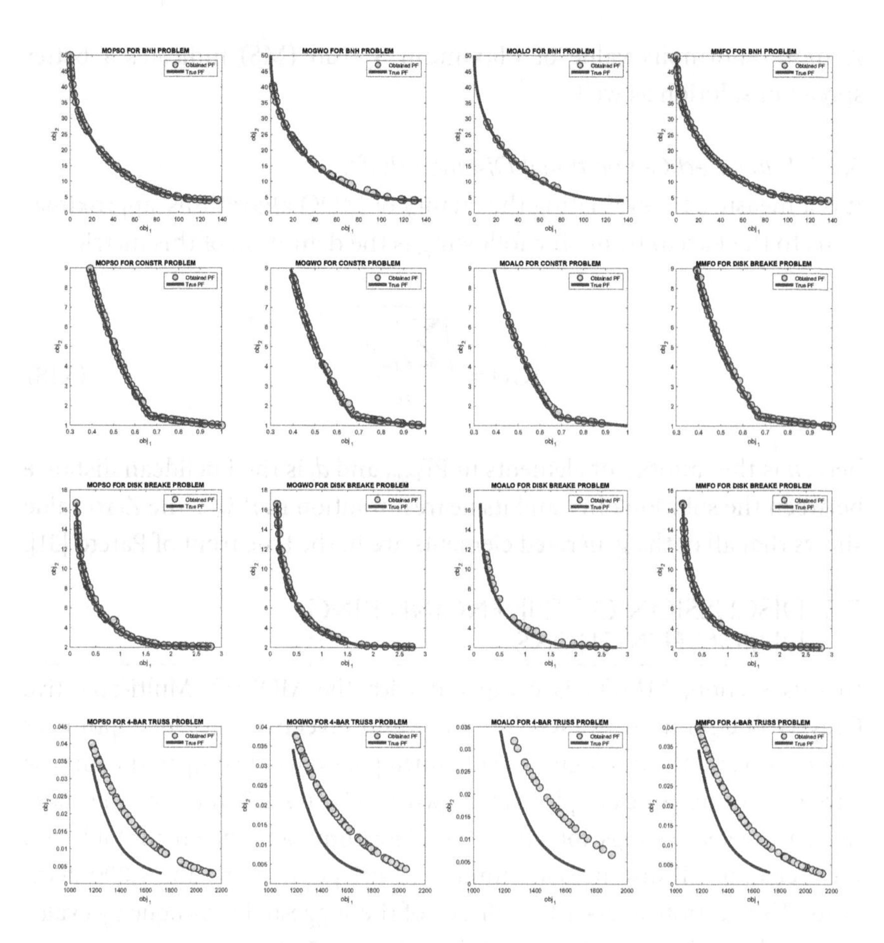

FIGURE 5.4 Estimated Pareto optimal solutions on BNH, CONSTR, DISK BREAKE, and 4-BAR TRUSS case studies.

TABLE 5.1 Parameters Setting of All Algorithms

Parameters	MOPSO	MOGWO	MOALO	MMFO
Mutation probability (P_w, or pro)	0.5	-		-
Population size (N_{pop})	100	100	100	100
Archive size (N_{rep} or TM)	100	100	100	100
Number of adaptive grid (N_{grid})	30	30	30	30
Personal learning coefficient (C_1)	1	-	-	-
Global learning coefficient (C_2)	2	-	-	-
Inertia weight (w)	0.4	-	-	-
Beta	4	4	4	4
Gamma	2	2	2	2

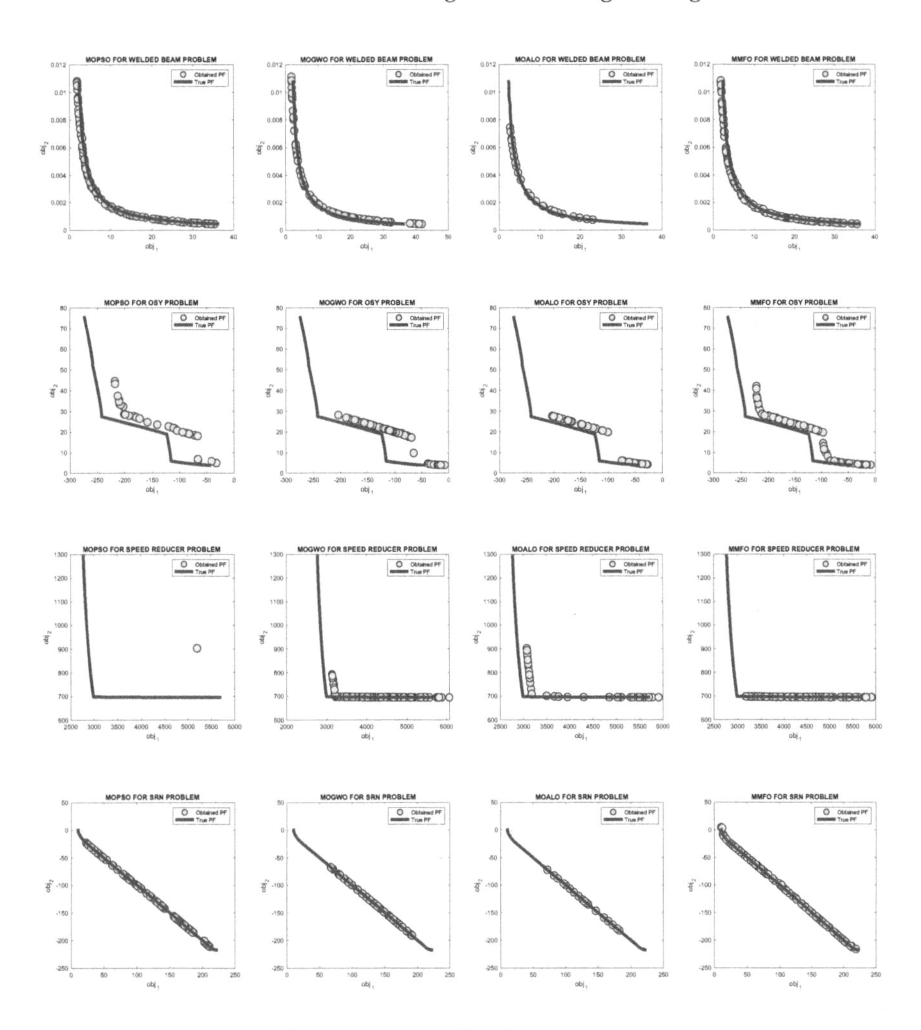

FIGURE 5.5 Estimated Pareto optimal solutions on WELDED BEAM, OSY, SPEED REDUCER, and SRN case studies.

for five out of eight problems: BNH, CONSTR, DISK BREAKE, SPEED REDUCER, and SRN. Thanks to this, the proposed method has competitive coverage in comparison with the other mentioned methods.

The statistical findings of engineering challenges for the maximum spread performance metric are shown in Table 5.4. The proposed MMFO acquired the most excellent results in seven out of eight test cases. The proposed MAOA produced the best results in most of the problems tested with low SD values, except Welded Beam, Speed Reducer, and OSY. The MOPSO and MOGWO obtained just one best results regarding the maximum spread performance metric.

TABLE 5.2 Benchmark Function Statistical Findings for IGD Performance Measure

		Algorithm			
Functions		**MOPSO**	**MOGWO**	**MOALO**	**MMFO**
BNH	Ave	9.6868E-04	3.3817E-03	1.2198E-02	9.5928E-04
	SD	1.7147E-04	1.1513E-03	3.6455E-03	8.7345E-05
CONSTR	Ave	5.1838E-04	7.4460E-04	2.5041E-03	5.1456E-04
	SD	5.5618E-05	1.7697E-04	9.6528E-04	4.8592E-05
DISK BREAKE	Ave	5.8831E-04	7.2091E-04	1.7399E-03	5.8422E-04
	SD	5.5995E-05	1.1206E-04	7.7421E-04	5.1654E-05
4-BAR TRUSS	Ave	2.0010E-02	2.1409E-02	2.2004E-02	1.9977E-02
	SD	3.9632E-05	1.5445E-04	1.1024E-03	3.2378E-05
WELDED BEAM	Ave	5.9705E-04	1.3441E-03	6.1770E-03	6.1991E-04
	SD	4.6341E-05	3.9377E-04	2.8159E-03	4.0206E-05
OSY	Ave	1.4663E-02	7.6220E-03	8.1016E-03	7.0087E-03
	SD	8.6917E-03	6.4006E-04	2.2613E-04	1.7325E-03
SPEED REDUCER	Ave	6.0305E-01	1.4243E-02	1.7737E-02	1.2731E-02
	SD	7.2130E-02	3.2032E-03	3.3878E-03	1.0569E-02
SRN	Ave	4.5146E-04	2.4823E-03	6.2137E-03	3.4377E-04
	SD	1.1656E-04	1.0614E-03	1.8810E-03	9.9166E-05

TABLE 5.3 Benchmark Function Statistical Findings for GD Performance Measure

		Algorithm			
Functions		***MOPSO***	***MOGWO***	***MOALO***	***MMFO***
BNH	Ave	3.3894E-02	6.2465E-02	5.3758E-02	3.3611E-02
	SD	2.1392E-03	1.5645E-02	1.7132E-02	1.9638E-03
CONSTR	Ave	8.3098E-04	8.9220E-04	1.5711E-03	8.0251E-04
	SD	3.3369E-05	1.0565E-04	7.7806E-04	1.0058E-04
DISK BREAKE	Ave	2.3045E-03	4.3311E-03	4.1887E-02	1.6933E-03
	SD	5.6273E-04	1.8074E-03	5.4951E-03	1.6256E-04
4-BAR TRUSS	Ave	1.4095E+01	1.1696E+01	2.5901E+00	1.4874E+01
	SD	5.1580E-01	2.3340E+00	2.2161E+00	1.0404E+00
WELDED BEAM	Ave	1.0846E-02	3.9207E-02	4.4399E-03	1.3382E-02
	SD	1.9956E-03	3.8097E-02	6.9956E-04	3.2405E-03
OSY	Ave	3.5765E+00	1.2736E+00	7.0760E-01	1.1503E+00
	SD	2.5250E+00	4.0179E-01	2.4932E-01	7.1836E-01
SPEED REDUCER	Ave	8.2516E+01	8.2889E+00	7.6817E+00	6.8962E+00
	SD	9.8695E+01	1.6129E+00	3.4697E+00	1.0599E+00
SRN	Ave	3.1617E-02	1.6743E-02	1.5798E-02	3.0736E-02
	SD	1.0695E-02	3.9924E-03	3.2892E-03	1.4179E-02

TABLE 5.4 Benchmark Function Statistical Findings for MS Performance Measure

Functions		Algorithm			
		MOPSO	*MOGWO*	*MOALO*	*MMFO*
BNH	Ave	1.0000E+00	8.7156E-01	5.4090E-01	1.0000E+00
	SD	0.0000E+00	6.6957E-02	1.0692E-01	0.0000E+00
CONSTR	Ave	9.9384E-01	9.7790E-01	7.7822E-01	9.9471E-01
	SD	6.8603E-03	2.0113E-02	7.6343E-02	5.8120E-03
DISK BREAKE	Ave	9.9928E-01	9.9582E-01	9.1779E-01	9.9950E-01
	SD	1.1417E-03	1.2807E-02	2.0904E-01	3.4466E-04
4-BAR TRUSS	Ave	1.4876E+00	1.3787E+00	8.4793E-01	1.4879E+00
	SD	5.4502E-04	4.6120E-02	1.2454E-01	4.9481E-04
WELDED BEAM	Ave	1.0072E+00	1.1107E+00	6.2463E-01	1.0322E+00
	SD	6.1053E-02	1.0194E-01	6.9354E-02	7.8287E-02
OSY	Ave	3.2390E-01	6.6490E-01	5.7530E-01	6.7654E-01
	SD	3.4284E-01	2.5118E-02	2.6351E-02	5.9364E-02
SPEED REDUCER	Ave	1.0408E-03	7.6707E-01	6.6868E-01	7.8149E-01
	SD	2.6703E-01	2.4182E-02	6.0939E-02	8.6327E-02
SRN	Ave	9.0900E-01	7.0014E-01	3.9177E-01	9.3385E-01
	SD	4.5463E-02	8.0177E-02	8.2165E-02	3.3579E-02

TABLE 5.5 Benchmark Function Statistical Findings for *S* Performance Measure

Functions		Algorithm			
		MOPSO	*MOGWO*	*MOALO*	*MMFO*
BNH	Ave	1.0901E+00	1.7477E+00	8.8402E-01	1.1807E+00
	SD	1.3631E-01	4.7352E-01	4.4648E-01	1.3487E-01
CONSTR	Ave	5.8740E-02	5.4894E-02	6.9071E-02	4.7608E-02
	SD	8.8936E-03	8.3095E-03	1.8623E-02	8.0015E-03
DISK BREAKE	Ave	1.1452E-01	1.3331E-01	1.4457E-01	1.0837E-01
	SD	1.3022E-02	1.9321E-02	1.1754E-02	2.4915E-02
4-BAR TRUSS	Ave	5.3605E+00	5.9045E+00	4.6988E+00	4.0172E+00
	SD	2.6169E-01	1.7462E+00	1.1721E+00	7.8121E-01
WELDED BEAM	Ave	2.3432E-01	4.4716E-01	2.2431E-01	2.1395E-01
	SD	2.5702E-02	2.1585E-01	1.3595E-01	2.3963E-02
OSY	Ave	1.1278E+00	1.6275E+00	1.4382E+00	1.1090E+00
	SD	1.4675E+00	4.7312E-01	4.8498E-01	2.1121E-01
SPEED REDUCER	Ave	3.6574E+02	2.0624E+01	3.6722E+01	2.0604E+01
	SD	1.8976E+01	3.1861E+00	9.3623E+00	5.5307E+00
SRN	Ave	2.2396E+00	2.6491E+00	1.7386E+00	1.1787E+00
	SD	4.7324E-01	1.2558E+00	8.6872E-01	2.9438E-01

To compare the diversity of MMFO with MOPSO, MOGWO, and MOALO algorithms with respect to the spacing metric, the statistical results are obtained from 30 individual runs with these benchmarks. Table 5.5 shows the results for the spacing performance metric. MMFO obtained the best results in terms of this metric for all problems, except the BNH. It is clear that the MMFO presented here will provide a reasonably reliable estimate of true Pareto optimal solutions.

5.6 CONCLUSION AND FUTURE DIRECTIONS

In this chapter, multi-objective version of MFO was proposed leveraging on an archive to store non-dominated solutions and a selection mechanism to choose "best" solutions for the MFO algorithm. In MMFO, a concept of archive is included to obtain the non-dominated Pareto optimal solutions. The leader selection feature recommended enabled the MMFO algorithm to reveal excellent coverage as well as convergence at the same time. The proposed MMFO method is analyzed on eight constrained engineering design problems to evaluate its suitability to find real-world engineering problems' solutions. Its performance is compared with several well-known multi-objective methods, including MOPSO, MOGWO, and MOALO. Several evaluation metrics are used to evaluate the obtained results, such as Generational Distance (GD), Inverted Generational Distance (IGD), MS, and Spacing (S). Therefore, this algorithm gives a highly competitive result in comparison with the mentioned algorithm. The examination functions used are of a different kind and also have varied Pareto optimal fronts. The results from the empirical analysis illustrated that the proposed MMFO method is better than other existing algorithms and has high convergence.

Future works are advised to use MMFO for various other engineering design problems such as truss structures and developing the structural health assessment. Furthermore, one can check out the performance of various stable handling approaches on MMFO that permit it to manage various kinds of unpredictability, which is crucial in solving real problems. Also, for this algorithm, it is worth investigating and determining the best-constrained handling strategy.

REFERENCES

1. K. Deb, A. Pratap, S. Agarwal, and T. Meyarivan, "A fast and elitist multiobjective genetic algorithm: NSGA-II," *IEEE Trans. Evol. Comput.*, vol. 6, no. 2, pp. 182–197, 2002.

2. A. K. Qin, V. L. Huang, and P. N. Suganthan, "Differential evolution algorithm with strategy adaptation for global numerical optimization," *IEEE Trans. Evol. Comput.*, vol. 13, no. 2, pp. 398–417, 2008.
3. M. Dorigo, M. Birattari, and T. Stutzle, "Ant colony optimization," *IEEE Comput. Intell. Mag.*, vol. 1, no. 4, pp. 28–39, 2006.
4. J. Kennedy and R. Eberhart, "Particle swarm optimization," in Proceedings of ICNN'95-International Conference on Neural Networks, Australia, 1995, vol. 4, pp. 1942–1948.
5. A. H. Gandomi, X.-S. Yang, and A. H. Alavi, "Cuckoo search algorithm: A metaheuristic approach to solve structural optimization problems," *Eng. Comput.*, vol. 29, no. 1, pp. 17–35, 2013.
6. A. Askarzadeh, "A novel metaheuristic method for solving constrained engineering optimization problems: Crow search algorithm," *Comput. Struct.*, vol. 169, pp. 1–12, 2016.
7. A. Kaveh, S. Talatahari, and N. Khodadadi, "Stochastic paint optimizer: Theory and application in civil engineering," *Eng. Comput.*, pp. 1–32, 2020.
8. S. Mirjalili, S. M. Mirjalili, and A. Lewis, "Grey wolf optimizer," *Adv. Eng. Softw.*, vol. 69, pp. 46–61, 2014.
9. A. Kaveh, N. Khodadadi, B. F. Azar, and S. Talatahari, "Optimal design of large-scale frames with an advanced charged system search algorithm using box-shaped sections," Eng. Comput., pp. 1–21, 2020.
10. S. Talatahari and M. Azizi, "Chaos game optimization: A novel metaheuristic algorithm," *Artif. Intell. Rev.*, vol. 54, no. 2, pp. 917–1004, 2021.
11. L. Abualigah, A. Diabat, S. Mirjalili, M. Abd Elaziz, and A. H. Gandomi, "The arithmetic optimization algorithm," *Comput. Methods Appl. Mech. Eng.*, vol. 376, p. 113609, 2021.
12. S. O. Degertekin, G. Y. Bayar, and L. Lamberti, "Parameter free Jaya algorithm for truss sizing-layout optimization under natural frequency constraints," *Comput. Struct.*, vol. 245, p. 106461, 2021.
13. A. Kaveh, S. Talatahari, and N. Khodadadi, "Hybrid invasive weed optimization-shuffled frog-leaping algorithm for optimal design of truss structures," *Iran. J. Sci. Technol. Trans. Civ. Eng.*, pp. 1–16, 2019.
14. Y. Yuan, L. Lv, X. Wang, and X. Song, "Optimization of a frame structure using the Coulomb force search strategy-based dragonfly algorithm," *Eng. Optim.*, vol. 52, no. 6, pp. 915–931, 2020.
15. A. Kaveh, S. Talatahari, and N. Khodadadi, "The hybrid invasive weed optimization-shuffled frog-leaping algorithm applied to optimal design of frame structures," *Period. Polytech. Civ. Eng.*, vol. 63, no. 3, pp. 882–897, 2019.
16. M. P. Saka and Z. W. Geem, "Mathematical and metaheuristic applications in design optimization of steel frame structures: an extensive review," *Math. Probl. Eng.*, vol. 2013, pp. 1–33, 2013.
17. M. Dehghani, M. Mardaneh, and O. P. Malik, "FOA:'Following'Optimization Algorithm for solving Power engineering optimization problems," *J. Oper. Autom. Power Eng.*, vol. 8, no. 1, pp. 57–64, 2020.

18. A. Kaveh, N. Khodadadi, and S. Talatahari, "A Comparative Study for the Optimal Design of Steel Structures Using Css and Acss Algorithms," *Iran Univ. Sci. Technol.*, vol. 11, no. 1, pp. 31–54, 2021.
19. S. Z. Mirjalili, S. Mirjalili, S. Saremi, H. Faris, and I. Aljarah, "Grasshopper optimization algorithm for multi-objective optimization problems," *Appl. Intell.*, vol. 48, no. 4, pp. 805–820, 2018.
20. H. Zille, H. Ishibuchi, S. Mostaghim, and Y. Nojima, "A framework for large-scale multiobjective optimization based on problem transformation," *IEEE Trans. Evol. Comput.*, vol. 22, no. 2, pp. 260–275, 2017.
21. V. Savsani and M. A. Tawhid, "Non-dominated sorting moth flame optimization (NS-MFO) for multi-objective problems," *Eng. Appl. Artif. Intell.*, vol. 63, pp. 20–32, 2017.
22. D. H. Wolpert and W. G. Macready, "No free lunch theorems for optimization," *IEEE Trans. Evol. Comput.*, vol. 1, no. 1, pp. 67–82, 1997.
23. S. J. Nanda, "Multi-objective moth flame optimization," in *2016 International* Conference on Advances in Computing, Communications and Informatics (ICACCI), India, 2016, pp. 2470–2476.
24. Z. Zhang et al., "Improved multi-objective moth-flame optimization algorithm based on R-domination for cascade reservoirs operation," *J. Hydrol.*, vol. 581, p. 124431, 2020.
25. Y. Soussi, N. Rokbani, A. Wali, and A. M. Alimi, "Multi-objective quantum moth flame optimization for clustering," in: A. E. Hassanien, A. Darwish, S. M. A. El-Kader, and D. A. Alboaneen (eds), Enabling Machine Learning Applications in Data Science. Springer: Berlin, Germany, 2021, pp. 193–205.
26. W. K. Li, W. L. Wang, and L. Li, "Optimization of water resources utilization by multi-objective moth-flame algorithm," *Water Resour. Manag.*, vol. 32, no. 10, pp. 3303–3316, 2018.
27. F. Y. Edgeworth, Mathematical psychics. McMaster University Archive for the History of Economic Thought, 1881.
28. V. Pareto, *Cours d'économie politique*, vol. 1. Librairie Droz, 1964.
29. P. Ngatchou, A. Zarei, and A. El-Sharkawi, "Pareto multi objective optimization," in Proceedings of the 13th International Conference on, Intelligent Systems Application to Power Systems, Morocco, 2005, pp. 84–91.
30. J. D. Knowles and D. W. Corne, "Approximating the nondominated front using the Pareto archived evolution strategy," *Evol. Comput.*, vol. 8, no. 2, pp. 149–172, 2000.
31. M. R. Sierra and C. A. C. Coello, "Improving PSO-based multi-objective optimization using crowding, mutation and ∈-dominance," in International Conference on Evolutionary Multi-Criterion Optimization, Mexico, 2005, pp. 505–519.
32. N. Khodadadi, M. Azizi, S. Talatahari, and P. Sareh, "Multi-objective crystal structure algorithm (MOCryStAl): Introduction and performance evaluation," *IEEE Access*, vol. 9, pp. 117795–117812, 2021.

CHAPTER 6

Accelerating Optimization Using Vectorized Moth-Flame Optimizer (vMFO)

AmirPouya Hemmasian and Kazem Meidani

Carnegie Mellon University

Seyedali Mirjalili

Torrens University Australia
Yonsei University

Amir Barati Farimani

Carnegie Mellon University

CONTENTS

6.1 Introduction 98
6.2 MFO Algorithm and the Vanilla Implementation 99
6.3 Vectorized MFO 102
6.4 Experiments 104
6.5 Conclusion 108
References 109

DOI: 10.1201/9781003205326-8

6.1 INTRODUCTION

Metaheuristic optimization algorithms, and particularly multi-agent systems, cover a wide range of algorithms with different sources of inspiration. Such algorithms, however, share common features and a similar optimization process. One of the main advantages of these nature-inspired algorithms is that they are derivative-free, meaning that there is no need for the full mathematical form of the objective function and its derivative. Therefore, they rely solely on the fitness evaluations to intelligently seek the global optima throughout the search space in an iterative manner.

The main assumption that has motivated the development of the current large number of optimization algorithms in this area is the limitation of computation time and the number of fitness evaluations. Improving the speed of the optimization process, i.e., reducing the computation time, can be achieved in two ways: First, by algorithmic improvement, where the goal is to provide an algorithm that has faster convergence rate with the same number of fitness evaluations. This viewpoint has been the main focus of the researchers in this field, and the reason for the proposal of a variety of nature-inspired algorithms, such as MFO [1], and their improved and hybrid versions [2–4]. The second viewpoint to accelerate the optimization process is to perform the same operations and calculations with efficient and parallel utilization of the computational resources and hardware [5–7]. This approach leverages on both efficient hardware utilization and software implementation techniques to enhance the computation efficiency of the optimization process.

Faster implementation techniques for the metaheuristic optimization methods have been previously discussed in the context of parallel computation. Parallel techniques aim to reduce the computation time by performing the tasks in a simultaneous manner for different stages of the optimization [8]. Such methods can be applied to different levels and extents to the optimization process [9,10].

The workflow of the multi-agent metaheuristic optimization algorithms is quite similar to each other. We consider MFO to showcase the steps of optimization. At each iteration, the search agents are first evaluated by the objective function to obtain their current values and to find the best agents in terms of their fitness. Next, the unique update rule of the algorithm is used to change the position of the agents in the search space in a stochastic yet intelligent manner. These two main steps along with parameter calculation and elite selection steps are repeated throughout the optimization to converge as fast as possible to a good optimum value (Figure 6.1).

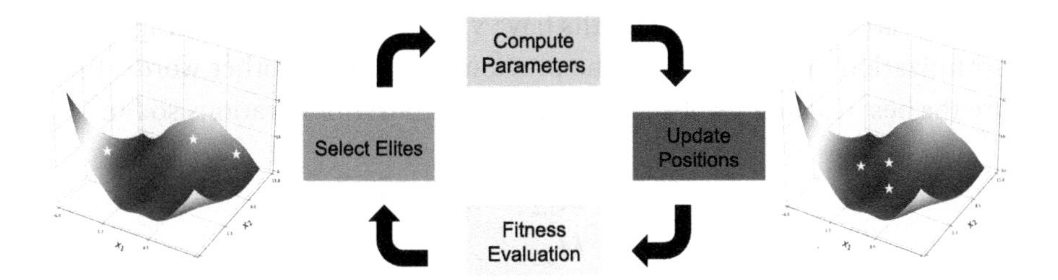

FIGURE 6.1 The general workflow of multi-agent metaheuristic algorithms based on swarm intelligence like Moth-Flame Optimization (MFO) algorithm. The steps at each iteration include evaluating the fitness of the agents, selecting the elites to follow, calculating parameters involved in the algorithm, and updating the positions of search agents in the space based on the algorithm's update rule.

The general workflow of multi-agent metaheuristic algorithms is based on swarm intelligence like Moth-Flame Optimization (MFO) algorithm. The steps at each iteration include (a) evaluating the fitness of the agents, (b) selecting the elites to follow, (c) calculating parameters involved in the algorithm, and (d) updating the positions of search agents in the space based on the algorithm's update rule.

The evaluation steps and the update step consist of various sub-steps which involve computations performed on the agents and in several dimensions of the search space. The implementation of loops over all the agents and the dimensions, however, employs a considerable portion of the computation time required in a single iteration. Here, we replace the time-consuming traditional implementation of MFO algorithm by a new version that employs vectorization to simultaneously compute the otherwise repetitive operations.

6.2 MFO ALGORITHM AND THE VANILLA IMPLEMENTATION

As the name suggests, the MFO algorithm is inspired by the spiral movement of moths around artificial lights (called flame). In this algorithm, we have n number of moths and n number of flames, all in the same d dimensional search space. The locations of the moths and the flames are stored in the matrices M and F, respectively, each with n rows and d columns. The element in the ith row and the jth column of M ($M_{i,j}$) is the value of the jth dimension of the ith moth, and the same holds for the flames in the matrix F as shown in Equation (6.1). The n flames are the best n locations in the search space that the moths have visited from the beginning of the optimization process up until the previous iteration. In other words, they are the best n solutions observed during the course of iterations so far.

$$M = \begin{bmatrix} M_{1,1} & M_{1,2} & \cdots & M_{1,d} \\ M_{2,1} & M_{2,2} & \cdots & M_{2,d} \\ \vdots & \vdots & \ddots & \vdots \\ M_{n,1} & M_{n,2} & \cdots & M_{n,d} \end{bmatrix},$$

$$F = \begin{bmatrix} F_{1,1} & F_{1,2} & \cdots & F_{1,d} \\ F_{2,1} & F_{2,2} & \cdots & F_{2,d} \\ \vdots & \vdots & \ddots & \vdots \\ F_{n,1} & F_{n,2} & \cdots & F_{n,d} \end{bmatrix} \tag{6.1}$$

We store the fitness values of the moths and flames in the vectors OM and OF, respectively, as shown in Equations (6.2a) and (6.2b), where f denotes the objective function:

$$\text{OM}_{n\times 1} = \left[f(M_1), f(M_2),\ldots, f(M_n) \right]^T \tag{6.2a}$$

$$\text{OF}_{n\times 1} = \left[f(F_1), f(F_2),\ldots, f(F_n) \right]^T \tag{6.2b}$$

In order to mathematically model the spiral movement of a moth around a flame, Equations (6.3a)–(6.3c) were proposed by Mirjalili [1]. In these equations, t is the current iteration, T indicates the total number of iterations, a shows the parameter which shrinks the spiral over time, and r is a random number sampled from a uniform distribution between 0 and 1. The parameter b is a constant that affects the shape of the spiral and is set to $b = 1$ in the default setting suggested by the author of this chapter.

It should be noted that each moth is appointed to the same index of the flame matrix during all of the optimization process, meaning that the moth i is going to move around the flame i in every iteration. Considering that the flame F_i might change during the optimization and be replaced with a better location in the search space, moth i would also change its center of spiral movement accordingly.

$$a = -1 - \frac{t}{T},\; r_{ij} \in [0,1],\; l_{i,j} = (a-1)r_{i,j} + 1 \tag{6.3a}$$

$$D_{i,j}(t) = \left|F_{i,j}(t) - M_{i,j}(t)\right| \tag{6.3b}$$

$$M_{i,j}(t+1) = D_{i,j}(t) \cdot e^{bl_{i,j}} \cdot \cos(2\pi l_{i,j}) + F_{i,j}(t) \tag{6.3c}$$

In order to make the whole population converge toward a single location as the optimization reaches the final iterations, the number of active flames (n_f) changes during the course of iterations according to Equation (6.4). All the moths with an index bigger than n_f are going to move around the flame with the index n_f because their flame is not active anymore. In the very final iterations, all of the moths are moving around the first flame F_1, i.e., the best location observed during the whole optimization, and performing local search and exploitation in its neighborhood. Therefore, in the case where $i > n_f$, the moths are going to move around the flame F_{n_f} according to Equation (6.5):

$$n_f = \text{round}\left(n - (n-1)\frac{t}{T}\right) \tag{6.4}$$

$$M_{i,j}(t+1) = D_{i,j}(t) \cdot e^{bl_{i,j}} \cdot \cos\left(2\pi l_{i,j}\right) + F_{n_f} \cdot j(t) \tag{6.5}$$

These equations are applied to each element of the matrix M one by one using two nested for loops, one for going through the moths and the other for going through the dimensions of each moth as shown in Algorithm 6.1.

Algorithm 6.1: Original Implementation of MFO [1]

Initialize the moth population matrix M
for $i = 1:n$ **do**
  compute the fitness value for M_i and store it in OM_i
sort OM to obtain the initial values for F and OF
while $t \leq T$ **do**
  Check if any search agent goes beyond the search space and amend it
  update a, n_f
    for $i = 1:n$ **do**
      for $j = 1:d$ **do**
        if $i \leq n_f$ **then**
          update $M_{i,j}$ using Equation (6.3)
        if $i > n_f$ **then**
          update $M_{i,j}$ using Equation (6.5)

for $i = 1 : n$ **do**

compute the fitness value for M_i and store it in OM_i

concatenate the vectors OM and OF, then sort the resulted combined vector and choose the first n ones as the new flames, storing their positions and fitness values in F and OF, respectively

$t = t + 1$

return F_1

MFO has a simple update equation and is easy to implement, and has shown magnificent performance especially in optimizing low-dimensional objective functions. However, there are some downsides to this algorithm because of its update scheme. In each iteration, a population consisting of $2n$ points needs to be sorted, which can significantly increase the computational efficiency of the algorithm for high number of search agents. Although this step is inevitable, we can save some time by making the other steps of the algorithm faster and more efficient. These steps include function evaluation, update equations, and random number sampling, which will be discussed more in the following section.

6.3 VECTORIZED MFO

When the same update equation is used to modify the values of all the elements in a matrix, there is a much more efficient way of changing those values than doing it one by one with two nested loops. The value of each element in the matrix only depends on the value of the same element of the matrix from the previous iteration and not on the other elements in the matrix. So, there is no need to compute the elements of the new matrix one by one, making it possible to compute them in parallel. Vectorization is the approach of performing the same calculations on some or all of the elements of a matrix by implementing the same update equation on whole matrices instead of their individual elements iteratively. The way that this line of code is interpreted by the computer is that every element of the matrix is supposed to be changed according to the same rules simultaneously and with efficient allocation of the memory. Vectorization is included in many programming frameworks like MATLAB® and the numpy library in Python. The platform used here is coded in python and the vectorized computations are performed with numpy [11].

In the case for Metaheuristics and MFO in this work, the serial updating of the matrices through the course of iterations is inevitable because they depend on the value of the matrices in the previous iteration. On the other hand, the sub-steps including the fitness evaluation of the agents and the update of agents' positions in each dimension can be performed simultaneously, due to the fact that they are independent of each other and there is no reason to wait to calculate one until you have calculated all the previous ones. Therefore, instead of the computationally expensive nested loops over all the dimensions and the search agents, we apply the update equations to matrices themselves, which leads to the exact same result but in a much shorter time. The fitness evaluation step is also implemented using the function numpy.apply_along_axis which enables us to apply the objective function to every row or every column of a matrix in one line and without any for loops. This line can also be replaced with an objective function which already has a vectorized implementation, meaning that it takes a whole M matrix and returns the vector OM. A comparison of the original and the vectorized updates in a single iteration is illustrated in Figure 6.2.

In MFO, there are two different update equations based on the number of the moth and the number of active flames. In the case with several update equations, we slice the matrix and implement each update equation on the relevant slice of the matrix. This way the elements which are supposed to be updated using the same equation are in the same slice, and

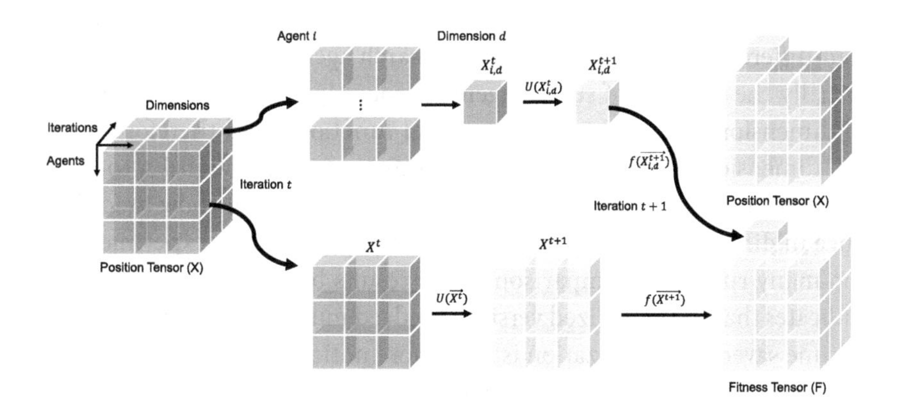

FIGURE 6.2 The scheme of updating position and fitness tensors in a single iteration via original serial implementation compared to the vectorized mapping. In the original version, the positions and function values are computed for each agent i and dimension d sequentially (top). In the proposed vectorization, we can compute the position update rule $U\left(\vec{X}\right)$ and fitness evaluation $f(X)$ simultaneously and significantly reduce the computation time.

again, a lot of computational time is going to be saved by implementing the updates with this approach and using vectorization.

One other important aspect of this algorithm is the random values $r_{i,j}$ and $l_{i,j}$ that are needed for calculating the next position of each moth. In the original implementation of MFO, these random values were sampled and calculated one by one for each element of the matrix. Similar to the update equations, we can sample a whole matrix of random values instead of sampling the elements one by one. This also helps to speed up the algorithm and save computational time. So, in each iteration, we sample the matrix r which has the same dimension as the matrix M and calculate the matrix l from it. Then, we use the whole matrix or slices from it for the vectorized calculations.

6.4 EXPERIMENTS

In order to quantitatively demonstrate the amount of the computational time saved by vectorization, we ran the vanilla MFO algorithm and the vectorized implementation (vMFO) on the 23 benchmark functions commonly used in the literature [12,13], containing several unimodal and multimodal landscapes. We recorded the wall-clock time spent for every run of the optimization. The processor used in our experiment is Intel(R) Core (TM) i9-9900K CPU with a rating of 3.6 GHz. We set the maximum number of iterations to 200 and conducted the experiments for 10 independent runs and reported the average time. The experiments were conducted for four different values of the number of search agents as shown in Table 6.1. Usually the number of search agents is roughly the same as the number of dimensions for the high-dimensional functions, and therefore, we set the dimension of the search space in functions F_1 through F_{13} to be the same as the number of search agents. The time does not show much variance in different runs, so there is no need to perform the experiments for too many runs. The comparison of the results between MFO and vMFO indicates that the vectorized version is indeed much faster, and the amount of time saved by vectorization is even more noticeable for large number of search agents and high-dimensional search spaces.

Algorithm 6.2: Vectorized Implementation of MFO

Initialize the moth population matrix M
obtain OM by performing vectorized function evaluation on M
sort the moth population to create the initial flames matrix F

TABLE 6.1 The Amount of Time (in Seconds) Needed to Execute an Optimization for 200 Iterations with MFO and vMFO, and the Speedup Ratio (SR) of Vectorization

	MFO	vMFO	SR	MFO	vMFO	SR	MFO	vMFO	SR	MFO	vMFO	SR
F1	1.83	0.06	32.68	7.24	0.11	63.58	16.09	0.19	84.33	29.55	0.28	105.10
F2	1.87	0.06	30.13	7.30	0.17	44.22	16.04	0.32	49.53	30.07	0.52	57.39
F3	2.36	0.54	4.35	9.34	2.20	4.24	20.47	4.77	4.29	38.42	9.05	4.25
F4	1.82	0.04	42.93	7.09	0.10	71.46	15.94	0.18	87.14	29.68	0.30	100.27
F5	1.90	0.08	22.79	7.24	0.17	42.59	16.41	0.27	59.95	29.78	0.40	75.24
F6	1.84	0.06	29.15	7.10	0.13	55.81	16.22	0.22	75.41	29.96	0.31	95.78
F7	1.82	0.13	14.47	7.18	0.31	23.11	16.11	0.56	28.58	29.48	0.87	33.98
F8	1.82	0.06	29.86	7.04	0.15	47.24	16.42	0.28	58.11	29.69	0.46	64.89
F9	1.78	0.08	21.38	6.90	0.18	39.22	15.58	0.30	52.44	28.70	0.44	64.93
F10	1.94	0.12	16.00	7.30	0.25	28.72	16.28	0.41	40.16	29.62	0.59	49.95
F11	1.92	0.13	14.67	7.29	0.32	23.03	16.93	0.58	29.33	30.01	0.91	33.01
F12	2.07	0.24	8.51	7.59	0.55	13.84	17.05	0.90	19.01	30.97	1.35	23.03
F13	2.13	0.27	7.78	7.76	0.62	12.53	17.04	1.01	16.91	31.07	1.51	20.58
F14	2.10	1.98	1.06	4.14	3.98	1.04	6.41	5.86	1.09	8.43	8.03	1.05
F15	0.38	0.12	3.13	0.74	0.23	3.22	1.13	0.33	3.43	1.62	0.45	3.59
F16	0.16	0.03	4.97	0.30	0.05	6.20	0.46	0.07	6.86	0.64	0.08	7.61
F17	0.16	0.04	4.50	0.31	0.06	5.41	0.47	0.08	5.86	0.65	0.10	6.23
F18	0.17	0.04	3.94	0.32	0.07	4.37	0.49	0.10	4.81	0.70	0.13	5.38
F19	0.41	0.21	1.96	0.82	0.42	1.93	1.25	0.61	2.06	1.70	0.86	1.99
F20	0.60	0.22	2.72	1.18	0.44	2.68	1.80	0.63	2.86	2.52	0.90	2.79
F21	1.14	0.85	1.34	2.35	1.74	1.35	3.52	2.56	1.38	4.63	3.52	1.32
F22	1.44	1.14	1.26	2.99	2.34	1.28	4.45	3.52	1.26	5.90	4.77	1.24
F23	1.91	1.63	1.17	3.90	3.25	1.20	5.83	4.98	1.17	7.69	6.59	1.17

Note: In high-dimensional functions (F1–F13), the dimension of the search space is equal to the number of search agents.

while $t \leq T$ **do**

Check if any search agent goes beyond the search space and amend it

Calculate n_f according to Equation (6.4)

update a, sample r, calculate l

update $M\left[1:n_f,:\right]$ using the flames $F\left[1:n_f,:\right]$

update $M\left[n_f+1:n,:\right]$ using the flame $F\left[n_f,:\right]$

obtain OM by performing vectorized function evaluation on M

concatenate the vectors OM and OF, then sort the resulted combined vector and choose the first ones as the new flames, storing their positions and fitness values in F and OF, respectively

$t = t+1$

return F_1

In order to provide a better insight about the amount of time needed for MFO for different objective functions and how it scales with the number of dimensions and search agents, we plot the time for different values of N and d and see how the computational time grows. As we can see in Figure 6.3, when we are performing optimization in high-dimensional search spaces like 100 dimensions, even just 200 iterations of MFO can take up to 30 seconds. For challenging optimization problems, we might need a lot more iterations in order to reach a satisfactory optimum. Moreover, when we want to evaluate modified versions of the algorithm and analyze their performance for research purposes, we need to conduct the experiments for many number of independent runs (usually 30) in order to reach statistically significant results. Therefore, using the vanilla implementation of MFO for many runs and functions might take up to several hours.

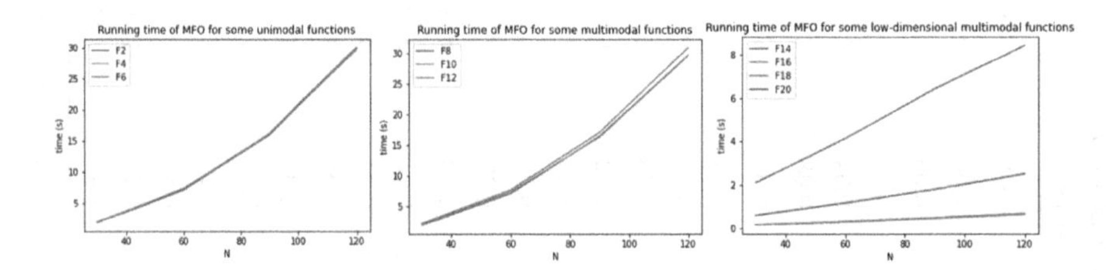

FIGURE 6.3 The average time spent by MFO to perform a full optimization with 200 iterations on several objective functions in python. Low-dimensional functions have different number of dimensions and diverse functional forms, therefore their computational time varies more than high-dimensional functions.

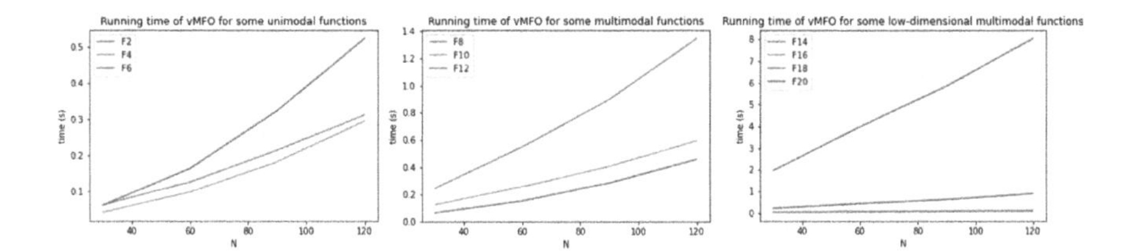

FIGURE 6.4 The average time spent by vMFO to perform a full optimization with 200 iterations on several objective functions in python. Compared to Figure 6.3, the difference for high-dimensional functions is very outstanding.

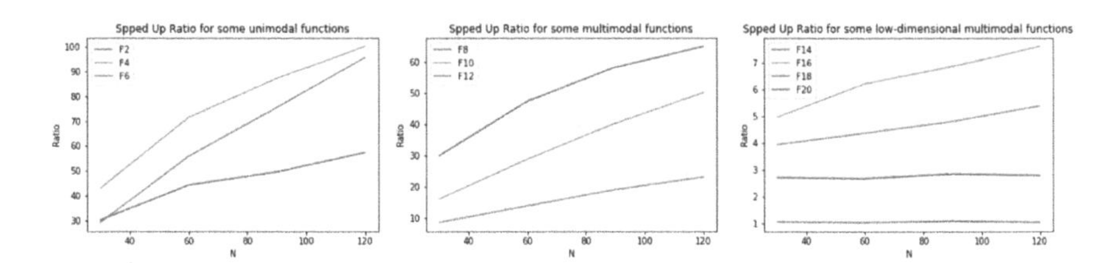

FIGURE 6.5 The ratio of the time spent by MFO to vMFO for optimizing different objective functions. We call this the Speedup Ratio. For high-dimensional functions, this ratio can even reach values around 100. The computational advantage is less significant for low-dimensional functions. Because of the diversity of the function forms and dimension of the search space, we see a rather different result for low-dimensional functions.

Now if we look at Figure 6.4, we observe that the computational time in the high-dimensional functions can be orders of magnitude lower than that of the vanilla implementation. Even for very large number of dimensions and search spaces, the computational time is much more scalable than the vanilla implementation. The Speedup Ratio (SR), which is the time spent by MFO divided by the time for vMFO, is plotted in Figure 6.5. We can see that for high-dimensional problems, the vectorized version can be up to 100 times faster, which means 100 times more experiments and more independent runs both for research purposes and for applications. Due to the randomness of metaheuristic algorithms, they are usually run for many times in order to find a good enough optima for practical purposes. The faster the algorithm, the more trials possible and the chance of getting a better optima in the same amount of time is highly increased. The difference for low-dimensional functions might not be as noticeable as the high-dimensional functions, but in large number of runs, even smaller SRs can make a difference. One important factor is the complexity of the objective function itself, meaning more complex functions need more time for evaluation and the time saved in vectorized update equations will consist a lower portion of the total time spent for the optimization. In the end, the vectorized implementation can be much faster and needs much less lines of code to write and program.

To summarize, vMFO performs the exact same calculations as MFO but with a vectorized implementation for function evaluations and update equations. This means that there is no for loop to go over the search agents and dimensions one by one, and everything is almost simultaneously calculated for all search agents and dimensions.

6.5 CONCLUSION

In this chapter, we introduced a more efficient approach of implementing the MFO algorithm using the feature of vectorization which is included in many programming frameworks like MATLAB® and the numpy library in python. We ran the vanilla algorithm and the vectorized implementation on our system with the processor Intel(R) Core (TM) i9-9900K CPU with a rating of 3.6 GHz and recorded the average wall-clock time spent on each run of the algorithm on several objective functions. The result of the comparison between the two implementations indicates that vectorization can probably reduce the computational time needed and increase the speed significantly, especially when dealing with large number of dimensions of the search space and the number of search agents. For high-dimensional functions like F_1, the vectorized version might be even 100 times faster than the vanilla implementation. In the end, vMFO performs the exact same update equations but in a very more efficient way than the vanilla implementation, allowing engineers to make the most out of their time and run the algorithm for their applications for more times and achieve better optimal solution. It also allows researchers to spend less time on trials while developing new variants of the algorithm and more time on designing the new algorithm itself. With this efficient implementation, there are now more possibilities to add new modifications and operations to improve other aspects of the MFO algorithm that were not practical or scalable before because of the computational cost of the original implementation. We encourage researchers to also apply their modifications in a vectorized implementation to decrease the wall-clock time needed for their experiments and running their algorithms for their applications. The vectorized version can also be applied with a much larger number of search agents and number of iterations, and be utilized to optimize functions with a very high-dimensional search space compared to the original implementation. Usually, the step which requires the most computational time is function evaluation. If this step can be vectorized, the algorithm is going to be very fast, allowing many opportunities as mentioned above.

REFERENCES

1. S. Mirjalili, Moth-flame optimization algorithm: A novel nature-inspired heuristic paradigm, *Knowledge-Based Systems* 89 (2015) 228–249.
2. D. Pelusi, R. Mascella, L. Tallini, J. Nayak, B. Naik, Y. Deng, An improved moth-flame optimization algorithm with hybrid search phase, *Knowledge-Based Systems* 191 (2020) 105277. doi: 10.1016/j.knosys.2019.105277. URL https://www.sciencedirect.com/science/article/pii/S0950705119305763.
3. G. I. Sayed, A. E. Hassanien, A hybrid SA-MFO algorithm for function optimization and engineering design problems, *Complex & Intelligent Systems* 4(3) (2018) 195–212. doi: 10.1007/s40747-018-0066-z.
4. M. Shehab, H. Alshawabkah, L. Abualigah, N. AL-Madi, Enhanced a hybrid moth-flame optimization algorithm using new selection schemes, *Engineering with Computers* 37(4) (2021) 2931–2956. doi:10.1007/s00366-020-00971-7.
5. A. Hassani, J. Treijs, An overview of standard and parallel genetic algorithms (2009).
6. E. Alba, P. Vidal, Systolic optimization on GPU platforms, in: R. Moreno-Díaz, F. Pichler, A. Quesada-Arencibia (Eds.), *Computer Aided Systems Theory: EUROCAST 2011*, Springer, Berlin, Heidelberg (2012), pp. 375–383.
7. E.-G. Talbi, V. Bachelet, Cosearch: A parallel cooperative metaheuristic, *Journal of Mathematical Modelling and Algorithms* 5(1) (2006) pp. 5–22. doi:10.1007/s10852-005-9029-7. doi: 10.1007/s10852-005-9029-7.
8. T. Crainic, *Parallel Metaheuristics and Cooperative Search*, Springer International Publishing, Cham (2019) pp. 419–451. doi: 10.1007/978-3-319-91086-4_13.
9. E. Alba, M. Tomassini, Parallelism and evolutionary algorithms, *IEEE Transactions on Evolutionary Computation* 6(5) (2002) 443–462. doi:10.1109/TEVC.2002.800880.
10. E. Alba, G. Luque, S. Nesmachnow, Parallel metaheuristics: Recent advances and new trends, *International Transactions in Operational Research* 20 (2013) 1–48.
11. C. R. Harris, K. J. Millman, S. J. van der Walt, R. Gommers, P. Virtanen, D. Cournapeau, E. Wieser, J. Taylor, S. Berg, N. J. Smith, R. Kern, M. Picus, S. Hoyer, M. H. van Kerkwijk, M. Brett, A. Haldane, J. F. del Río, M. Wiebe, P. Peterson, P. Gérard-Marchant, K. Sheppard, T. Reddy, W. Weckesser, H. Abbasi, C. Gohlke, T. E. Oliphant, Array programming with NumPy, *Nature* 585(7825) (2020) 357–362. doi:10.1038/s41586-020-2649-2.
12. S. Mirjalili, S. M. Mirjalili, A. Lewis, Grey wolf optimizer, *Advances in Engineering Software* 69 (2014) 46–61.
13. J. Digalakis, K. Margaritis, On benchmarking functions for genetic algorithms, *International Journal of Computer Mathematics* 77(4) (2001) 481–506. doi: 10.1080/00207160108805080.

CHAPTER 7

A Modified Moth-Flame Optimization Algorithm for Image Segmentation

Sanjoy Chakraborty

Iswar Chandra Vidyasagar College

National Institute of Technology Agartala

Sukanta Nama

Maharaja Bir Bikram University

Apu Kumar Saha

National Institute of Technology Agartala

Seyedali Mirjalili

Torrens University Australia

Yonsei University

CONTENTS

7.1 Introduction 112

7.2 Problem formulation of multilevel thresholding 114

7.2.1 Kapur's entropy method 115

7.2.2 Segmentation performance measure 116

7.3 Moth-flame optimization 116

7.4 Proposed modified m-MFO algorithm 119

7.5 Comparison of numerical results 120

DOI: 10.1201/9781003205326-9

7.5.1 Comparison with evaluated benchmark function results 120
7.5.2 Comparison of Image segmentation results 120
7.6 Convergence analysis 123
7.7 Conclusion 125
References 126

7.1 INTRODUCTION

Image segmentation is a crucial cycle of image preprocessing in which a given image is divided into several important homogenous sub-areas [1]. The similitude technique based on pixel intensity is used by the majority of image segmentation algorithms. It divides using the similarity between the image objects and a predefined standard. Thresholding, region increasing, region splitting, and region merging are several well-known similitude procedures. In terms of precision, simplicity, and heartiness, which are dependent on the image histogram, the thresholding approach outperforms all other segmentation algorithms. Thresholding divides the image into several portions by applying a variety of gray level limits. If there is just one gray level value, the image is divided into two localities; this is a reference to bi-level thresholding. Multiple thresholding is a bi-level thresholding extension in which more than one gray level value is used to divide the image into different sub-areas. To find the threshold value in a picture, the histogram-based technique is most usually utilized. Kapur's entropy [2] is one of the most common one-dimensional thresholding approach cited in the literature that uses the histogram of an image.

These days multilevel thresholding is firmly suggested over the bi-level thresholding for genuine pictures since bi-level thresholding does not give proper execution [3]. The multilevel image thresholding problem may be viewed as an optimization problem in which nature-inspired approaches are employed to discover the best answer. In recent years, researchers have expanded their investigation of multilevel image thresholding based on advancement methodologies. PSO and ABC algorithms are used in Refs. [4] and [5] for multilevel least cross-entropy rule and they enhance the most extreme cross-entropy standard. In Ref. [6], the authors proposed multilevel thresholding method based on Harmony Search leveraging on the work done by Otsu and Kapur. In Ref. [7], the authors improved beta differential development algorithm to enhance Otsu's variance-based measures. Other similar works can be found in Refs. [8] and [9].

The Moth-flame optimization (MFO) [10] method is a novel algorithm proposed by Mirjalili which has performed great capacities in a few applications. It was developed by mirroring the winding fly of moths around artificial lights and it performs very well in exploiting the search space. Still the lack of diversity in solution may be trapped into local solution [11]. Considering the weakness in MFO, numerous researches have been done to upgrade its performance and solve several issues. Few of those are in Ref. [12] in which the authors developed BMMFOA (modified moth-flame binary-coded optimization algorithm) to solve unit commitment (UC) problem. MMFOA (modified MFO algorithm) was used to increase the exploitation and lessen the number of flames. The productivity of the proposed BMMFOA has been verified on four single commitment choices. DELMFO [13] was introduced with two evolutionary learning strategies, viz., dynamic flame guidance (DFG) and DE flame generation (DEFG). It was compared with the other six MFO variants, ten metaheuristic algorithms, and nine optimization algorithms using various benchmark suits. E-MFO was presented by Kaur et al. [14]. The authors modified the algorithm by dividing the iterations, introducing a Cauchy distribution function, altering the influence of best flame and adaptive step size, and maintaining the balance between exploration and exploitation. Abd Elaziz et al. [15] introduced OMFODE by modifying the classical MFO algorithm with the OBL (opposition-based learning) strategy and DE. The new method was used to enhance ten UCI datasets for feature selections. An improved variant of MFO was developed by Ma et al. [16], where the objective was to alleviate slow convergence and premature convergence in MFO. Inertia weight with diversity feedback control and a probability mutation factor were introduced. MFO has been updated and used to solve several problems like segmentation of color image [17], parameter identification of single-phase inverter [18], and photovoltaic power generation [19].

One of the common trends of metaheuristic study is improving and modifying the existing algorithms for better performance. In the above paragraph, we have seen several variants of MFO which have been suggested by researchers. Likewise, some modifications on other metaheuristic algorithms have been made by many scholars. Few recent modifications on algorithms, such as SOS, BOA, and WOA, can be found in Refs. [20–26]. It is apparent from the above discussion that modification of algorithms and also their application to solve various physical world issues is a common phenomenon.

In this study, first, MFO has been modified by introducing a new parameter C and a weight W. The parameter C varies non-linearly from 2 to 0 and takes a random value between the range. The value of C is used while finding the distance of the moth from its corresponding flame position. The use of C enhances the exploration process in the algorithm. Weight W is used to exploit the nearing value of a flame position that also increases the convergence speed. Thus, C and W in the basic MFO algorithm balance the algorithm from an exploration and exploitation perspective. The proposed algorithm's efficacy is tested by solving 20 benchmark functions and comparing them to other approaches. The images are also segmented using the updated algorithm and Kapur's entropy approach. Finally, convergence analysis is performed to assess the algorithm's convergence speed.

The remainder of the chapter is organized as follows: Section 7.2 contains the picture of segmentation formulation with multilayer thresholding. Section 7.3 describes the fundamental MFO algorithm. Section 7.4 presents the newly adjusted m-MFO algorithm. Section 7.5 provides a comparison of the evaluated findings using various methodologies. Convergence analysis is covered in Section 7.6. Finally, Section 7.7 brings the research to a close.

7.2 PROBLEM FORMULATION OF MULTILEVEL THRESHOLDING

As discussed, thresholding can be bi-level or multilevel. In bi-level thresholding, only one threshold value t and two classes C^0 and C^1 are created. But in multilevel thresholding, n number of threshold values $\{t_1, t_2, \ldots, t_n\}$ are used that splits the image, Im, into $n+1$ classes of $\{C^0, C^1, C^2, \ldots, C^n)$.

In an image, Im, of g gray levels, bi-level thresholding can be written as follows:

$$C^0 = \{f(x,y) \in \mathrm{Im} \mid 0 \le f(x,y) \le t_h - 1\} \tag{7.1}$$

$$C^1 = \{f(x,y) \in \mathrm{Im} \mid t_h \le f(x,y) \le g - 1\}$$

where $f(x,y)$ denotes the intensity of the pixels of image Im.

For multilevel image thresholding, the above equations can be extended to

$$C^0 = \{f(x,y) \in \mathrm{Im} \mid 0 \le f(x,y) \le t_1 - 1\}$$

$$C^1 = \{f(x,y) \in \text{Im} \mid t_1 \le f(x,y) \le t_2 - 1\}$$

$$C^3 = \{f(x,y) \in \text{Im} \mid t_2 \le f(x,y) \le t_3 - 1\}$$

$$C^n = \{f(x,y) \in \text{Im} \mid t_n \le f(x,y) \le g - 1\}$$

7.2.1 Kapur's entropy method

Kapur's function estimates class distinctness and ascertains entropy estimation depending on the likelihood dispersion of the image gray level values. The threshold's optimal values are gained whenever entropy measure in segmented classes has the highest value. Accordingly, the point is to track down the most elevated entropy fitness value, which returns the best threshold value. The fundamental Kapur's entropy was produced for bi-level thresholding of pictures. Further, it very well may be stretched out to multilevel thresholding. For bi-level thresholding, the fitness function can be composed as

$$f_1(t_i) = et_0 + et_1 \tag{7.2}$$

where

$$et_0 = -\sum_{k=0}^{t_i-1} \frac{p_k}{\omega_0} \ln\left(\frac{p_k}{\omega_0}\right), \ \omega_0 = \sum_{k=0}^{t_i-1} p_k$$

$$et_1 = -\sum_{k=t_i}^{g-1} \frac{p_k}{\omega_1} \ln\left(\frac{p_k}{\omega_1}\right), \ \omega_1 = \sum_{i=t_i}^{g-1} p_k$$

In the above-written equations, et_0 and et_1 represent entropies; ω_0 and ω_1 imply class probabilities of the segmented classes C^0 and C^1, respectively. p_k is the probability of gray level k which is calculated as follows:

$$p_k = \frac{hg(k)}{\sum_{k=0}^{g-1} hg(k)}, \ i = 0,1,\ldots\ldots g-1$$

where $hg(k)$ signifies the histogram value of the pixel in the kth position.

For multilevel thresholding, the thought can be reached out to $(n+1)$ classes. This way the target function of multilevel thresholding can be composed as

$$f_1(t_1,t_2,\ldots,t_n)=et_0+et_1+\cdots+et_n \tag{7.3}$$

where

$$et_0=-\sum_{k=0}^{t_1-1}\frac{p_k}{\omega_0}\ln\left(\frac{p_k}{\omega_0}\right),\quad \omega_0=\sum_{i=0}^{t_1-1}p_k$$

$$et_1=-\sum_{k=t_1}^{t_2-1}\frac{p_k}{\omega_1}\ln\left(\frac{p_k}{\omega_1}\right),\quad \omega_1=\sum_{k=t_1}^{t_2-1}p_k$$

$$et_n=-\sum_{i=t_n}^{g-1}\frac{p_k}{\omega_n}\ln\left(\frac{p_k}{\omega_n}\right),\quad \omega_n=\sum_{i=t_n}^{g-1}p_k$$

$et_0,et_1,\ldots,et_n$ addresses entropies and $\omega_0,\omega_1,\ldots,\omega_n$ demonstrates class probabilities of the segmented classes $C^0,C^1,\ldots,C^n$ individually.

7.2.2 Segmentation performance measure

Numerous exhibition files are accessible to gauge multilevel image thresholding performance. Here, we have utilized pick signals to noise ratio (PSNR) for estimating execution in our examination. PSNR estimates the level of fragmented picture quality in decibels (DB). Numerically, it may be addressed as follows:

$$\text{PSNR}=10\log_{10}\left(\frac{255^2}{\text{MSE}}\right) \tag{7.4}$$

where MSE is the mean square error. MSE is calculated as follows:

$$\text{MSE}=\frac{i}{mn}\sum_{i=1}^{m}\sum_{j=1}^{n}\left[mg(i,j)-mg'(i,j)\right]^2$$

Variables m and n are the size of the images. Im and Im′ represent the original and segmented images, respectively.

7.3 MOTH-FLAME OPTIMIZATION

In classical MFO, all moths are expressed as solution vectors. Moths can fly in any dimension 1D, 2D, 3D,…,ND. Moths are represented in the solution space by the following matrix:

1
2

$$mo = \begin{bmatrix} mo_{1,1} & mo_{1,2} & \cdots & mo_{1,D-1} & mo_{1,D} \\ mo_{2,1} & \cdots & \cdots & \cdots & mo_{2,D} \\ \vdots & \cdots & \cdots & \cdots & \vdots \\ mo_{n-1,1} & \cdots & \cdots & \cdots & mo_{n-1,D} \\ mo_{n,1} & mo_{n,2} & \cdots & mo_{n,D-1} & mo_{n,D} \end{bmatrix}$$

where $mo_i = [mo_{i,1}, mo_{i,2}, \ldots, mo_{i,D}], i \in \{1,2,\ldots,n\}$.

n is the number of moths in the population, and D signifies the dimension of the solution. The fitness of each moth in the solution can be represented as follows:

$$\text{fit}[mo] = \begin{bmatrix} \text{fit}[mo_1] \\ \text{fit}[mo_2] \\ \vdots \\ \text{fit}[mo_n] \end{bmatrix}$$

3

Another critical part of the MFO method is flames. A grid-like moth matrix is considered as follows:

$$fm = \begin{bmatrix} fm_{1,1} & fm_{1,2} & \cdots & fm_{1,D-1} & fm_{1,D} \\ fm_{2,1} & \cdots & \cdots & \cdots & fm_{2,D} \\ \vdots & \cdots & \cdots & \cdots & \vdots \\ fm_{n-1,1} & \cdots & \cdots & \cdots & fm_{n-1,D} \\ fm_{n,1} & fm_{n,2} & \cdots & fm_{n-1} & fm_{n,D} \end{bmatrix}$$

Also, the fitness vector of the flame matrix is stored in the following matrix:

$$\text{fit}[fm] = \begin{bmatrix} \text{fit}[fm_1] \\ \text{fit}[fm_2] \\ \vdots \\ \text{fit}[fm_n] \end{bmatrix}$$

MFO has two essential components: one is the moth and the other is the flame, where the moth moves through the respective flame to achieve suitable outcomes, and the best outcomes acquired by the moth are known as flame. The moth physically moves following a spiral path. The movement of the moths mathematically designed using logarithmic spiral and is represented using the following equation:

$$mo_i^{it+1} = \begin{cases} \bar{D}_i \cdot e^{bt} \cdot \cos(2\pi t) + fm_i^{it}, & i \le \text{nfm} \\ \bar{D}_i \cdot e^{bt} \cdot \cos(2\pi t) + fm_{nfm}^{it}, & i \ge \text{nfm} \end{cases} \tag{7.5}$$

where $\bar{D} = |mo_i^k - fm_i^k|$ represents the distance of moth at the ith place and its specific flame (fm_i). The distance between the ith moth is mo_i and its particular flame, b, is a constant used to specify the spiral movement and t be any arbitrary number somewhere in the range of 0 to 1, alluding to how much nearer the moth is to its particular flame. A movable strategy has been recommended to diminish the variable t value over the redundancy, which improves the viability of both exploration and exploitation in the first and last iteration individually.

$$a_1 = -1 + it\left(\frac{-1}{\text{maxit}}\right) \tag{7.6}$$

$$t = (a_1 - 1) \times rnd + 1 \tag{7.7}$$

where maxit represents the number of maximum iterations, and a_1 is the convergence constant that decreases from (–1) to (–2) linearly, proving that both diversification and intensification occur in the MFO algorithm.

The flame position for the current and last iteration is collected and arranged according to the fitness value for the global and local search in every iteration. Only the best nfm flames are preserved and other flames are wiped away leading to imperfection. Both first and final flames are the best fitness and the worst fitness. Then, depending on the same order, the moths came to capture the flames one by one. The last flame will always be captured by same- and lower-ordered moths over the number of iterations. The following formula can obtain the number of flames (nfm) that has been reduced over the iteration:

$$\text{nfm} = \text{round}\left(mo - it * \frac{(mo-1)}{\text{maxit}}\right) \tag{7.8}$$

7.4 PROPOSED MODIFIED m-MFO ALGORITHM

The MFO algorithm is a population-based metaheuristic which is reliant on a component called transverse orientation. In this component, the moths will, in general, keep a fixed point regarding the moon. MFO suffers from the degeneration of the global search capacity and convergence speed [27]. In MFO, the flames are produced by arranging the best moths. Then again, the moths updated themselves using the flames. As a result, one of the shortcomings of MFO is the limited population variation, which makes it difficult to discover promising solutions. Considering these defects in MFO in this study, a new variant of MFO called m-MFO is introduced. A new parameter C and a weight W are employed to modify the algorithm. The parameter C is designed to vary non-linearly from 2 to 0. It also takes a random value between the range. The mathematical equation of the parameter is written as follows:

$$C = 2 - \cos(\text{rnd}) * \left(\frac{\text{it} - 1}{\text{maxit} - 1} \right) \tag{7.9}$$

where rnd is an arbitrary number within the limit [0,1], and it and maxit signify the present iteration and the maximum number of iterations, respectively. The value of C is multiplied by the distance $\bar{D}$ to shift the focus of the process from the particular flame. Thus, the inclusion of C diversifies the solution. On either side, a weight value is calculated using the equation

$$W = 0.3 + \text{rnd} * 0.3 \tag{7.10}$$

In the above equation, rnd specifies a random number between [0,1]. The value of W is multiplied with the flame position when updating the position of the moth. Therefore, the nearby value of the corresponding flame position is traversed. Weight W helps the search process in local search. Now after modification, Equation (7.5) is changed to

$$mo_i^{it+1} = \begin{cases} \bar{D}_i \cdot e^{bt} \cdot \cos(2\pi t) + W * fm_i^{it}, & i \leq \text{nfm} \\ \bar{D}_i \cdot e^{bt} \cdot \cos(2\pi t) + W * fm_{nfm}^{it}, & i \geq \text{nfm} \end{cases} \tag{7.11}$$

$$\text{and} \quad \bar{D} = C * \left| mo_i^{\text{it}} - fm_i^{\text{it}} \right| \tag{7.12}$$

The pseudo-code of the proposed m-MFO is given in Figure 7.1.

```
Initialize a population of moths
Until the satisfaction of an end condition repeat the following
Use Eq. 8 to update nfm (number of flames)
Update the flame set
For every moth
Use Eq. 6 and 7 to update a1 and t
Fig 1- Pseudo code of the proposed m-MFO.
Use Eq. (12) to update D̄
Use Eq. (11) to update the set of moths
After terminating the end condition, return to the best flame
```

FIGURE 7.1 Pseudo-code of the MFO algorithm.

7.5 COMPARISON OF NUMERICAL RESULTS

This section compares the results obtained by the proposed m-MFO using 20 classical benchmark functions and image segmentation using multi-level thresholding with various other well-known algorithms. A computer system with an Intel I3 processor, 8 GB RAM, and Windows 10 operating system and MATLAB® 2015a software is used to evaluate the results of all the algorithms.

7.5.1 Comparison with evaluated benchmark function results

The suggested m-MFO is employed to evaluate 20 classical benchmark functions. Description of all the functions can be found in Refs. [28,29]. F1 to F11 are the unimodal functions. Rest are the multimodal functions. The mean and standard deviation results of 30 individual runs using a population size of 30 and the maximum iteration of 500 are illustrated in Table 7.1, which are then used for performance measure. The algorithms considered here for comparison are WOA [30], MFO [31], BOA [32], DE [33], PSO [34], and SCA [35], while evaluation parameters for the employed algorithms are set as suggested in their original study. Analysis of Table 7.1 exposes that m-MFO found superior results than all other algorithms in F1, F2, F3, F4, F5, F7, F9, F10, F11, F13, F14, F15, F16, F17, F18, and F20. The proposed m-MFO shares similar optimal results with few other algorithms in F6 and F12. Among the comparison algorithms, WOA evaluates better result than m-MFO in F8. WOA, MFO, DE, and PSO estimate better result than m-MFO in F19.

7.5.2 Comparison of Image segmentation results

Results are evaluated using images airport (a) and couple (b). The images are taken from the USC-SIPI image database and are given in Figure 7.2. Results are compared with the evaluated results of MPBOA [36], HBO [37],

TABLE 7.1 Comparison of Results with State-of-the-Art Algorithms

Algorithm	F1 Mean	F1 SD	F2 Mean	F2 SD	F3 Mean	F3 SD	F4 Mean	F4 SD	F5 Mean	F5 SD
m-MFO	**4.12E−116**	2.26E−115	**3.55E−64**	1.39E−63	**8.07E−96**	3.34E−95	**6.22E−65**	2.05E−64	**1.10E−02**	2.10E−02
WOA	3.23E−74	1.40E−73	7.43E−50	3.31E−49	4.51E+04	1.18E+04	4.96E+01	2.97E+01	2.81E+01	4.52E−01
MFO	3.20E+01	7.64E+01	1.82E+01	1.92E+01	2.55E+04	1.35E+04	5.70E+01	8.70E+00	3.46E+04	4.60E+04
BOA	1.67E−01	8.63E−03	2.24E−01	9.92E−02	1.98E−01	1.32E−02	5.89E−01	2.59E−02	2.89E+01	3.02E−02
DE	9.13E−01	2.51E+00	2.05E−02	7.80E−02	4.29E+02	2.23E+02	1.80E+01	5.06E+00	8.52E+02	2.85E+03
PSO	3.95E−03	3.40E−03	9.54E−03	5.46E−03	4.83E+01	1.91E+01	7.78E−01	1.28E−01	4.91E+01	4.33E+01
SCA	1.72E−65	8.73E−65	1.20E−34	6.49E−34	1.67E−36	9.16E−36	1.13E−35	4.51E−35	9.58E+00	1.22E+01

Algorithm	F6 Mean	F6 SD	F7 Mean	F7 SD	F8 Mean	F8 SD	F9 Mean	F9 SD	F10 Mean	F10 SD
m-MFO	**0.00E+00**	0.00E+00	**1.88E−04**	1.35E−04	6.97E−160	3.82E−159	**8.94E−206**	0.00E+00	**1.36E−117**	7.10E−117
WOA	**0.00E+00**	0.00E+00	4.64E−03	5.41E−03	**3.79E−184**	0.00E+00	1.02E−80	5.26E−80	1.93E−66	5.97E−66
MFO	9.42E+01	1.17E+02	2.84E−01	1.10E−01	8.91E−72	4.15E−71	2.71E−91	1.20E−90	3.73E+08	2.43E+08
BOA	**0.00E+00**	0.00E+00	4.83E−04	2.81E−04	8.62E−03	3.67E−03	1.97E−02	4.93E−03	4.05E−01	2.41E−02
DE	7.20E+00	5.62E+00	5.27E−02	3.05E−02	2.97E−95	9.76E−95	3.59E−103	8.78E−103	1.27E+08	6.60E+08
PSO	2.67E-01	5.83E−01	2.66E−02	8.85E−03	2.41E−35	1.13E−34	1.69E−43	8.17E−43	9.47E+04	6.46E+04
SCA	**0.00E+00**	0.00E+00	7.15E−02	5.74E−02	1.45E−94	6.19E−94	4.19E−73	2.30E−72	3.45E−60	1.05E−59

Algorithm	F11 Mean	F11 SD	F12 Mean	F12 SD	F13 Mean	F13 SD	F14 Mean	F14 SD	F15 Mean	F15 SD
m-MFO	**5.31E−106**	2.91E−105	**0.00E+00**	0.00E+00	**8.88E−16**	0.00E+00	**0.00E+00**	0.00E+00	**1.45E−06**	2.65E−06
WOA	2.44E−05	1.62E−05	**0.00E+00**	0.00E+00	4.20E−15	2.63E−15	5.07E-03	2.78E-02	3.81E−02	4.59E−02
MFO	2.33E−05	2.01E−05	1.67E+02	3.70E+01	1.21E+01	8.71E+00	4.18E+00	1.64E+01	1.10E+01	7.13E+00
BOA	2.58E−02	8.45E−03	2.32E-01	2.14E-02	7.39E−01	2.69E−02	1.01E+00	9.19E−03	8.38E−01	1.77E−01
DE	4.84E−49	1.13E−48	1.65E+02	3.36E+01	7.30E−01	8.92E−01	1.29E−01	1.74E−01	7.64E−01	1.03E+00
PSO	3.22E−07	3.15E−07	3.62E+01	1.50E+01	1.27E−02	4.54E−03	2.95E−02	3.81E−02	4.32E−01	5.15E−01
SCA	4.85E−48	2.66E−47	**0.00E+00**	0.00E+00	8.88E−16	0.00E+00	0.00E+00	0.00E+00	3.28E−03	3.28E−03

(Continued)

TABLE 7.1 (*Continued*) Comparison of Results with State-of-the-Art Algorithms

Algorithm	F16		F17		F18		F19		F20	
	Mean	SD	Mean	SD	Mean	SD	Mean	SD	Mean	SD
m-MFO	**1.39E–05**	2.02E–05	**1.18E+00**	3.76E–01	**4.03E–04**	3.70E–05	–2.98E+00	1.26E-01	**–1.01E+01**	8.21E–02
WOA	4.96E–01	2.70E–01	2.77E+00	2.94E+00	7.01E–04	4.69E–04	–3.19E+00	1.85E–01	–7.86E+00	2.69E+00
MFO	8.03E+01	1.81E+02	2.67E+00	2.48E+00	2.45E–03	5.06E–03	–3.23E+00	6.32E–02	–6.55E+00	3.34E+00
BOA	3.24E+00	2.55E–01	3.73E+00	2.11E+00	1.02E–02	2.28E–02	–2.35E+00	4.34E–01	–4.28E+00	1.28E+00
DE	9.43E+02	4.17E+03	9.98E–01	0.00E+00	1.04E–03	3.65E–03	**–3.30E+00**	4.51E–02	–8.82E+00	2.76E+00
PSO	2.71E–03	5.44E–03	1.86E+00	1.41E+00	4.30E–04	3.45E–04	–3.27E+00	5.92E–02	–5.06E+00	3.04E+00
SCA	4.64E–02	4.65E–02	4.53E+00	5.08E+00	9.85E–04	8.05E–04	–2.36E+00	5.37E–01	–8.08E+00	2.17E+00

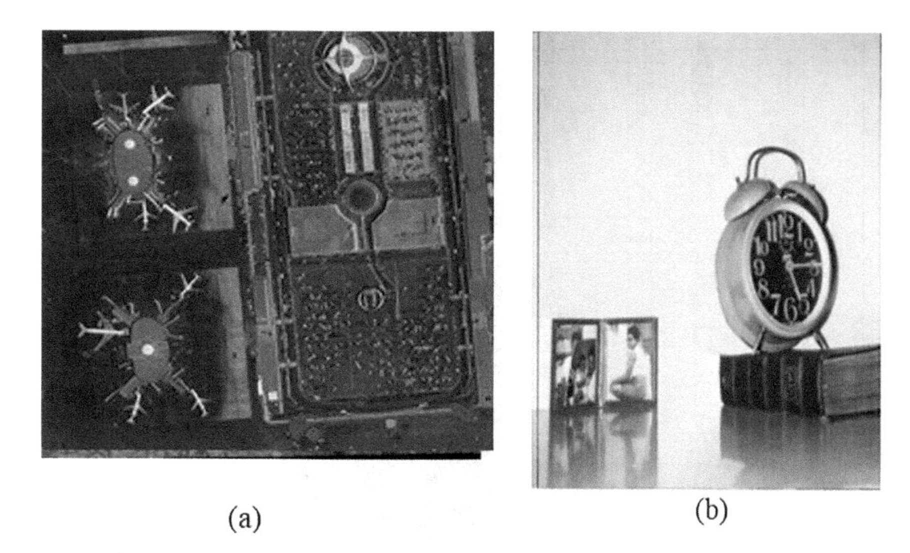

(a) (b)

FIGURE 7.2 Benchmark test images: (a) airport and (b) couple.

and SMA [38]. Population size, maximum iteration, and independent runs considered here are 50, 100, and 30, respectively. It is observed that the algorithms' results are almost the same for a small threshold value. Therefore, in this study, images are segmented using threshold values 4, 5, and 6. Fitness graph, segmented image, and segmented histogram for images a and b in different gray levels are shown in Figures 7.3 and 7.4, respectively. In maximization problems, if the fitness value is greater, the result of the optimization is better. Tables 7.2 and 7.3 represent the mean fitness value, PSNR value, and best value evaluated b for the algorithms. It is clear from the table data that m-MFO estimates maximum fitness value among the algorithms in every gray level for both the considered images.

7.6 CONVERGENCE ANALYSIS

Figure 7.5 presents some of the convergence graphs of m-MFO with its component algorithms MFO. The best solution's fitness value in each iteration is considered to draw convergence curves in these figures. From these figures, it can be noted that the proposed m-MFO algorithm converges faster for both types of functions, which establishes that the exploration and exploitation capacities of the proposed algorithm are more balanced than MFO.

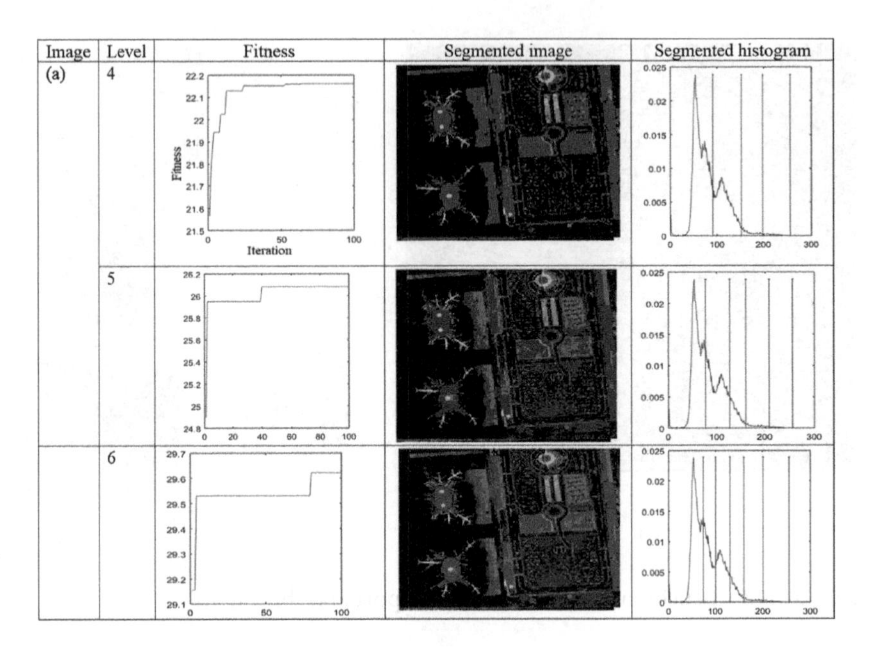

FIGURE 7.3 Optimized results after applying m-MFO using Kapur's method on image airport.

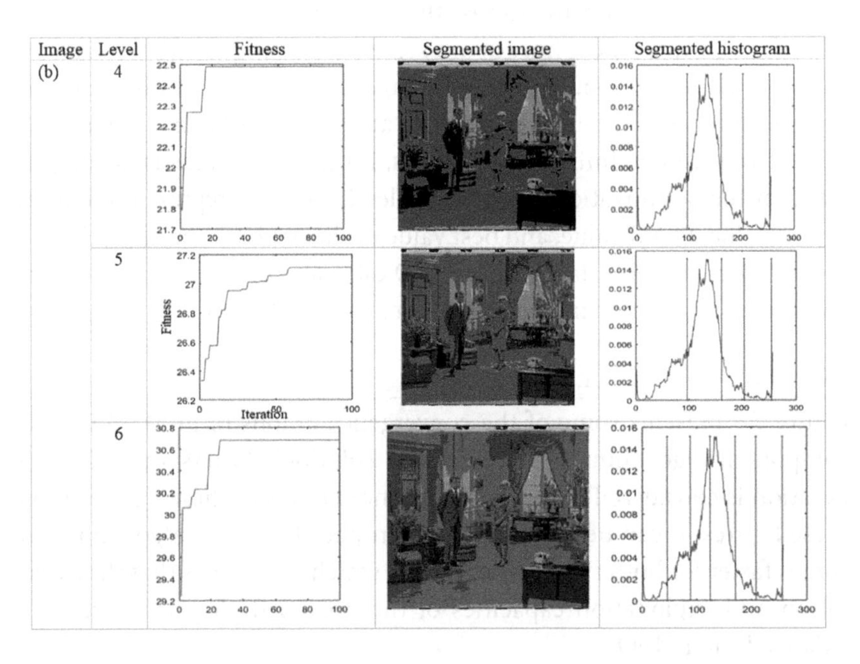

FIGURE 7.4 Optimized results after applying m-MFO using Kapur's method on image couple.

TABLE 7.2 Optimization Experimental Results on Benchmark Image Airport

Algorithms	**Image**	**Level**	**Intensity**	**Mean**	**PSNR**	**Best**
m-MFO	(a)	4	91 153 198 255	22.0192	13.3350	22.1476
MPBOA			91 150 197 250	21.8031	13.3531	22.0664
HBO			12 35 205 229	11.5874	13.1404	14.4335
SMA			125 152 184 256	21.0437	10.7769	21.5167
m-MFO	(a)	5	77 127 161 208 255	26.0458	14.8629	26.2136
MPBOA			75 117 157 197 256	25.7737	15.3172	26.2355
HBO			18 128 167 175 191	14.1584	12.6636	17.2298
SMA			79 143 205 219 256	24.2354	14.3169	24.9926
m-MFO	(a)	6	74 100 131 161 201 255	29.6992	15.7739	29.8752
MPBOA			75 109 140 165 192 256	29.3495	15.5154	29.7832
HBO			71 73 81 83 167 210	17.1270	15.3933	20.3645
SMA			116 159 175 189 231 256	27.4752	11.4224	28.0722

TABLE 7.3 Optimization Experimental Results on Benchmark Image Couple

Algorithms	**Image**	**Level**	**Intensity**	**Mean**	**PSNR**	**Best**
m-MFO	(b)	4	97 162 204 254	**22.5307**	15.0044	22.6199
MPBOA			63 112 182 253	22.1972	18.0043	22.5742
HBO			123 125 130 226	12.4030	11.7827	14.8519
SMA			52 110 207 256	21.1498	16.9063	21.7837
m-MFO	(b)	5	59 111 161 199 255	26.8427	18.9296	27.0787
MPBOA			68 110 167 203 248	26.3555	18.5428	26.8332
HBO			47 135 171 222 223	15.2328	14.3796	18.1667
SMA			67 117 128 192 256	24.6479	18.9060	25.4796
m-MFO	(b)	6	46 88 125 171 208 255	30.6332	20.8377	30.8352
MPBOA			52 81 113 168 207 249	30.1160	20.1270	30.6584
HBO			141 144 164 182 209 246	18.5393	8.9917	21.8935
SMA			42 49 119 161 204 243	27.9306	17.7526	28.8969

7.7 CONCLUSION

The MFO study reveals that the method suffers from poor exploration during the search phase, as well as a slow convergence speed. To address the inherent shortcomings of MFO, a newly updated method (m-MFO) is suggested, which includes a unique non-linear parameter C and a weight W. The parameter C is utilized to improve exploration capacity and the parameter W aids in exploitation. The comparison of m-MFO with other metaheuristic family algorithms employing benchmark functions and picture segmentation with multilayer thresholding problem indicates the improved performance of the suggested method.

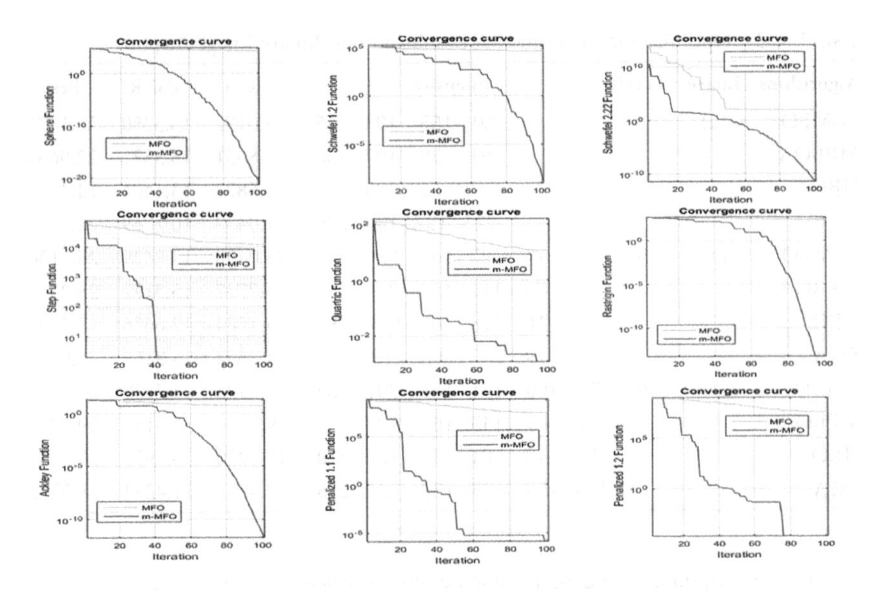

FIGURE 7.5 Convergence graphs of m-MFO with MFO in randomly selected functions.

REFERENCES

1. Ayala, H. V. H., dos Santos, F. M., Mariani, V. C., & dos Santos Coelho, L. (2015). Image thresholding segmentation based on a novel beta differential evolution approach. *Expert Systems with Applications,* 42(4), 2136–2142.
2. Kapur, J. N., Sahoo, P. K., & Wong, A. K. (1985). A new method for gray-level picture thresholding using the entropy of the histogram. *Computer Vision, Graphics, and Image Processing,* 29(3), 273–285.
3. Bhandari, A. K., Kumar, A., & Singh, G. K. (2015). Tsallis entropy based multilevel thresholding for colored satellite image segmentation using evolutionary algorithms. *Expert Systems with Applications,* 42(22), 8707–8730.
4. Yin, P. Y. (2007). Multilevel minimum cross entropy threshold selection based on particle swarm optimization. *Applied Mathematics and Computation,* 184(2), 503–513.
5. Horng, M. H. (2011). Multilevel thresholding selection based on the artificial bee colony algorithm for image segmentation. *Expert Systems with Applications* 38(11), 13785–13791. doi: 10.1016/j.eswa.2011.04.180.
6. Oliva, D., Cuevas, E., Pajares, G., Zaldivar, D., & Perez-Cisneros, M. (2013). Multilevel thresholding segmentation based on harmony search optimization. *Journal of Applied Mathematics,* 2013, 1–24.
7. Ayala, H. V. H., dos Santos, F. M., Mariani, V. C., & dos Santos Coelho, L. (2015). Image thresholding segmentation based on a novel beta differential evolution approach. *Expert Systems with Applications,* 42(4), 2136–2142.
8. Farshi, T. R. (2019). A multilevel image thresholding using the animal migration optimization algorithm. *Iran Journal of Computer Science,* 2(1), 9–22.

9. Wunnava, A., Naik, M. K., Panda, R., Jena, B., & Abraham, A. (2020). An adaptive Harris hawks optimization technique for two dimensional grey gradient based multilevel image thresholding. *Applied Soft Computing*, 95, 106526.
10. Mirjalili, S. (2015). Moth-flame optimization algorithm: A novel nature-inspired heuristic paradigm. *Knowledge-Based Systems*, 89, 228–249.
11. Khalilpourazari, S., & Khalilpourazary, S. (2019). An efficient hybrid algorithm based on water cycle and moth-flame optimization algorithms for solving numerical and constrained engineering optimization problems. *Soft Computing*, 23(5), 1699–1722.
12. Reddy, S., Panwar, L. K., Panigrahi, B. K., & Kumar, R. (2018). Solution to unit commitment in power system operation planning using binary coded modified moth flame optimization algorithm (BMMFOA): A flame selection based computational technique. *Journal of Computational Science*, 25, 298–317.
13. Li, C., Niu, Z., Song, Z., Li, B., Fan, J., & Liu, P. X. (2018). A double evolutionary learning moth-flame optimization for real-parameter global optimization problems. *IEEE Access*, 6, 76700–76727.
14. Kaur, K., Singh, U., & Salgotra, R. (2020). An enhanced moth flame optimization. *Neural Computing and Applications*, 32(7), 2315–2349.
15. Abd Elaziz, M., Ewees, A. A., Ibrahim, R. A., & Lu, S. (2020). Opposition-based moth-flame optimization improved by differential evolution for feature selection. *Mathematics and Computers in Simulation*, 168, 48–75.
16. Ma, L., Wang, C., Xie, N., Shi, M., Ye, Y., & Wang, L. (2021). Moth-flame optimization algorithm based on diversity and mutation strategy. *Applied Intelligence*. doi: 10.1007/s10489-020-02081-9.
17. Jia, H., Ma, J., & Song, W. (2019). Multilevel thresholding segmentation for color image using modified moth-flame optimization. *IEEE Access*, 7, 44097–44134.
18. Wu, Z., Shen, D., Shang, M., & Qi, S. (2019). Parameter identification of single-phase inverter based on improved moth flame optimization algorithm. *Electric Power Components and Systems*, 47(4–5), 456–469.
19. Lin, G. Q., Li, L. L., Tseng, M. L., Liu, H. M., Yuan, D. D., & Tan, R. R. (2020). An improved moth-flame optimization algorithm for support vector machine prediction of photovoltaic power generation. *Journal of Cleaner Production*, 253, 119966.
20. Saha, A., Nama, S., & Ghosh, S. (2019). Application of HSOS algorithm on pseudo-dynamic bearing capacity of shallow strip footing along with numerical analysis. *International Journal of Geotechnical Engineering*, 15, 1298–1311.
21. Nama, S., Saha, A. K., & Sharma, S. (2020). A novel improved symbiotic organisms search algorithm. *Computational Intelligence*, 1–31.
22. Nama, S. (2021). A modification of I-SOS: Performance analysis to large scale functions. *Applied Intelligence*, 51, 7881–7902.
23. Sharma, S., & Saha, A. K. (2020). m-MBOA: A novel butterfly optimization algorithm enhanced with mutualism scheme. *Soft Computing*, 24(7), 4809–4827.

24. Sharma, S., Saha, A. K., Majumder, A., & Nama, S. (2021). MPBOA-A novel hybrid butterfly optimization algorithm with symbiosis organisms search for global optimization and image segmentation. *Multimedia Tools and Applications*, 80(8), 12035–12076.
25. Chakraborty, S., Saha, A. K., Sharma, S., Mirjalili, S., & Chakraborty, R. (2021). A novel enhanced whale optimization algorithm for global optimization. *Computers & Industrial Engineering*, 153, 107086.
26. Chakraborty, S., Saha, A. K., Sharma, S., Chakraborty, R., & Debnath, S. (2021). A hybrid whale optimization algorithm for global optimization. *Journal of Ambient Intelligence and Humanized Computing*, 1–37.
27. Pelusi, D., Mascella, R., Tallini, L., Nayak, J., Naik, B., & Deng, Y. (2020). An improved moth-flame optimization algorithm with hybrid search phase. *Knowledge-Based Systems*, 191, 105277.
28. Banaie-Dezfouli, M., Nadimi-Shahraki, M. H., & Beheshti, Z. (2021). R-GWO: Representative-based grey wolf optimizer for solving engineering problems. *Applied Soft Computing*, 106, 107328.
29. Nadimi-Shahraki, M. H., Taghian, S., Mirjalili, S., & Faris, H. (2020). MTDE: An effective multi-trial vector-based differential evolution algorithm and its applications for engineering design problems. *Applied Soft Computing*, 97, 106761.
30. Mirjalili S., & Lewis, A. (2016). The whale optimization algorithm. *Advances in Engineering Software*, 95, 51–67. doi: doi: 10.1016/j.advengsoft.2016.01.008.
31. Mirjalili, S. (2015). Moth-flame optimization algorithm: A novel nature-inspired heuristic paradigm. *Knowledge-Based Systems*, 89, 228–249.
32. Arora, S, & Singh, S. (2018). Butterfly optimization algorithm: a novel approach for global optimization. *Soft Computing* 23,715. doi: 10.1007/s00500-018-3102-4.
33. Storn, R., & Price, K. (1997). Differential evolution: A simple and efficient heuristic for global optimization over continuous spaces. *Journal of Global Optimization*, 11(4), 341–359. doi: 10.1023/a:1008202821328.
34. Kennedy, J., & Eberhart, R. (1995). Particle swarm optimization. *Proceedings of ICNN'95: International Conference on Neural Networks*, 4, 1942–1948. doi: 10.1109/ICNN.1995.488968.
35. Mirjalili, S. (2016). SCA: A sine cosine algorithm for solving optimization problems. *Knowledge-Based Systems*, 96, 120–133.
36. Sharma, S., Saha, A. K., Majumder, A., & Nama, S. (2021). MPBOA-A novel hybrid butterfly optimization algorithm with symbiosis organisms search for global optimization and image segmentation. *Multimedia Tools and Applications*, 80(8), 12035–12076.
37. Askari, Q., Saeed, M., & Younas, I. (2020). Heap-based optimizer inspired by corporate rank hierarchy for global optimization. *Expert Systems with Applications*, 161, 113702.
38. Li, S., Chen, H., Wang, M., Heidari, A. A., & Mirjalili, S. (2020). Slime mould algorithm: A new method for stochastic optimization. *Future Generation Computer Systems*, 111, 300–323.

CHAPTER 8

Moth-Flame Optimization-Based Deep Feature Selection for Cardiovascular Disease Detection Using ECG Signal

Arindam Majee and Shreya Biswas
Jadavpur University

Somnath Chatterjee
Future Institute of Engineering and Management

Shibaprasad Sen
University of Engineering and Management

Seyedali Mirjalili
Torrens University Australia
Yonsei University

Ram Sarkar
Jadavpur University

DOI: 10.1201/9781003205326-10

CONTENTS

8.1 Introduction 130
8.2 Related Work 132
8.3 Motivation 134
8.4 Dataset Used 134
8.4.1 Pre-Processing 135
8.5 Methodology 136
8.5.1 VGG16: A Brief Overview 137
8.5.2 Moth-Flame Optimization 137
8.6 Experimental Results and Discussion 142
8.7 Conclusion 148
References 148

8.1 INTRODUCTION

Cardiovascular diseases (CVDs), such as Coronary, Cerebrovascular, Peripheral, Ischemic, Hypertensive, Congenital, Rheumatic, and Non-rheumatic valvular, are considered as the main reason of fatality with almost 17.9 million deaths each year. It accounts for almost 31% of all deaths worldwide [1]. In the United States, affecting most ethnic groups, it is one of the main diseases present [2]. The electrocardiogram (ECG) is used as a major tool in diagnosing CVDs. Automated analysis of standard ECG gained paramount importance in order to save time and effort, as the world transitioned from analog to digital. ECG is usually performed when a patient experiences acute chest pain following which the treatment can be immediately determined [3].

However, the problem is that many physicians solely rely on elemental diagnostic analysis of ECG results whereas they always require the understanding and confirmation by a trained technician [4].

On the other hand, Deep Learning (DL) has achieved remarkable success in medical diagnosis tasks [5,6] and has the potential to improve health care and clinical practice on a large scale [7]. Though an expert's confirmation is probably required in many clinical settings, DL can help an expert in the clinical environment. Studies show that Supervised Learning can perform better than a human specialist in medical testing and diagnosis [8,9]. However, efficient training of Deep Neural Networks (DNNs) requires a big dataset which, for medical applications, is very scarce—mainly due to confidentiality issues [10].

The ECG is the most common evaluation technique of the heart providing a proper evaluation of the patient's heart activity - including the cardiac rhythm, repolarization, arrhythmias, coronary syndromes, and effects of drugs. Thus, an automatic DL approach interpreting ECGs can be useful for better diagnosis of the patients.

ECG analysis has been done using DL in recent works [11], in which the authors trained DNNs on a fairly large-sized dataset consisting of 91,232 single-lead ECGs achieving a ROC area of 0.97. In Ref. [12], the authors used DNN to train the large publicly available PhysioNet Challenge dataset and achieved better performance when compared to that of cardiologists. Out of a total of 75 teams that entered the challenge, 4 teams won with an F1 score of 0.83.

The standard short-duration 12-lead ECGs are often performed in healthcare units, often with no specialists to analyze the ECG signals. Also, doctors during their training may lack a complete understanding of these tracings [13]. The automatic yet precise interpretation of ECG is the need of the hour with CVDs increasing at an alarming rate—owing to the unhealthy lifestyles that most people are incorporating nowadays [14]. In countries where maximum deaths are related to CVDs and people often do not have access to trained cardiologists, this can be a successful alternative.

The benefits of DNN usage for ECG evaluation are still largely unexploited due to the shortage of medically accurate digital ECG databases [15]. According to the authors of Ref. [16], there are very few databases with a large number of ECG tracings and their standardized explanation, hence limiting their effectiveness as training datasets for the DL methods which follow the supervised learning approach. Despite this disadvantage, the results that DL models produce are better than the analysis of most physicians.

In light of the above facts, under this procedure, we have initially applied a pre-trained Deep Convolutional Neural Network (DCNN) model—VGG16 for arrhythmia classification. Though this gives decent outcomes, the usage of DL models involves large computation cost and requires huge training data so as to generalize over new samples. However, we have used Transfer Learning (TL) in order to extract main features from the data for further processing. But the sizable features so produced may have some redundant and unessential features. Therefore, these obtained features need to be optimized by some means to reduce the computation time and storage requirement apart from the reduction of the redundancy of the features.

For dimension reduction of the feature vector, we have applied a swarm intelligence-based metaheuristic algorithm, the Moth-Flame Optimization (MFO) algorithm, which was originally proposed in Ref. [17]. This optimal subset of the feature set produced by tuning the parameters of MFO was then fed into a Support Vector Machine (SVM) classifier [18] to detect arrhythmia accurately.

Research highlights of this work are listed below:

1. Combining a DCNN model with a nature-inspired metaheuristic feature selection algorithm to classify ECG signals for identification of the CVDs—hence creating a two-staged approach architecture.
2. To manage the overfitting problem, we have used TL. The initial weights are generalized to get the full benefit of TL. Another use of this is to reduce the training time. Training a large model like VGG16 from scratch would have taken a lot of time.
3. For dimension reduction of the feature set that we got from the original DCNN model, a popular metaheuristic, called the MFO algorithm, has been used.
4. The proposed model gained state-of-the-art results when evaluated on the publicly available MIT-BIH Arrhythmia database. The optimized feature subset results in better classification accuracy than the non-optimized feature set produced by VGG16.

The remaining portion of this chapter has been divided into the following sections: Section 8.2 contains the related work in this field of ECG classification and diagnosis using DL methods. Section 8.3 mentions the motivation behind this work. Section 8.4 provides information about the dataset we have used in this work. Section 8.5 explains the pre-processing we have applied in the MIT-BIH dataset to obtain the image data. Section 8.6 contains the methodology and Section 8.7 explains the results we have obtained and also compares our work with some state-of-the-art works in this area. Section 8.8 concludes this chapter by discussing some future prospects to work on.

8.2 RELATED WORK

Here, we briefly discuss some past works related to the said research topic. In Ref. [19], the authors proposed an algorithm to classify noisy ECG signals from the MIT-BIH Arrhythmia database [20] using a one-dimensional CNN. The authors had applied two approaches. In one approach, a five-class classification from the 1D ECG data is obtained using 1D CNN, achieving about 97.4% accuracy. In another approach, they converted the 1D ECG data into 2D gray-scale image data and then used an eight-class classification on those images using a 2D CNN, achieving a classification accuracy of 99.02%.

In Ref. [21], the authors used Marine Predators Algorithm (MPA) with CNN, naming the method MPA-CNN, to classify four different types of arrhythmia. They worked on three separate datasets—the MIT-BIH, the European ST-T dataset (EDB) [22], and the St Petersburg INCART [23]. They obtained precisions of 99.31%, 99.76%, and 99.47% on the datasets, respectively.

The authors of the work reported in Ref. [24] proposed hybrid feature extraction of T-wave for arrhythmia classification on the MIT-BIH database. They used the windowing technique, feature extraction, followed by classification for the purpose, and obtained an accuracy of 98.3%, specificity of 98.0%, and sensitivity of 98.6%. The authors of Ref. [25] used cross-correlation feature extraction. A Least-Square SVM classifier was used for a three-class classification on the MIT-BIH database. The accuracy achieved was about 96%.

In Ref. [26], the authors proposed a 1D DNN model to correctly classify three different arrhythmias on the MIT-BIH database, achieving a 97.44% accuracy. In Ref. [27], the authors presented a classification system designed for detecting Ventricular Ectopic Beats. Accuracy–sensitivity performances of the system were 98.3%–84.6% and 97.4%–63.5%, respectively.

The authors of the work [28] used a DCNN model to classify five different arrhythmias and Myocardial Infarction (MI) from the MIT-BIH and PTB Diagnostics datasets. They obtained 93.4% and 95.9% accuracy on each dataset. The authors of Ref. [29] used Particle Swarm Optimization (PSO) to enhance the classification results of the SVM classifier using data from the MIT-BIH dataset. PSO gave an 89.72% accuracy.

The authors of Ref. [30] used wavelet-transformation on waveforms from 18 files of the MIT-BIH database. They generated the feature set for classification and obtained an accuracy of 95.16%–96.82%. In the work [31], the authors used a deep 2D CNN model for arrhythmia classification, achieving 99.05% accuracy and 97.85% sensitivity.

Recently, many Swarm intelligence optimization techniques (inspired by animal groups and insect colonies, mimicking their behavior) have been proposed and tested for general and medical diagnosis [32]. Robustness, flexibility, and the ability to quickly find the optimal solution to a particular problem are some of the usefulness of these metaheuristic algorithms [33]. Genetic Algorithm [34], Ant Colony Optimization [35], PSO [36], Artificial Bee colony optimization [37], and Cuckoo Search [38] are a few of them. Some of the applications of swarm intelligence-based optimization algorithms in the medical field deal with cancer screening [39,32], the fusion of MRI and CT scans [40], endocrinology [41], tumor classification [42], and so on.

8.3 MOTIVATION

MIT-BIH dataset has been used in this study because of its large amount of data as a DL model needs to properly train in order to correctly predict ECG classes. We have chosen a DL-based model for initial feature extraction as it has been seen that DL models perform better than machine learning-based models as the former can learn complex patterns automatically from the raw inputs. The choice of the MFO algorithm as the chosen metaheuristic algorithm is because it is a population-based algorithm with a local search strategy. This helps to find out the best possible solution using global and local exploitation [43]. MFO has few parameters, is flexible and easy to implement, and has faster convergence. Hence, it is being used to solve many problems such as parameter estimation [44], classifications [18], medical diagnoses [45], and image processing [46]. Other applications of MFO have been elucidated in Ref. [47].

8.4 DATASET USED

We have used the Massachusetts Institute of Technology-Beth Israel Hospital (MIT-BIH) Arrhythmia Database [20] that has 48 properly annotated records (each with a duration of 30 minutes and sampled at 360 samples per second with 11-bit resolution over a 10 mV voltage range). This dataset gives a proper view of all important waves present in an ECG signal, including P-waves, Q-waves, R-waves, S-waves, and T-waves. There are almost 15 labels, including 1 normal class. Among these labels, some are unclassified and some are normally not used in Arrhythmia detection. We have chosen to work with the MIT-BIH database because it provides a properly annotated collection of normal and several different types of arrhythmia to train our classifier.

For this work, we have collected the required arrhythmia recordings from the database which has more than 110,000 ECG beats with almost 16 different types of arrhythmia and 1 normal. As we have already discussed in Section 8.2, past authors have done arrhythmia classification tasks with different classes. After surveying previous works, we have decided to move with an eight-class classification as this contains normal and the most common Arrhythmia classes. From the MIT-BIH database, we have considered eight different ECG beats. These are normal beats (NOR) and seven different types of ECG arrhythmias. These seven different types of Arrhythmia are as follows: Premature Ventricular Contraction (PVC), Paced Beat (PAB), Right and Left Bundle Branch Block Beat (RBB and LBB), Atrial Premature Contraction (APC), Ventricular Flutter Wave (VFW), and Ventricular Escape Beat (VEB). We have ignored some less important beats in ECG arrhythmia classification studies like non-conducted P-wave and unclassifiable beats.

8.4.1 Pre-Processing

ECG signal is a time-based signal. To use these ECG 1D data in a 2D CNN model, we need to convert these into image data. We have followed a pre-processing strategy for this conversion. We plotted every ECG signal as an individual 128×128 gray-scale image and transformed them into corresponding ECG images. The MIT-BIH dataset is sampled with a frequency of 360 Hz, and the label is specified during the heartbeat's Q peak. So, each signal is segmented based on the Q-wave peak time and converted into gray-scale images after plotting. Thus, each ECG beat's Q-wave peak is kept at the center and the first and last 20 sampled values from the previous and afterward Q-wave peak signals are excluded as the duration of the Q-wave peak is 0.03 s or less. So, in the time domain, we can define the range of a single ECG beat as follows;

$$T\Big(\big(Q_{\text{peak}}\big)(n-1)+20\Big) \leq T(n) \leq \Big(\big(Q_{\text{peak}}\big)(n+1)-20\Big)$$

Following the above formula, we have excluded the first and last ECG beats. Figure 8.1 shows some images obtained after pre-processing. As a result, we have obtained a total of 1,07,620 images belonging to eight different classes. Then, we have divided this dataset into training, validation, and testing set in a ratio of 70%, 15%, and 15%. The number of images in every class is presented in Table 8.1.

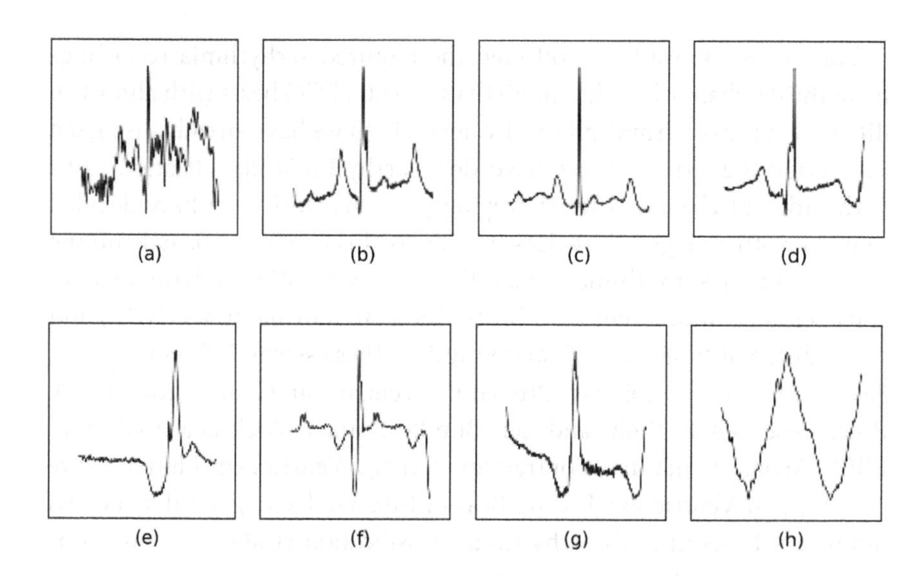

FIGURE 8.1 ECG signal Images formed after pre-processing. A sample from all eight classes is shown here: (a) APC, (b) LBB, (c) normal beat (NOR), (d) RBB, (e) PAB, (f) PVC, (g) VFW, and (h) VEB.

TABLE 8.1 Division of the Dataset—A Total of Eight Classes—into Train Set, Test Set, and Validation Set

	No. of Samples in		
Class	**Training Data**	**Validation Data**	**Testing Data**
APC	1,760	382	402
LBB	5,630	1,204	1,238
Normal	52,509	11,250	11,257
PAB	4,937	1,036	1,051
PVC	4,983	1,097	1,050
RBB	5,095	1,088	1,073
VEB	78	16	12
VFW	342	70	60
Total	75,334	16,143	16,143

8.5 METHODOLOGY

Before using the MFO algorithm, we have first extracted the features from the images. We have employed the VGG16 model pre-trained on the ImageNet dataset as our feature extractor model. After freezing the top layers up to block 5, we have fine-tuned the model on our ECG dataset. The pre-trained VGG16 model accepts three channel input images. So, we

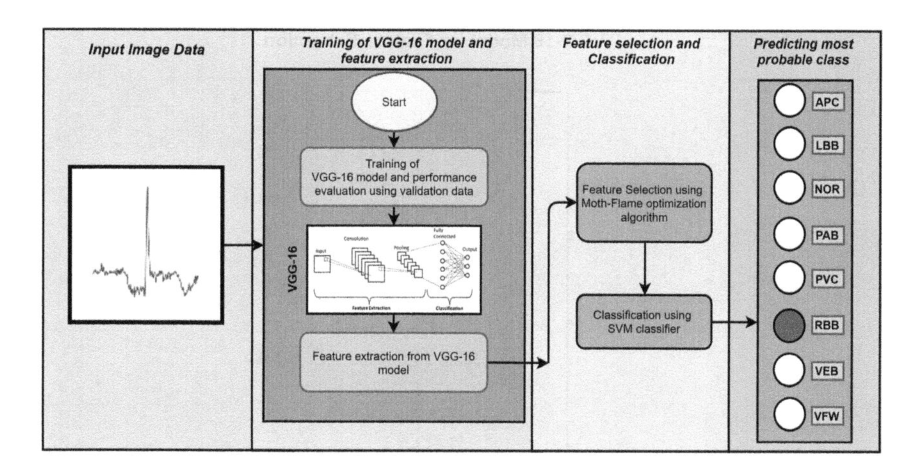

FIGURE 8.2 Block diagram representing the proposed methodology.

have converted the gray-scale images obtained after pre-processing into three-channel (RBG) images. The block diagram depicting the flow of our proposed methodology can be seen in Figure 8.2.

8.5.1 VGG16: A Brief Overview

In the first stage of the procedure, we have applied VGG16 for feature extraction on our dataset. The initial layers of this model are initialized with ImageNet weights [48], whereas the final layers are made trainable to modify their weights according to the images in our dataset. VGG [49], proposed in 2014, has its main aspect in its small kernel size of (3×3). This feature enables VGG to apprehend provincial features, thus improving the performance. Figure 8.3 shows the VGG16 model we have used in this work after inserting a few supplementary layers for achieving better results fine-tuned especially for our purpose.

8.5.2 Moth-Flame Optimization

MFO is a nature-inspired metaheuristic algorithm simulating the movement of moths, and this movement is commonly known as the Transverse Orientation. While traveling at night, a moth maintains a fixed angle with the Moon (a bright source of light) as a result of which it stays in a straight line. But in most cases, the moths keep on spiraling to the source until they get exhausted. This is because these moths are fooled by the presence of artificial lights and they try to follow a similar movement method around them. Being extremely close to the Moon, these lights resulted in a spiral path for the moths. Figure 8.4 shows diagrammatically the spiral motion.

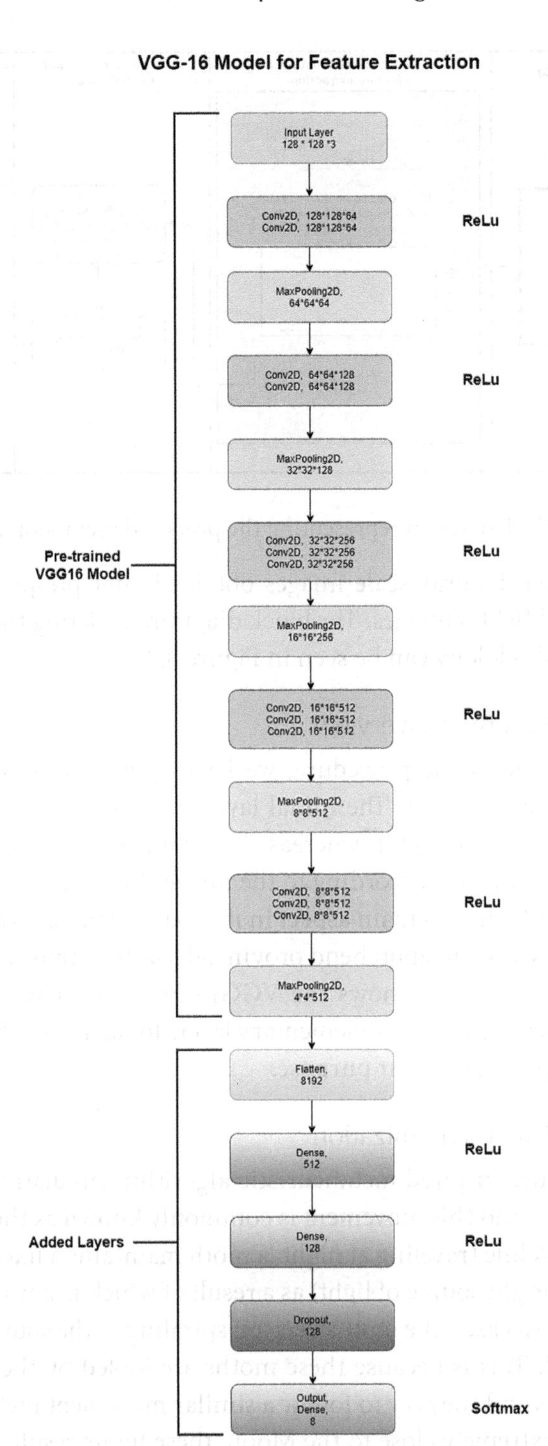

FIGURE 8.3 VGG16 model used for feature extraction from ECG records.

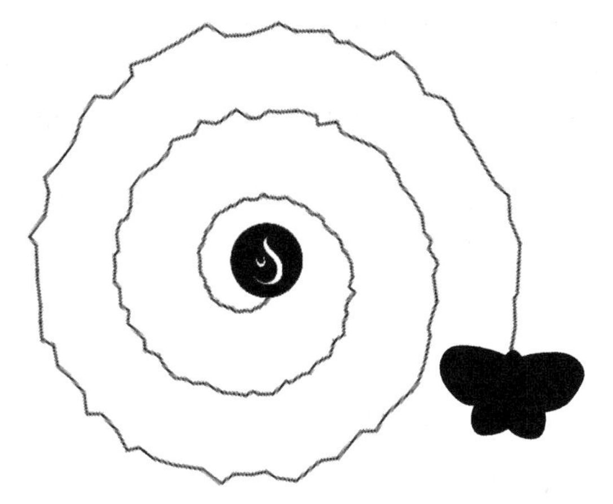

FIGURE 8.4 Movement of moths around an artificial light following the transverse orientation ultimately leading to a spiraling motion.

The MFO algorithm can solve many complex optimization problems. For this, the moths and the artificial light sources (flames) are assumed to be the solutions to the problem, and the position of the moth in space becomes the variable in any number of dimensions. We represent the moth population as a matrix M:

$$M = \begin{bmatrix} m_{1,1} & m_{1,2} & \cdots & \cdots & m_{1,d} \\ m_{2,1} & m_{2,2} & \cdots & \cdots & m_{2,d} \\ \vdots & \vdots & \vdots & \vdots & \vdots \\ m_{n,1} & m_{n,2} & \cdots & \cdots & m_{n,d} \end{bmatrix} \tag{8.1}$$

where n is the total number of moths and d the number of dimensions in which the moth travels.

A problem-specific fitness function is defined and an array OM stores the corresponding fitness values. As there are n number of moths, there are n fitness values in the array, as shown below:

$$\text{OM} = \begin{bmatrix} \text{OM}_1 \\ \text{OM}_2 \\ \vdots \\ \text{OM}_n \end{bmatrix} \tag{8.2}$$

Another matrix F, with dimensions same as that of F, stores the best-obtained solutions so far as flames, and a matrix OF stores the fitness values of the individual flames as shown in the below two equations:

$$F = \begin{bmatrix} F_{1,1} & F_{1,2} & \cdots & \cdots & F_{1,d} \\ F_{2,1} & F_{2,2} & \cdots & \cdots & F_{2,d} \\ \vdots & \vdots & \vdots & \vdots & \vdots \\ F_{n,1} & m_{n,2} & \cdots & \cdots & F_{n,d} \end{bmatrix} \tag{8.3}$$

$$\text{OF} = \begin{bmatrix} \text{OF}_1 \\ \text{OF}_2 \\ \vdots \\ \text{OF}_n \end{bmatrix} \tag{8.4}$$

The MFO algorithm is defined as follows:

$$\text{MFO} = (I, P, T) \tag{8.5}$$

Here, the function I generates a random population of moths and their corresponding fitness values:

$$I : \varnothing \rightarrow \{M, \text{OM}\} \tag{8.6}$$

The function P returns the updated positions of the moths:

$$P : M \rightarrow M \tag{8.7}$$

The function T returns true if the stopping condition is satisfied:

$$T : M \rightarrow \{\text{True}, \text{False}\} \tag{8.8}$$

S is the spiral function (usually a logarithmic spiral) using which the position of a moth with respect to a flame is updated and stored in M.

$$M_i = S(M_i, F_j) = D_i \cdot e^{bt} \cdot \cos(2\pi t) + F_j \tag{8.9}$$

where D_i indicates the distance of the ith moth for the jth flame, b is a constant, and t is a random number in [–1,1].

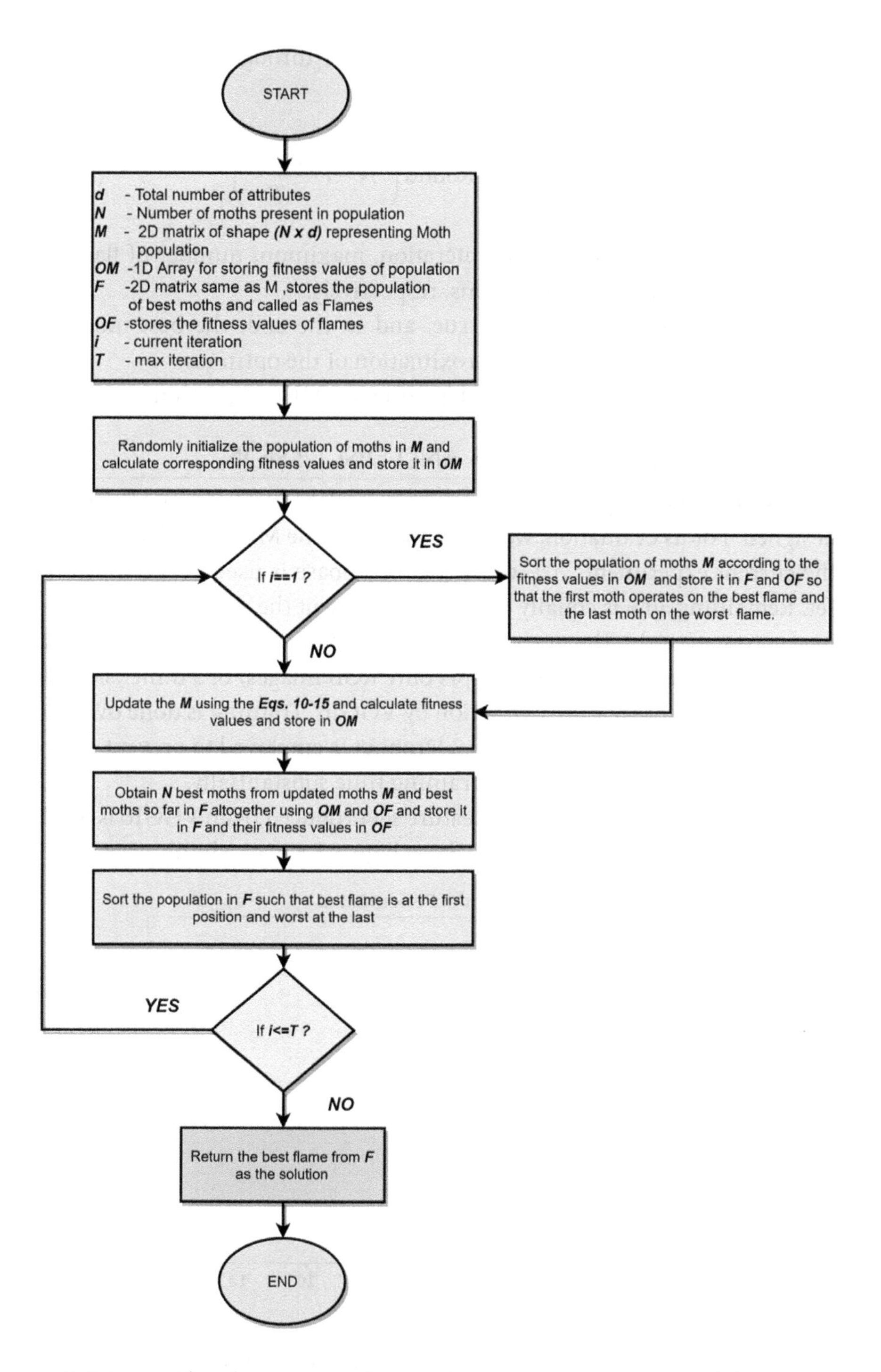

FIGURE 8.5 Flowchart representing the feature optimization using MFO.

The number of flames adaptively decreases throughout iterations as shown:

$$\text{Flame number} = \text{Round}\left(N - l * \frac{N-1}{T} \right) \tag{8.10}$$

where l, N, and T denote current iteration, maximum number of flames, and maximum number of iterations, respectively.

This continues till T returns true, and in the end, the best moth is returned as the best-obtained approximation of the optimum.

Figure 8.5 shows the flowchart depicting the MFO algorithm.

8.6 EXPERIMENTAL RESULTS AND DISCUSSION

In this study, a bi-stage model for the classification of ECG beats has been designed. For its evaluation, we have considered the MIT-BIH database. In the current experiment, 70% of the total database is used as the training set. Remaining 30% is equally divided to represent the validation and testing set, respectively. The gray-scale sample images are firstly transformed into RGB images (since VGG accepts only RGB images) of a dimension of $128 \times 128 \times 3$ before feature extraction by VGG16. Training is done over 20 epochs with a batch size of 64. A 0.2 Dropout is employed to prevent overfitting. Also, TL helps reduce the training time substantially.

In the present experiment, the feature set produced from VGG16 is used in the next stage of the framework. From Figure 8.6, it can be observed that

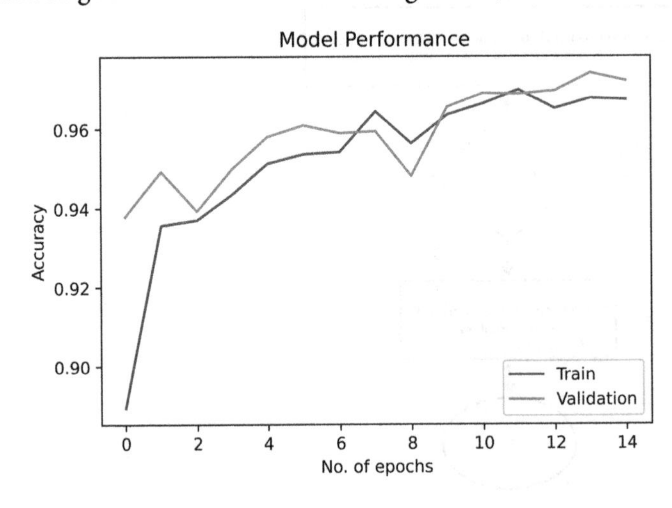

FIGURE 8.6 The train and validation accuracy curves with respect to the number of epochs for the VGG-16 classifier.

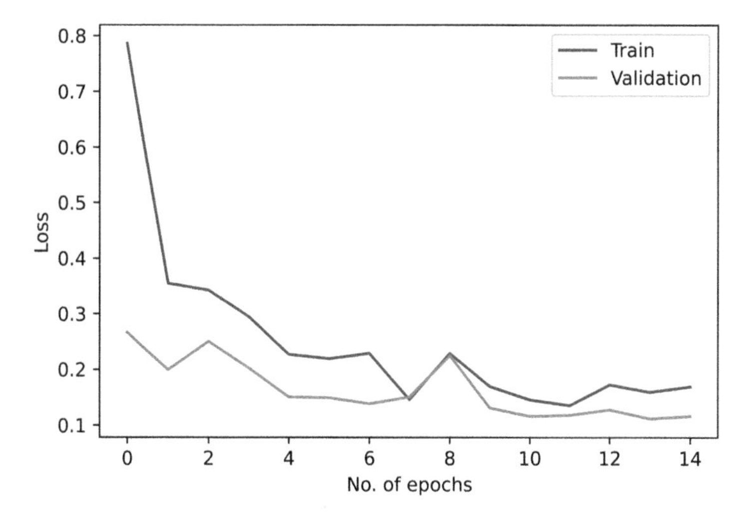

FIGURE 8.7 The train and validation loss curves with respect to the number of epochs for the VGG-16 classifier.

the VGG16 model has a good performance and achieves a 97.36% accuracy on the test set. Figures 8.7, and 8.8 represent the loss curves and confusion matrix, respectively. In the next stage of our framework, the MFO algorithm has been used to optimize the feature vector achieved from the VGG16 model. The mechanism eliminates the irrelevant features and thus helps in reducing the dimension of the feature set without compromising the classification performance. The results of the VGG16 model before applying the MFO algorithm are given in Table 8.2.

From the VGG16 model, we have extracted 128 features from the penultimate layer of the model. Convolutional and Pooling blocks extract the generic features, while the Dense layer in the model distinguishes those special features of the images which makes the classification tasks easier. So, we have used the features of the second last layer for optimization as they contain most of the information related to the dataset.

It has been observed that the MFO algorithm can reduce the feature set's dimension from 128 to 58, i.e., almost a 55% reduction without any significant loss in performance. Instead, the performance of the overall classification task improves marginally by 0.5%. The results are detailed in Table 8.3.

After applying the MFO-based feature selection approach, the confusion matrix is shown in Figure 8.9. It is clear from this figure that after optimizing the feature set, the performance has improved.

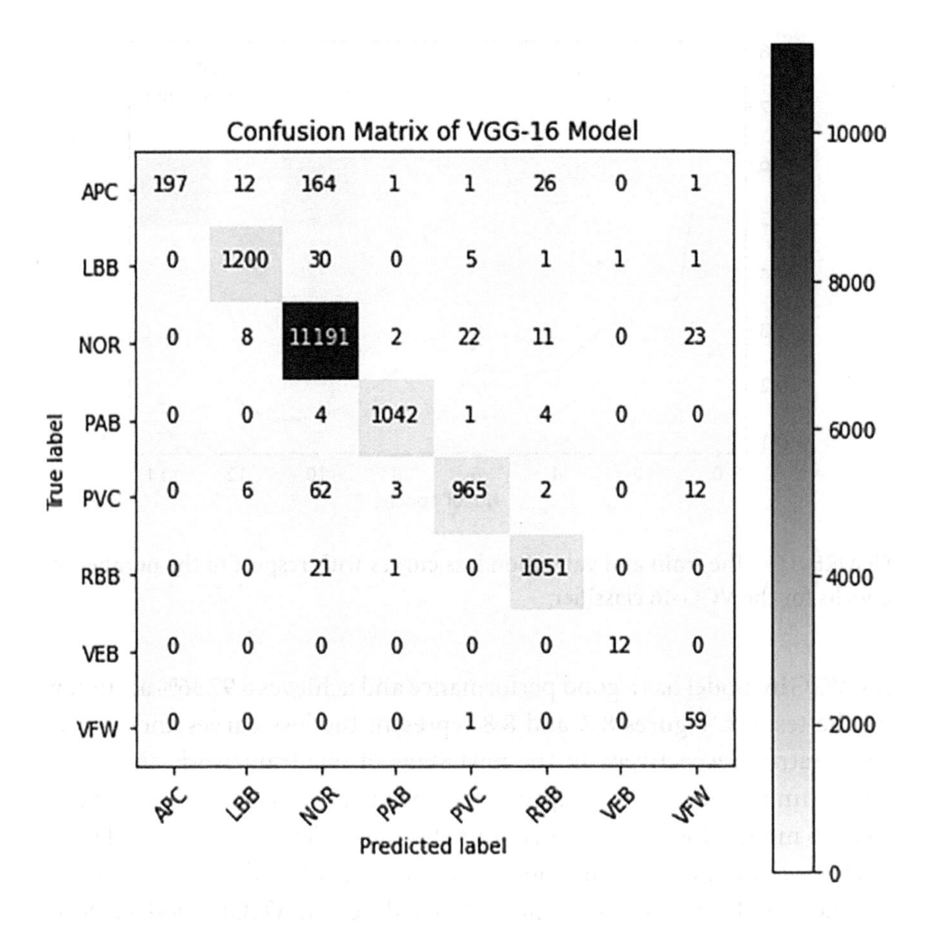

FIGURE 8.8 Confusion matrix of the VGG-16 classifier model.

TABLE 8.2 Performance of the Pre-Trained VGG16 Classifier for Extraction of the Feature Set

Model	Accuracy (%)	Precision	Recall	F_1 Score
VGG16	97.36	0.97	0.97	0.97

TABLE 8.3 Classification Performance of ECG Samples Before and After Application of the MFO Algorithm

Method	Accuracy (%)	Dimension	Precision	Recall	F1 Score
Before feature selection	97.36	128	0.97	0.97	0.97
After applying MFO-based feature selection approach	97.86	58	0.98	0.98	0.98

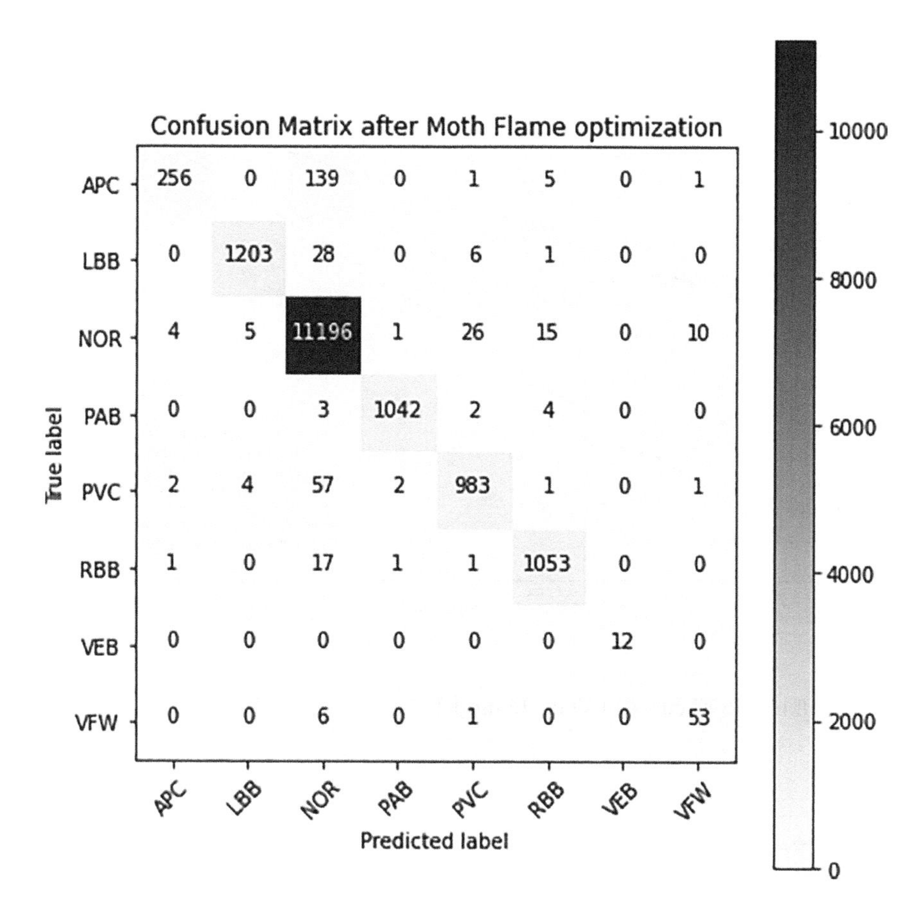

FIGURE 8.9 Confusion matrix after applying MFO algorithm and SVM classifier.

The ROC curves before and after using the MFO algorithm are given in Figures 8.10 and 8.11, respectively.

We have used 20 moths for optimization, the fitness and dimension of each moth are given in Table 8.4. This represents the dimensions of each moth in the population after the final iteration along with the classification accuracy.

Table 8.3 justifies that the proposed method is assuring with respect to performance and efficiency. We have tried to provide a comparative study with the other existing approaches evaluated on the same dataset in Table 8.5. In Ref. [50], the authors claim a classification accuracy of 98.71% on eight classes with a test set of size 4,900. First, they have used Independent Component Analysis (ICA) to separate independent sources from mixed ECG signals components. After that, they used neural

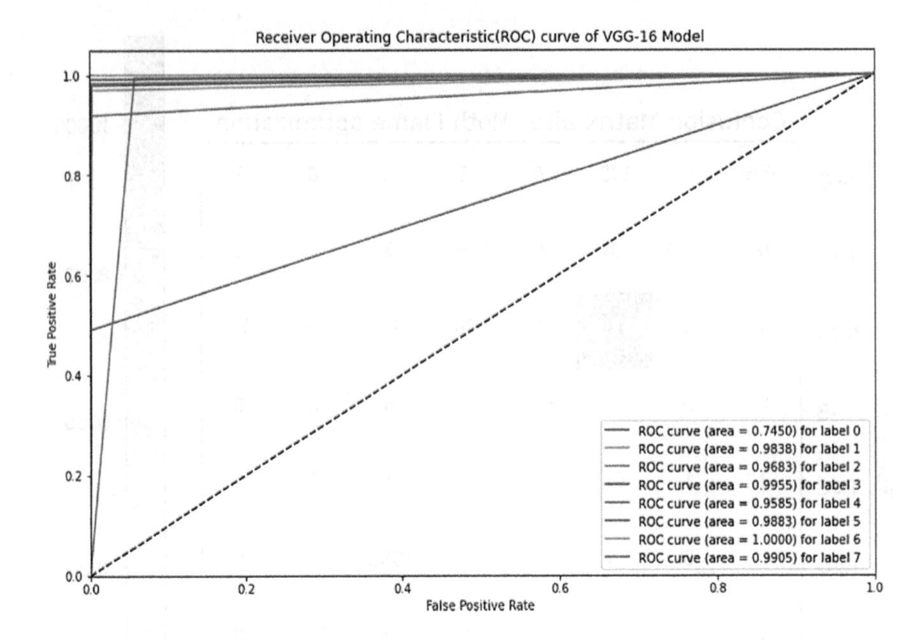

FIGURE 8.10 ROC curve of VGG-16 model.

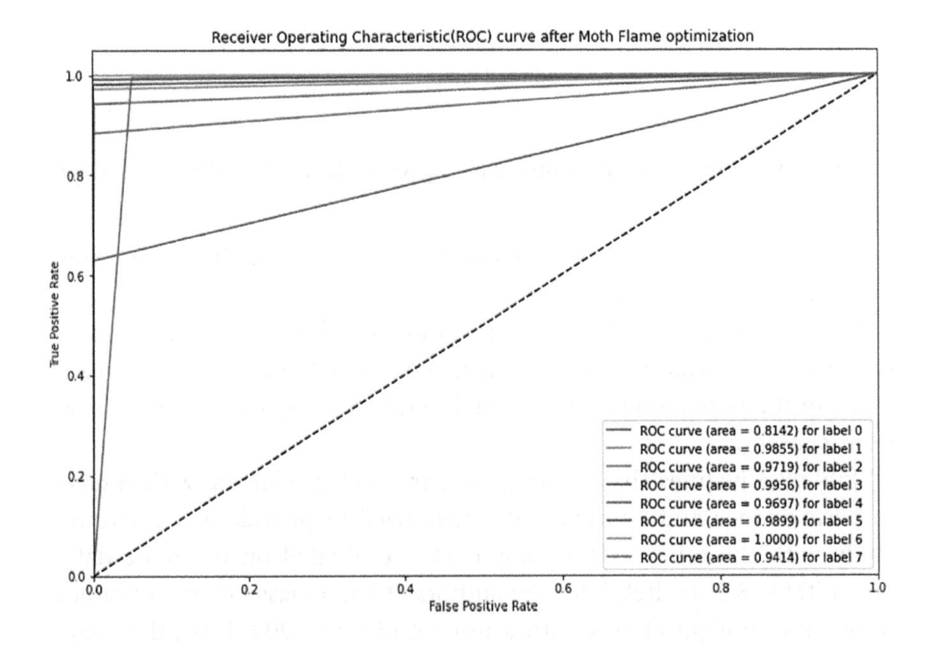

FIGURE 8.11 ROC curve after Moth-Flame Optimization.

TABLE 8.4 Fitness and Feature Dimensions of the Population After the Final Iteration

Moth No.	Accuracy (%)	Dimension	Moth No.	Accuracy (%)	Dimension
1	97.86	58	11	97.83	76
2	97.86	65	12	97.83	67
3	97.85	71	13	97.83	79
4	97.85	71	14	97.82	67
5	97.84	87	15	97.82	80
6	97.84	72	16	97.82	74
7	97.84	65	17	97.82	85
8	97.84	63	18	97.82	57
9	97.83	70	19	97.81	72
10	97.83	72	20	97.81	67

TABLE 8.5 A Comparison between the Proposed Approach and Other Existing Approaches

Work Ref.	Method Used	No. of Data Samples in the Test Set	Classification Accuracy (in %)
Yu et al. [51]	Integration of ICA and classification using neural networks	4,900	98.71
Yu et al.[50]	Novel independent components arrangement approach for feature extraction and classification using SVM	4,900	98.70
Wang et al. [52]	They used PCA and LDA for ECG feature set minimization. Then, a probabilistic neural network was used to classify the reduced feature set	4,900	99.71
Proposed Method	A two-staged network involves optimizing the features of ECG scans using MFO and classifying it using SVM	16,143	97.86

networks to classify the ECG beats. The dataset used in Ref. [51] is a small subset of the original dataset MIT-BIH. They had selected two samples at random from each record corresponding to the eight classes which produced a dataset of size 9,800. A similar kind of random sub-sampling is used by Yu et al. [50] and Wang et al. [52]. The size of the mentioned datasets in these said papers [50,52] is <10% of what we have considered in the present study. Therefore the classification accuracies provided in these papers [50,52] may not act as benchmarking results rather motivate us to achieve comparable results.

Table 8.5 shows that the proposed method in this work is comparable with past works in ECG classification and produces satisfactory results.

8.7 CONCLUSION

This work proposes a bi-stage framework for arrhythmia classification using ECG signals. In the first level, pixel-level features are extracted from the 2D images (constructed from the 1D ECG signals) using the VGG16 model. After this, to optimize this feature set, we have applied a meta-heuristic MFO algorithm on it. The obtained performance shows that this method successfully reduces the size of the feature set and also improves the classification results. This method can be hence used as an aid in arrhythmia diagnosis in humans. As future work, this framework can be evaluated on other domains to measure its robustness. Again, in the future, an ensemble of different DCNN models or the stacking of different feature sets can be tried out to have better recognition accuracy. Besides, the MFO algorithm can be modified and/or blended with other optimization algorithms for better optimization of the feature set produced by some classification models at the initial stage.

REFERENCES

1. https://www.who.int/health-topics/cardiovascular-diseases/.
2. Heron, M., Deaths: Leading causes for 2017. *Nat. Vital Stat. Rep.*, 68(6), 77 (2019).
3. American Heart Association, Electrocardiogram (ECG or EKG) (2015). https://www.heart.org/en/health-topics/heart-attack/diagnosing-a-heart-attack/electrocardiogram-ecg-or-ekg.
4. Kligfield, P., The centennial of the Einthoven electrocardiogram. *J. Electrocardiol.*, 35(Suppl), 123–129 (2002).
5. Bakator, M., & Radosav, D., Deep learning, and medical diagnosis: A review of literature. *Multimodal Technol. Interact.*, 2, 47 (2018). doi: 10.3390/mti2030047.
6. Stead, W.W., Clinical implications and challenges of artificial intelligence and deep learning. *JAMA*, 320, 1107–1108 (2018).
7. Naylor, C., On the prospects for a (deep) learning health care system. *JAMA*, 320, 1099–1100 (2018).
8. Bejnordi, B. E. et al., Diagnostic assessment of deep learning algorithms for detection of lymph node metastases in women with breast cancer. *JAMA*, 318, 2199 (2017).
9. De Fauw, J. et al., Clinically applicable deep learning for diagnosis and referral in retinal disease. *Nat. Med.*, 24, 1342–1350 (2018).
10. Beck, E. J., Gill, W. & De Lay, P. R., Protecting the confidentiality and security of personal health information in low- and middle-income countries in the era of SDGs and Big Data. *Glob. Health Action*, 9, 32089 (2016).
11. Hannun, A. Y. et al., Cardiologist-level arrhythmia detection and classification in ambulatory electrocardiograms using a deep neural network. *Nat. Med.*, 25, 65–69 (2019).

12. Clifford, G. D. et al., AF classification from a short single lead ECG recording: The PhysioNet/Computing in Cardiology Challenge 2017. *Comput. Cardiol.*, 44, 1–4 (2017).
13. Cook, D. A., Oh, S., & Pusic, M. V., Accuracy of physicians' electrocardiogram interpretations: A systematic review and meta-analysis. *JAMA Internet Med.*, 180(11), 1461–1471 (2020). doi: 10.1001/jamainternmed.2020.3989.
14. Robertson, S., Unhealthy lifestyle raises heart disease risk more than genetics. *News-Medical* (2019) viewed 12 July 2021, https://www.news-medical.net/news/20190903/Unhealthy-lifestyle-raises-heart-disease-risk-more-than-genetics.aspx.
15. Sassi, R. et al., PDF-ECG in clinical practice: a model for long-term preservation of digital 12-lead ECG data. *J. Electrocardiol.*, 50, 776–780 (2017).
16. Lyon, A., Mincholé, A., Martínez, J. P., Laguna, P. & Rodriguez, B., Computational techniques for ECG analysis and interpretation in light of their contribution to medical advances. *J. R. Soc. Interface*, 15, 20170821 (2018).
17. Mirjalili, S., Moth-flame optimization algorithm: A novel nature-inspired heuristic paradigm. *Knowledge-Based Syst.*, 89, 228–249 (2015). doi: 10.1016/j.knosys.2015.07.006.
18. Zawbaa, H.M., Emary, E., Parv, B., & Sharawi, M., Feature selection approach based on the moth-flame optimization algorithm. *In 2016 IEEE Congress on Evolutionary Computation (CEC)*. IEEE, Canada, pp. 4612–4617 (2016).
19. Ullah, A., Rehman, S. U., Tu, S., Mehmood, R.M., Fawad, & Ehatisham-ul-Haq, M., A hybrid deep CNN model for abnormal arrhythmia detection based on cardiac ECG signal. *Sensors*, 21, 951 (2021). doi: 10.3390/s21030951.
20. Moody, G. B. & Mark, R. G., The impact of the MIT-BIH arrhythmia database. *IEEE Eng. Med. Biol.*, 20(3), 45–50 (2001). doi: 10.13026/C2F305.
21. Houssein, E. H., Abdelminaam, D. S., Ibrahim, I. E., Hassaballah, M., & Wazery, Y. M., A hybrid heartbeats classification approach based on marine predators algorithm and convolution neural networks. *IEEE Access*, 9, 86194–86206 (2021). doi: 10.1109/ACCESS.2021.3088783.
22. Taddei, A., Distante, G., Emdin, M., Pisani, P., Moody, G. B., Zeelenberg, C., & Marchesi, C., The European ST-T database: Standard for evaluating systems for the analysis of ST-T changes in ambulatory electrocardiography. *Eur. Heart J.*, 13, 1164–1172 (1992). doi: 10.13026/C2D59Z.
23. Goldberger, A., Amaral, L., Glass, L., Hausdorff, J., Ivanov, P. C., Mark, R., & Stanley, H. E., PhysioBank, PhysioToolkit, and PhysioNet: Components of a new research resource for complex physiologic signals. *Circulation*, 101(23), e215–e220 (2000). doi: 10.13026/C2V88N.
24. Raghu, N., Arrhythmia detection based on hybrid features of T-wave in electrocardiogram. In: J. Joshua Thomas, et al. (eds.), *Deep Learning Techniques and Optimization Strategies in Big Data Analytics*. IGI Global, pp. 1–20 (2020). doi: 10.4018/978-1-7998-1192-3.ch001.
25. Dutta, S., Chatterjee, A., & Munshi, S., Correlation technique and least square support vector machine combine for frequency domain based ECG beat classification. *Med. Eng. Phys.*, 32(10), 1161–1169 (2010). doi: 10.1016/j.medengphy.2010.08.007.

26. Parveen, A., Vani, R. M, Hunagund, P.V., & Masroor, F., Deep learning: 1-D convolution neural network for ECG signal. *Int. J. Ind. Electron. Electr. Eng. (IJIEEE)*, 8(6), 1–17 (2020).
27. Ince, T., Kiranyaz, S., & Gabbouj, M., A generic and robust system for automated patient-specific classification of ECG signals. *IEEE Trans. Biomed. Eng.*, 56(5), 1415–1426 (2009). doi: 10.1109/TBME.2009.2013934.
28. Kachuee, M., Fazeli, S., & Sarrafzadeh, M., ECG heartbeat classification: A deep transferable representation, *2018 IEEE International Conference on Healthcare Informatics (ICHI)*, pp. 443–444 (2018). doi: 10.1109/ICHI.2018.00092.
29. Melgani, F. & Bazi, Y., Classification of electrocardiogram signals with support vector machines and particle swarm optimization. *IEEE Trans. Inf. Technol. Biomed.*, 12(5), 667–677 (2008). doi: 10.1109/TITB.2008.923147.
30. Inan, O. T., Giovangrandi, L., & Kovacs, G. T. A., Robust neural-network-based classification of premature ventricular contractions using wavelet transform and timing interval features. *IEEE Trans. Biomed. Eng.*, 53(12), 2507–2515 (2006). DOI: 10.1109/TBME.2006.880879.
31. Jun, T., Nguyen, H.M., Kang, D., Kim, D., Kim, D., & Kim, Y., ECG arrhythmia classification using a 2-D convolutional neural network. ArXiv, abs/1804.06812 (2018).
32. Pereira, D. C., Ramos, R. P., & do Nascimento, M.Z., Segmentation and detection of breast cancer in mammograms combining wavelet analysis and genetic algorithm. *Comput. Methods Programs Biomed.*, 114(1), 88–101 (2014). doi: 10.1016/j.cmpb.2014.01.014.
33. Blum, C. & Li, X., Swarm intelligence in optimization. In: *Swarm Intelligence.* Springer: Berlin, Heidelberg, pp. 43–85. (2008).
34. Man, K.F., Tang, K.S., & Kwong, S., Genetic algorithms: Concepts and applications in engineering design. *IEEE Trans. Ind. Electr.*, 43(5), 519–534 (1996). doi: 10.1109/41.538609.
35. Dorigo, M., Birattari, M., & Stutzle, T., Ant colony optimization. *IEEE Comput. Intell. Mag.*, 1(4), 28–39 (2006). doi: 10.1109/MCI.2006.329691.
36. Kennedy, J. & Eberhart, R., Particle swarm optimization. *Proceedings of ICNN'95- International Conference on Neural Networks*, pp. 1942–1948, vol. 4 (1995). DOI: 10.1109/ICNN.1995.488968.
37. Karaboga, D., & Basturk, B., Artificial Bee Colony (ABC) optimization algorithm for solving constrained optimization problems. In: Melin, P., Castillo, O., Aguilar, L.T., Kacprzyk, J., & Pedrycz, W. (eds) *Foundations of Fuzzy Logic and Soft Computing.* IFSA 2007. Lecture Notes in Computer Science, vol. 4529. Springer: Berlin, Heidelberg (2007). doi: 10.1007/978-3-540-72950-1_77.
38. Mohamad, A., Zain, A., Bazin, N.E.N., & Udin, A., Cuckoo search algorithm for optimization problems: A literature review. *Appl. Mech. Mater.* (2013). doi: 10.4028/www.scientific.net/AMM.421.502.
39. Ghaheri, A., Shoar, S., Naderan, M., & Hoseini, S.S. The applications of genetic algorithms in medicine. *Oman. Med. J.*, 30(6), 406–416 (2015). doi: 10.5001/omj.2015.82.

40. Valsecchi, A., Damas, S., & Santamaria, J. (eds.), An image registration approach using genetic algorithms. *2012 IEEE Congress on Evolutionary Computation (CEC)*. IEEE, Australia (2012).
41. Ling, S.S., & Nguyen, H.T., Genetic-algorithm-based multiple regression with fuzzy inference system for detection of nocturnal hypoglycemic episodes. *IEEE Trans. Inf. Technol. Biomed.*, 15(2), 308–15 (2011).
42. Kumar, A., Ashok, A., & Ansari, M. A., Brain tumor classification using hybrid model of PSO and SVM classifier, *2018 International Conference on Advances in Computing, Communication Control and Networking (ICACCCN)*, pp. 1022–1026 (2018). DOI: 10.1109/ICACCCN.2018.8748787.
43. Jangir, N., Pandya, M.H., Trivedi, I.N., Bhesdadiya, R., Jangir, P., & Kumar, A., Moth-flame optimization algorithm for solving real challenging constrained engineering optimization problems. *In 2016 IEEE Students' Conference on Electrical, Electronics and Computer Science (SCEECS)*. IEEE, pp. 1–5 (2016).
44. Hazir, E., Erdinler, E.S., & Koc, K.H., Optimization of CNC cutting parameters using design of experiment (doe) and desirability function. *J. For. Res.*, 29(5), 1423–1434 (2018).
45. Wang, M., Chen, H., Yang, B., Zhao, X., Hu, L., Cai, Z., Huang, H., & Tong, C., Toward an optimal kernel extreme learning machine using a chaotic moth-flame optimization strategy with applications in medical diagnoses. *Neurocomputing*, 267, 69–84 (2017).
46. El Aziz, M.A., Ewees, A.A., & Hassanien, A.E., Whale optimization algorithm and moth-flame optimization for multilevel thresholding image segmentation. *Expert. Syst. Appl.*, 83, 242–256 (2017).
47. Muangkote, N., Sunat, K., & Chiewchanwattana, S., Multilevel thresholding for satellite image segmentation with moth-flame based optimization. *In 2016 13th International Joint Conference on Computer Science and Software Engineering (JCSSE)*. IEEE, pp. 1–6 (2016).
48. Russakovsky, O., Deng, J., Su, H., Krause, J., Satheesh, S., Ma, S., Huang, Z., Karpathy, A., Khosla, A., Bernstein, M., & Berg, A. C., Imagenet large scale visual recognition challenge. *Int. J. Comput. Vision*, 115(3), 211–252 (2015).
49. Liu, S., & Deng, W., Very deep convolutional neural network-based image classification using small training sample size. *In 2015 3rd IAPR Asian Conference on Pattern Recognition (ACPR)*, IEEE, Malaysia, pp. 730–734 (2015).
50. Yu, S.-N., & Chou, K.-T., Selection of significant independent components for ECG beat classification. *Expert Syst. Appl.*, 36, 2088–2096 (2009). doi:10.1016/j.eswa.2007.12.016.
51. Yu, S.-N., & Chou, K.-T., Integration of independent component analysis and neural networks for ECG beat classification. *Expert Syst. Appl.*, 34, 2841–2846 (2008). doi: 10.1016/j.eswa.2007.05.006.
52. Wang, J.-S., Chiang, W.-C., Hsu, Y.-L., & Yang, Y.-T.C., ECG arrhythmia classification using a probabilistic neural network with a feature reduction method. *Neurocomputing*, 116, 38–45 (2013). doi: 10.1016/j.neucom.2011.10.045.

III

Hybrids and Improvements of Moth-Flame Optimization Algorithm

CHAPTER 9

Hybrid Moth-Flame Optimization Algorithm with Slime Mold Algorithm for Global Optimization

Sukanta Nama
Maharaja Bir Bikram University

Sanjoy Chakraborty
Iswar Chandra Vidyasagar College
National Institute of Technology Agartala

Apu Kumar Saha
National Institute of Technology Agartala

Seyedali Mirjalili
Torrens University Australia
Yonsei University

DOI: 10.1201/9781003205326-12

CONTENTS

9.1 Introduction 156
9.2 Overview of Component Algorithms 158
9.2.1 MFO Algorithm 158
9.2.2 Overview of SMA 158
9.3 The Proposed hMFOSMA 158
9.3.1 Motivation of the Work 158
9.3.2 Framework of hMFOSMA 159
9.4 Performance Results and Analysis 161
9.4.1 Comparison of the Proposed hMFOSMA With Other Algorithm 162
9.4.2 Convergence Analysis 162
9.4.3 Statistical Analysis 170
9.5 Application of hMFOSMA on Engineering Problems 171
9.5.1 EP.1: Gas Transmission Compressor Design Problem 171
9.5.2 EP.2: Optimal Capacity of Gas Production Facilities 172
9.6 Conclusion and Future Directions 173
References 175

9.1 INTRODUCTION

Optimization is a method to identify the best way to solve a certain problem. Many difficulties in the actual world may be seen as challenges of optimization. As challenges get increasingly complicated than before, it becomes clearer that new optimization approaches are needed. In recent decades, several techniques for solving complex problems have been suggested and significant progress has been made. For example, prior to the invention of heuristic optimization, mathematical optimization approaches were the sole approach for the solution of optimization problem. These techniques should nevertheless know the feature of the problems of optimization, such as continuity or differentiation. A lot of heuristic optimization techniques have been created in recent years. These include Genetic Algorithm (GA) [1], Particle Swarm Optimization (PSO) [2], Slime Mold Algorithm (SMA) [3], Moth-Flame Optimization (MFO) [4], Differential Evolution (DE) [5], Sine Cosine Algorithm (SCA) [6], Artificial Bee Colony (ABC) [7], Butterfly Optimization Algorithm (BOA) [8], Cuckoo Search (CS) [9], and Bat Algorithm (BA) [10]. The same objective has to be achieved across all potential inputs, with the optimal output (Global optimum). To do this, two main features need to be provided by a heuristic algorithm to make sure that global optimum is achieved. Exploration and exploitation are the two primary features [11].

Exploration is an algorithm's capacity to search for entire sections of problem areas, while exploitation is the convergence capable of providing the optimal solution. Effective balance of operational and exploratory capability to optimize the overall global outcome is the vital objective of all heuristic optimization algorithms. The exploration and exploitation of evolutionary algorithms are not clear, according to Ref. [12], because of the commonly recognized impression of the lake. In contrast, the other one will be weak and vice versa by boosting one capability.

The present optimization heuristic algorithms are able to solve finite problems because of the aforementioned points. Also, no one can ensure that they are optimally suited to all optimization challenges despite the success of traditional and newer MAs. The No-Free-Lunch (NFL) idea has logically proved that [13]. This theorem prompted many scientists to create a new algorithm and to tackle problems more effectively. The combination of optimization algorithms is a technique to balance exploration with exploitation.

The MFO is a new technique of metaheuristic optimization by emulating the moth navigation mechanism in nature known as transverse orientation. The algorithm is used to control moth fire. Moth and flames are the two solutions in this algorithm. Seyedali Mirjalili, the developer of this method, has proven that the algorithm may provide extremely competitive outcomes in comparison to other advanced metaheuristic optimization techniques. The MFO method is nonetheless now in the investigation stage and this algorithm may be enhanced further with convergence speed and computation accuracy.

Because of its simplicity, convergence rate, and search capability, MFO is one of the most often utilized evolutionary algorithms for hybrid techniques. Some of the works that studied MFO with various algorithms can be found in Refs. [14–18]. These hybrid algorithms are designed to reduce the chance of trapping optimally at the local level. A new heuristic optimization approach termed SMA has just been proposed. We know that SMA has a significant capacity to enhance global search and tackle local trapping. We suggest hMFOSMA to take advantage of the good performance of SMA. The additional benefits of MFO and SMA allow for a quicker, more robust approach using the suggested algorithm. Fifty-six benchmarking functions and two engineering problems are tested for the suggested method.

The rest of this chapter is organized as follows: A brief introduction of MFO and SMA is presented in Section 9.2. In Section 9.3, an enhanced MFO adaption dubbed hMFOSMA is proposed. Section 9.4 depicts the experimental findings of test functions as well as the engineering design problem. Section 9.5 finally finishes the work in some direction for the future.

9.2 OVERVIEW OF COMPONENT ALGORITHMS

In this section, basics of MFO and SMA have been discussed.

9.2.1 MFO Algorithm

The MFO algorithm mimics the traverse orientation of moths in nature. A moth maintains a constant angle with a light source when migrating. When a source of light is in far distance, this leads to a straight fly path. When the light source is close, however, this results in a spiral movement. The details of the mathematical models of MFO can be found in Ref. [4].

9.2.2. Overview of SMA

SMA is a recent nature-based metaheuristics that mimics the intelligence of slime molds in nature [3]. The search process in this algorithm includes approaching a food source, wrap a food source, and grabble a good source. The details of the mathematical models of MFO can be found in Ref. [3].

9.3 THE PROPOSED hMFOSMA

This section introduces the suggested ensemble algorithm called hMFOSMA. The hMFOSMA is motivated in Section 9.3.1, and the quest approach is addressed in Section 9.3.2.

9.3.1 Motivation of the Work

The recent trend in work consists of hybridization of two or more variations in order to recognize higher functional quality and solutions to global optimization problem. Ensemble or hybrid algorithms become the subject of curiosity since their great determination is also boosted by their intriguing features rather than methods. The aim of this hybrid algorithm design is to increase the connection between execution and investigation, preserve the broad spectrum of solutions in the simulation process, and increase algorithm stability and speedy convergence [19–23].

The MFO algorithm is a new heuristic paradigm inspired by nature. The primary moth's navigation in nature called a transversal orientation is the major motivation of this approach. Moths fly at night at a constant angle on the Moon, a highly effective technique over vast distances traveling on a straight line. But these beautiful insects are confined around artificial lights on a spiral route. This leads to slow convergence and limited accuracy of MFO algorithms [15]. First, MFO's exploratory capacity has been claimed to be poorer than its operations exploitation [14]. But in ordinary MFOs, it is sufficiently demonstrated by this research mechanism in the traditional MFO that low exploitation, real solutions, and unassociated harmony between exploitation and exploration are skipped.

On the other hand, the newly formed SMA is based on an oscillation function of the natural slime mold. SMA needs to develop a more efficient approach via exploitation [24] and to expand the diversity of options to improve exploratory trends [25]. A hybrid SMA strategy-based variant of the MFO algorithm is suggested called hMFOSMA. The SMA mechanism can boost population diversity against early convergence and improve the efficiency of the algorithm. This method is useful for achieving a better balance between exploring and exploiting MFO capabilities, therefore making hMFOSMA quicker and more robust. The framework of proposed hMFOSMA will be as follows.

9.3.2 Framework of hMFOSMA

This section details the techniques in hMFOSMA for enhancing the search effectiveness of traditional MFOs in further detail. The precious description of the techniques suggested increasing the accuracy of MFO research is as follows:

1. In the air, the slime mold may come up to food via smell. This operator is responsible for leading the set of candidate solutions, increasing their scan capabilities, and conducting searches based on local search criteria.
2. Wrap food processes are built such that the early iterations of an algorithm may be exploited enough and a balance between exploitation and exploration is maintained.

The following is a summary of the proposed hMFOSMA algorithm flow:

Step 1: Initialize parameters including number of solution N, maximum number of iteration T, search dimension D, and lower bound LB and upper bound UB of domain space.

Step 2: Generate randomly a population set in search space as the initial solution and calculate the fitness value of the corresponding solution.

Step 3: Determine each individual objective function value in the present set of individuals and find the location with the best fitness value as the optimal location X_b.Step 4: Utilize MFO's following equation for spiral movement:

$$x_i^{K+1} = \begin{cases} \delta_i \cdot e^{bt} \cdot \cos(2\pi t) + Fm_i(k), & i \leq N \cdot Fm \\ \delta_i \cdot e^{bt} \cdot \cos(2\pi t) + Fm_{N.FM}(k), & i \geq N \cdot Fm \end{cases} \tag{9.1}$$

where $\delta_i = \left| x_i^K - Fm_i \right|$ represents the distance of moth at the ith place and its specific flame (Fm_i) is the distance between the ith moth M_i and its specific flame further; b is a constant used to recognize the shape of the search for spiral flight shape. And, $t \in (-1,1)$, indicating how close the moth is to its unique flame. It has been proposed that a customizable course of action be used to reduce the variable (t) value over time, so as to enhance the efficiency of both exploration and exploitation in the beginning and end iterations, respectively.

Step 5: Calculate the W using SMA's equations:

$$\overrightarrow{W(\text{SmellIndex}(i))} = \begin{cases} 1 + r \cdot \log\left(\dfrac{bF - S(i)}{bF - wF} + 1\right), & \text{condition} \\ 1 - r \cdot \log\left(\dfrac{bF - S(i)}{bF - wF} + 1\right), & \text{others} \end{cases} \tag{9.2}$$

$$\text{Smell Index} = \text{sort}(S) \tag{9.3}$$

where condition indicates that $S(i)$ ranks first half of the population, r defines within the interval $[0,1]$, bF illustrates the value of optimal fitness achieved in the current iteration procedure, wF signifies the current worst function value attained during the iterative procedure, and SmellIndex denotes the sorted order of objective function value (ascends in the minimization optimization problem).

Step 6: Update p, vb, vc:

$$p = \tanh\left| S(i) - DF \right| \tag{9.4}$$

where $i \in 1,2,\ldots,n$; $S(i)$ signifies the objective function value of $\vec{X}$and DF represents the best objective function value acquired in all iterations.

$$\overrightarrow{vb} = [-a, a] \tag{9.5}$$

$$a = \operatorname{arctanh}\left(-\left(\frac{t}{\max\_t}\right)+1\right) \tag{9.6}$$

Step 7: Update position using position updated in wrap food mechanism of SMA as follows:

$$\overrightarrow{X^*} = \begin{cases} r\text{and} \cdot (UB - LB) + LB, & r\text{and} < z \\ \overrightarrow{X_b(t)} + \overrightarrow{vb} \cdot \left(W \cdot \overrightarrow{X_A(t)} - \overrightarrow{X_B(t)}\right), & r < p \\ \overrightarrow{vc} \cdot \overrightarrow{X(t)}, & r \geq p \end{cases} \tag{9.7}$$

where LB and UB define the search range's lower and upper limits, respectively, and r denotes the random value in [0, 1].

Step 8: Output the optimal individual if the number of contemporary iterations reaches T. Otherwise, return to Step 3.

9.4 PERFORMANCE RESULTS AND ANALYSIS

In this discipline, it is usual to measure the algorithm performance by a set of mathematical features that have been known globally as optimum. This is the same procedure by using the literature [26–28] as comparable test beds for 56 typical benchmark functions. To assess the efficiency of the hMFOSMA algorithm from diverse viewpoints, three sets of benchmarking functions with distinct features have been selected. These criteria are grouped into three categories, as indicated in the appendix: unimodal functions, multimodal functions, and multimodal functions in a fixed dimension. Unimodal function is appropriate to measure the use and convergence of an optimization method since it has one ideal global and no optimal local value. On the other hand, there are several optimal multimodal functions such that they are more difficult than one single function. One of the optimum points is known as the global optimum and the remainder as local optimum. An optimization method should evade approaching and approximating any local optima. Therefore, multi-modal functions may be used to scan and avoid methods locally. Appendix provides the mathematical description of the used benchmark functions.

The authors chose some of the most recent and well-known algorithms in the field as follows: SMA, MFO, PSO, DE, SCA, ABC, BOA, CS, and BA in order to examine the efficiency of the proposed hMFOSMA method. All control parameter algorithms utilized for the analysis are of the same values as the original study. Note that for each method, 30 numbers of search agents and an evaluation of 9,000 functions are used.

9.4.1 Comparison of the Proposed hMFOSMA With Other Algorithm

To obtain the performance results on 56 test function, the number of solution size is measured as 30. And in each execution, maximum 9,000 function evaluations are calculated. The obtained outcome from the proposed hMFOSMA is compared with the conventional algorithms SMA, MFO, PSO, DE, SCA, ABC, BOA, CS, and BA. Tables 9.1–9.6 exhibit the results of the test problem under consideration, correspondingly. The mean and standard deviation of the fitness value in 30 runs are exposed in those tables. Table 9.7 shows the counting number of superior, inferior, and equal to hMFOSMA than the compared algorithm. From Table 9.7, it is clear that hMFOSMA is superior to SMA, MFO, PSO, DE, SCA, ABC, BOA, CS, and BA on 30, 47, 46, 44, 50, 49, 50, 56, and 50 test functions; hMFOSMA is inferior to SMA, MFO, PSO, DE, SCA, ABC, BOA, CS, and BA on 14, 5, 7, 8, 0, 5, 5, 0, and 6 test functions; hMFOSMA is equal to SMA, MFO, PSO, DE, SCA, ABC, BOA, CS, and BA on 12, 4, 3, 4, 6, 2, 1, 0, and 0 test functions. The schematic views of the results are shown in Figure 9.1. The comparison results show that the SMA operator improves the solution execution process to assist the optimizer in alleviating local minima and generally converge to the optimal solution. The hMFOSMA considerably improved the performance of the MFO method and obtained competitive results when equated to other state-of-the-art optimization methods according to the results.

9.4.2 Convergence Analysis

The convergence graphs are shown in Figure 9.2 among proposed hMFOSMA, conventional SMA, MFO, PSO, DE, SCA, ABC, BOA, CS, and BA algorithms. The functions for the convergence analysis are considered as F1, F2, F3, F9, F10, and F11. It can be seen in Figure 9.2 that the proposed ensemble of SMA and MFO (hMFOSMA) convergence is robust compared with SMA, MFO, PSO, DE, SCA, ABC, BOA, CS, and BA.

TABLE 9.1 Performance Results of hMFOSMA, SMA, MFO, PSO, DE, and SCA on F1–F20 Test Functions

Function	hMFOSMA		SMA		MFO		PSO		DE		SCA	
	Mean	Std	Mean	Std	Mean	Std	Mean	Std	Mean	Std	Mean	Std
1	2.33E−129	1.27E−128	3.55E−48	1.94E−47	1.21E+03	3.01E+03	7.05E+01	3.40E+01	1.30E+01	4.78E+01	1.62E+02	1.91E+02
2	4.77E−71	2.50E−70	1.58E−51	6.20E−51	3.90E+01	1.76E+01	2.80E+00	8.80E−01	3.08E−01	1.71E−01	3.73E−01	4.22E−01
3	2.77E−87	1.50E−86	1.30E−35	7.14E−35	2.67E+04	1.21E+04	1.43E+04	2.89E+03	1.37E+03	7.66E+02	1.39E+04	5.75E+03
4	7.14E−57	3.85E−56	4.92E−40	2.61E−39	6.85E+01	8.13E+00	2.94E+01	4.84E+00	1.83E+01	4.94E+00	4.57E+01	1.26E+01
5	1.70E+01	1.17E+01	5.13E+00	7.48E+00	2.74E+06	1.47E+07	1.63E+04	1.38E+04	3.29E+03	5.48E+03	1.05E+06	1.96E+06
6	0.00E+00	0.00E+00	0.00E+00	0.00E+00	2.47E+03	4.11E+03	9.55E+01	6.65E+01	1.07E+01	1.56E+01	1.65E+02	2.16E+02
7	6.65E−04	5.63E−04	5.16E−04	5.44E−04	4.21E+00	7.21E+00	2.49E−01	8.82E−02	7.40E−02	2.48E−02	2.75E−01	2.67E−01
8	−1.26E+04	2.22E+01	−1.26E+04	2.12E+00	−8.50E+03	8.79E+02	−1.04E+04	5.28E+02	−4.71E+03	4.26E+02	−3.55E+03	2.31E+02
9	0.00E+00	0.00E+00	0.00E+00	0.00E+00	1.74E+02	4.21E+01	1.11E+02	2.53E+01	1.98E+02	1.96E+01	6.38E+01	3.09E+01
10	8.88E−16	0.00E+00	8.88E−16	0.00E+00	1.60E+01	5.36E+00	3.76E+00	4.65E−01	1.19E+00	5.62E−01	1.27E+01	8.15E+00
11	0.00E+00	0.00E+00	0.00E+00	0.00E+00	3.25E+01	4.91E+01	1.73E+00	4.68E−01	8.17E−01	3.53E−01	3.21E+00	4.16E+00
12	1.79E−02	2.08E−02	3.24E−03	5.19E−03	1.24E+04	3.13E+04	2.66E+01	3.93E+01	2.78E+00	3.95E+00	1.99E+06	6.28E+06
13	3.35E−02	3.84E−02	3.45E−02	4.66E−02	7.55E+04	1.52E+05	3.09E+03	5.34E+03	3.05E+03	7.25E+03	3.18E+06	6.13E+06
14	6.40E−04	2.33E−04	6.04E−04	3.38E−04	1.97E−03	3.73E−03	6.06E−03	9.14E−03	3.38E−04	7.02E−05	1.08E−03	3.35E−04
15	3.98E−01	6.18E−08	3.98E−01	7.27E−08	3.98E−01	0.00E+00	3.98E−01	0.00E+00	3.98E−01	0.00E+00	4.00E−01	1.99E−03
16	3.00E+00	8.27E−11	3.00E+00	1.17E−08	3.00E+00	2.23E−15	3.00E+00	4.28E−15	3.00E+00	2.11E−15	3.00E+00	3.04E−04
17	−3.00E−01	2.26E−16	−3.00E−01	2.26E−16	−3.00E−01	2.26E−16	−2.63E−01	2.79E−02	−3.00E−01	1.59E−14	−3.00E−01	2.26E−16
18	−1.02E+01	1.22E−03	−1.02E+01	9.84E−04	−6.22E+00	3.40E+00	−6.37E+00	2.86E+00	−9.32E+00	2.22E+00	−2.42E+00	1.74E+00
19	−1.04E+01	7.74E−04	−1.04E+01	5.26E−04	−6.70E+00	3.38E+00	−8.39E+00	3.21E+00	−9.96E+00	1.69E+00	−2.35E+00	1.57E+00
20	−1.05E+01	6.47E−04	−1.05E+01	9.51E−04	−7.55E+00	3.53E+00	−8.53E+00	2.95E+00	−1.05E+01	1.36E−15	−3.62E+00	1.35E+00

TABLE 9.2 Performance Results of hMFOSMA, ABC, BOA, CS, and BA on F1–F20 Test Functions

Function	hMFOSMA		ABC		BOA		CS		BA	
	Mean	Std	Mean	Std	Mean	Std	Mean	Std	Mean	Std
1	2.33E-129	1.27E-128	5.57E-02	9.59E-02	1.81E-05	5.18E-06	6.68E+04	7.88E+03	4.27E+04	5.86E+03
2	4.77E-71	2.50E-70	8.63E-02	6.98E-02	1.03E-04	1.13E-04	4.24E+12	1.27E+13	1.84E+03	7.55E+03
3	2.77E-87	1.50E-86	2.58E+04	5.18E+03	1.33E-05	3.88E-06	1.45E+05	3.64E+04	5.00E+04	1.06E+04
4	7.14E-57	3.85E-56	6.10E+01	6.54E+00	1.52E-03	2.82E-04	8.75E+01	2.98E+00	6.92E+01	5.05E+00
5	1.70E+01	1.17E+01	3.91E+02	7.63E+02	2.88E+01	3.73E-02	2.61E+08	3.80E+07	1.91E+07	8.85E+06
6	0.00E+00	0.00E+00	1.67E+00	1.52E+00	0.00E+00	0.00E+00	6.78E+04	6.34E+03	4.97E+04	6.74E+03
7	6.65E-04	5.63E-04	4.58E-01	1.26E-01	1.01E-02	3.75E-03	1.16E+02	2.14E+01	1.46E-01	5.15E-02
8	−1.26E+04	2.22E+01	−1.07E+04	3.01E+02	−4.04E+03	3.14E+02	−2.23E+03	4.71E+02	−3.86E+03	3.31E+02
9	0.00E+00	0.00E+00	1.73E+01	4.36E+00	1.02E+02	9.81E+01	4.43E+02	3.41E+01	1.50E+02	2.27E+01
10	8.88E-16	0.00E+00	2.71E+00	5.80E-01	1.10E-03	2.53E-04	2.07E+01	1.85E-01	1.87E+01	4.87E-01
11	0.00E+00	0.00E+00	3.25E-01	2.12E-01	2.97E-05	1.32E-05	6.16E+02	6.87E+01	4.80E+02	7.62E+01
12	1.79E-02	2.08E-02	1.37E-02	3.08E-02	3.77E-01	1.02E-01	5.99E+08	1.15E+08	6.35E+07	3.06E+07
13	3.35E-02	3.84E-02	5.55E-03	6.24E-03	2.61E+00	4.47E-01	1.13E+09	2.84E+08	1.88E+08	5.87E+07
14	6.40E-04	2.33E-04	1.01E-03	3.92E-04	5.35E-04	1.20E-04	2.14E-01	2.08E-01	2.67E-03	5.62E-03
15	3.98E-01	6.18E-08	3.98E-01	0.00E+00	3.99E-01	8.69E-04	2.22E+00	1.59E+00	3.98E-01	1.15E-12
16	3.00E+00	8.27E-11	3.00E+00	1.37E-02	3.01E+00	1.34E-02	5.53E+01	5.11E+01	3.00E+00	1.21E-10
17	−3.00E-01	2.26E-16	−3.00E-01	2.26E-16	−3.00E-01	5.30E-04	−1.12E-02	2.09E-02	−2.45E-01	4.03E-02
18	−1.02E+01	1.22E-03	−9.97E+00	9.24E-01	−5.14E+00	1.10E+00	−6.01E-01	4.15E-01	−5.22E+00	2.97E+00
19	−1.04E+01	7.74E-04	−1.02E+01	9.65E-01	−4.51E+00	9.16E-01	−6.63E-01	2.26E-01	−4.27E+00	2.64E+00
20	−1.05E+01	6.47E-04	−1.03E+01	9.81E-01	−4.42E+00	8.94E-01	−1.00E+00	5.33E-01	−5.30E+00	3.33E+00

TABLE 9.3 Performance Results of hMFOSMA, SMA, MFO, PSO, DE, and SCA on F21–F40 Test Functions

Function	hMFOSMA		SMA		MFO		PSO		DE		SCA	
	Mean	Std	Mean	Std	Mean	Std	Mean	Std	Mean	Std	Mean	Std
21	0.00E+00	0.00E+00	0.00E+00	0.00E+00	0.00E+00	0.00E+00	0.00E+00	0.00E+00	0.00E+00	0.00E+00	0.00E+00	0.00E+00
22	0.00E+00	0.00E+00	0.00E+00	0.00E+00	0.00E+00	0.00E+00	0.00E+00	0.00E+00	0.00E+00	0.00E+00	0.00E+00	0.00E+00
23	0.00E+00	0.00E+00	0.00E+00	0.00E+00	2.88E−14	1.58E−13	1.85E−18	1.01E−17	0.00E+00	0.00E+00	0.00E+00	0.00E+00
24	2.87E−211	0.00E+00	6.56E−119	3.59E−118	1.08E−15	5.89E−15	3.34E−22	6.23E−22	3.11E−58	6.76E−58	2.88E−35	1.52E−34
25	2.85E−45	1.56E−44	1.03E−89	5.63E−89	3.94E+02	8.77E+01	1.27E+02	2.96E+01	5.81E+01	2.21E+01	5.95E+01	2.57E+01
26	2.19E−45	1.20E−44	2.61E+04	1.39E+05	1.02E+11	1.57E+11	2.43E+09	1.44E+09	1.00E+08	8.09E+07	7.21E+09	1.59E+10
27	5.23E−121	2.25E−120	2.08E−77	1.14E−76	1.04E−66	5.67E−66	8.70E−17	4.17E−16	1.18E−48	6.11E−48	8.87E−08	4.40E−07
28	2.98E−08	4.25E−08	1.31E−07	1.77E−07	0.00E+00	0.00E+00	5.67E−23	1.59E−22	0.00E+00	0.00E+00	2.24E−03	1.61E−03
29	4.80E−02	9.14E−02	2.64E−03	4.28E−03	1.45E+00	1.96E+00	7.87E−02	7.45E−02	3.65E−01	1.03E+00	1.83E+00	1.04E+00
30	6.91E−06	6.33E−06	6.00E−05	6.37E−05	1.91E−02	1.04E−01	2.32E−12	5.55E−12	9.21E−25	3.31E−24	2.91E−01	2.74E−01
31	5.95E−182	0.00E+00	1.54E−135	5.85E−135	2.88E−05	3.17E−05	2.67E−06	3.06E−06	2.86E−22	1.57E−21	1.99E−07	7.15E−07
32	6.79E−03	7.77E−03	9.44E−03	1.08E−02	1.82E−02	2.34E−02	7.90E−03	1.07E−02	4.67E−03	1.08E−02	2.39E+00	4.97E+00
33	1.45E+01	0.00E+00	1.45E+01	0.00E+00	1.45E+01	1.77E−03	1.45E+01	4.67E−03	1.45E+01	2.47E−03	1.45E+01	2.62E−03
34	−2.37E+03	7.18E+02	−1.37E+03	8.85E+01	3.25E+05	2.00E+05	9.06E+04	4.25E+04	7.25E+03	6.33E+03	7.03E+04	9.45E+04
35	2.44E−07	6.30E−07	2.86E−06	4.32E−06	1.20E−02	1.94E−02	8.53E−08	1.70E−07	2.52E−03	5.41E−03	8.19E−04	1.18E−03
36	9.60E−09	1.83E−08	7.84E−08	1.24E−07	9.64E−11	5.15E−10	1.58E−20	4.39E−20	0.00E+00	0.00E+00	6.08E−04	5.81E−04
37	−1.00E+00	5.55E−07	−9.93E−01	4.06E−02	−1.00E+00	0.00E+00	−9.33E−01	2.54E−01	−1.00E+00	0.00E+00	−9.96E−01	5.02E−03
38	9.98E−01	1.77E−11	9.98E−01	3.18E−12	2.37E+00	2.95E+00	9.98E−01	2.26E−16	1.32E+00	1.78E+00	1.86E+00	9.96E−01
39	−2.35E+03	1.14E−01	−2.35E+03	8.93E−02	−2.04E+03	6.09E+01	−2.11E+03	7.17E+01	−2.18E+03	1.22E+02	−1.15E+03	8.68E+01
40	0.00E+00	0.00E+00	1.44E−306	0.00E+00	5.38E−03	5.87E−03	8.15E−04	1.27E−03	1.27E−04	2.03E−04	2.75E−01	5.00E−01

TABLE 9.4 Performance Results of hMFOSMA, ABC, BOA, CS, and BA on F21–F40 Test Functions

Functions	hMFOSMA		ABC		BOA		CS		BA	
	Mean	Std	Mean	Std	Mean	Std	Mean	Std	Mean	Std
21	0.00E+00	0.00E+00	0.00E+00	0.00E+00	1.04E−08	1.87E−08	5.27E+02	4.65E+02	2.43E+00	7.23E−01
22	0.00E+00	0.00E+00	2.22E−17	3.13E−17	6.44E−09	2.01E−08	7.81E+02	6.78E+02	6.25E−01	4.57E−01
23	0.00E+00	0.00E+00	2.79E−05	5.06E−05	7.09E−10	1.52E−09	1.40E+02	1.10E+02	8.51E−01	6.96E−01
24	2.87E−211	0.00E+00	3.61E−05	4.19E−05	4.66E−14	1.23E−13	4.18E−01	3.87E−01	3.51E−14	3.00E−14
25	2.85E−45	1.56E−44	3.27E+02	4.42E+01	4.85E−06	1.90E−06	1.96E+09	2.60E+09	1.03E+02	3.44E+01
26	2.19E−45	1.20E−44	1.18E+06	1.84E+06	2.54E−05	1.06E−05	2.03E+12	1.86E+11	1.20E+12	1.43E+11
27	5.23E−121	2.25E−120	4.93E−04	5.87E−04	6.19E−06	2.74E−05	7.05E−02	1.15E−01	2.74E−08	8.21E−08
28	2.98E−08	4.25E−08	3.52E−14	1.12E−13	2.48E−04	2.14E−04	1.24E+01	1.53E+01	1.19E−12	1.12E−12
29	4.80E−02	9.14E−02	1.79E+00	1.83E+00	7.46E−01	6.11E−01	4.86E+03	4.35E+03	1.14E+00	1.36E+00
30	6.91E−06	6.33E−06	8.15E−05	1.20E−04	3.61E−03	3.89E−03	2.17E+03	2.39E+03	1.31E−01	2.66E−01
31	5.95E−182	0.00E+00	3.58E−01	5.81E−01	1.24E−10	1.49E−10	2.08E+02	1.47E+02	2.18E−06	1.27E−06
32	6.79E−03	7.77E−03	9.17E−02	8.88E−02	3.91E−01	3.23E−01	8.90E+01	1.03E+02	4.68E−03	9.22E−03
33	1.45E+01	0.00E+00	1.46E+01	4.42E−02	1.45E+01	4.25E−03	1.48E+01	1.35E−01	1.46E+01	6.53E−02
34	−2.37E+03	7.18E+02	2.90E+04	1.61E+04	−5.16E+01	3.63E+01	4.54E+06	8.11E+05	3.18E+06	3.51E+05
35	2.44E−07	6.30E−07	1.34E−02	1.73E−02	7.93E−03	1.49E−02	8.28E−01	8.44E−01	7.24E−12	7.32E−12
36	9.60E−09	1.83E−08	6.17E−04	2.50E−03	2.24E−03	5.17E−03	2.25E+00	4.35E+00	6.48E−13	5.40E−13
37	−1.00E+00	5.55E−07	−9.13E−01	2.59E−01	−1.54E−01	7.54E−01	−1.36E−50	7.47E−50	−2.00E−01	4.07E−01
38	9.98E−01	1.77E−11	9.98E−01	1.72E−16	1.31E+00	6.95E−01	1.38E+02	1.61E+02	5.37E+00	4.09E+00
39	−2.35E+03	1.14E−01	−2.28E+03	2.80E+01	−7.73E+03	5.20E+03	−9.08E+02	1.19E+02	−1.98E+03	6.28E+01
40	0.00E+00	0.00E+00	5.69E−10	2.93E−09	2.51E−08	1.59E−08	1.04E+02	2.11E+01	5.87E−25	7.83E−25

TABLE 9.5 Performance Results of hMFOSMA, SMA, MFO, PSO, DE, and SCA on F41–F56 Test Functions

Functions	hMFOSMA		SMA		MFO		PSO		DE		SCA	
	Mean	Std	Mean	Std	Mean	Std	Mean	Std	Mean	Std	Mean	Std
41	0.00E+00	0.00E+00	0.00E+00	0.00E+00	9.22E−01	7.18E−01	4.80E−01	2.79E−01	1.21E−01	1.37E−01	6.52E−02	1.49E−01
42	4.78E−30	2.29E−29	3.70E−31	2.01E−30	5.80E+00	3.06E+00	3.10E+00	4.55E−01	1.45E+00	3.34E−01	2.30E+00	1.09E+00
43	1.13E−258	0.00E+00	1.82E−194	0.00E+00	1.16E+01	1.63E+01	1.30E−03	1.60E−03	2.59E−04	3.74E−04	1.50E−02	3.09E−02
44	8.70E−62	4.76E−61	7.88E+01	4.26E+02	1.14E+10	1.62E+10	2.27E+08	1.36E+08	1.94E+07	5.81E+07	5.30E+08	7.77E+08
45	1.00E+00	0.00E+00	1.00E+00	0.00E+00	1.00E+00	0.00E+00	1.00E+00	1.22E−14	1.00E+00	0.00E+00	1.00E+00	0.00E+00
46	3.14E−04	6.03E−04	1.08E−108	5.87E−108	1.73E+01	1.97E+01	3.35E−02	1.73E−02	5.75E−02	9.75E−02	4.99E−02	7.72E−02
47	4.08E−07	9.83E−07	6.28E−04	3.30E−03	4.55E−02	1.20E−01	3.05E−06	5.81E−06	7.42E−04	4.04E−03	2.62E−03	2.91E−03
48	−2.50E+04	6.70E+02	−2.28E+04	9.17E+02	−1.55E+04	1.20E+03	−1.17E+04	1.99E+03	−5.95E+03	4.55E+02	−6.43E+03	7.15E+02
49	3.25E+06	1.11E+06	4.23E+06	6.52E−05	4.23E+06	3.79E−09	4.23E+06	8.18E+02	4.23E+06	3.05E−09	4.23E+06	4.57E+02
50	6.74E+01	6.96E+00	7.14E+01	2.89E−14	7.14E+01	2.89E−14	7.15E+01	2.72E−03	7.14E+01	2.89E−14	7.14E+01	2.89E−14
51	−1.00E+00	0.00E+00	−1.00E+00	0.00E+00	−9.77E−01	3.12E−02	−1.00E+00	0.00E+00	−9.98E−01	1.16E−02	−1.00E+00	0.00E+00
52	7.34E−249	0.00E+00	5.51E−185	0.00E+00	3.78E+07	1.05E+08	1.04E+04	1.49E+04	1.76E+02	6.08E+02	2.87E+04	5.21E+04
53	5.69E−145	3.12E−144	8.68E−121	4.75E−120	4.69E−02	5.38E−02	1.44E−02	1.10E−02	3.70E−03	1.14E−02	9.27E−01	6.55E−01
54	6.79E+04	2.49E+02	7.14E+04	4.93E+03	1.70E+09	3.82E+09	3.56E+08	3.58E+08	2.86E+08	5.05E+08	1.91E+10	5.03E+10
55	8.61E−246	0.00E+00	1.71E−175	0.00E+00	3.31E−55	1.39E−54	3.50E−21	7.93E−21	1.46E−56	7.57E−56	1.99E−37	7.26E−37
56	3.98E+04	4.15E+01	3.98E+04	8.87E−01	4.34E+04	9.58E+02	4.34E+04	7.84E+02	4.59E+04	3.69E+02	4.60E+04	2.91E+02

TABLE 9.6 Performance Results of hMFOSMA, ABC, BOA, CS, and BA on F41–F56 Test Functions

Functions	hMFOSMA		ABC		BOA		CS		BA	
	Mean	Std	Mean	Std	Mean	Std	Mean	Std	Mean	Std
41	0.00E+00	0.00E+00	1.87E–02	1.64E–02	1.13E–06	5.46E–07	9.64E+00	7.89E–01	1.93E+00	4.82E–01
42	4.78E–30	2.29E–29	6.43E+00	9.86E–01	4.00E–01	8.83E–02	2.67E+01	1.37E+00	2.38E+01	1.20E+00
43	1.13E–258	0.00E+00	5.22E–02	2.25E–01	8.30E–11	5.69E–11	1.27E+03	2.46E+03	3.76E+00	1.52E+00
44	8.70E–62	4.76E–61	5.37E+04	4.80E+04	2.58E–05	6.25E–06	1.83E+11	1.82E+10	1.09E+11	1.72E+10
45	1.00E+00	0.00E+00	1.00E+00	4.12E–17	1.00E+00	1.34E–12	8.11E+03	7.15E+03	1.72E+02	2.32E+02
46	3.14E–04	6.03E–04	5.26E–04	1.14E–03	4.53E–06	1.59E–06	3.77E+08	8.52E+08	1.05E–08	1.97E–09
47	4.08E–07	9.83E–07	2.85E–02	4.12E–02	6.60E–04	7.03E–04	3.36E+02	4.12E+02	4.17E–01	1.66E+00
48	–2.50E+04	6.70E+02	–1.65E+04	6.05E+02	–6.70E+03	7.10E+02	–3.46E+03	7.46E+02	–6.56E+03	7.46E+02
49	3.25E+06	1.11E+06	4.23E+06	3.79E–09	4.23E+06	9.80E+01	4.69E+06	2.58E+05	4.29E+06	3.90E+04
50	6.74E+01	6.96E+00	7.14E+01	2.89E–14	7.15E+01	1.22E–01	7.20E+01	2.96E–01	7.15E+01	4.67E–02
51	–1.00E+00	0.00E+00	–1.00E+00	8.05E–06	–3.00E+303	6.55E+04	–6.00E–01	1.75E–01	–9.03E–01	6.69E–02
52	7.34E–249	0.00E+00	1.80E–04	3.54E–04	3.90E–08	2.12E–08	4.74E+09	1.12E+09	1.83E+09	4.85E+08
53	5.69E–145	3.12E–144	2.04E–04	1.87E–04	3.96E–06	1.15E–06	2.99E+01	7.36E+00	5.85E–09	1.21E–09
54	6.79E+04	2.49E+02	6.81E+04	3.92E+02	1.33E+05	4.55E+03	6.24E+12	9.99E+11	3.52E+12	8.21E+11
55	8.61E–246	0.00E+00	1.98E–17	1.78E–17	5.14E–14	1.28E–13	1.25E+05	1.05E+05	5.86E+02	1.10E+03
56	3.98E+04	4.15E+01	4.05E+04	1.61E+02	4.62E+04	2.56E+02	4.79E+04	3.54E+02	4.36E+04	4.54E+02

TABLE 9.7 Schematic View of Superior, Inferior, and Equal to hMFOSMA With Other Compared Algorithm

	SMA	MFO	PSO	DE	SCA	ABC	BOA	CS	BA
Superior	30	47	46	44	50	49	50	56	50
Inferior	14	5	7	8	0	5	5	0	6
Equal	12	4	3	4	6	2	1	0	0

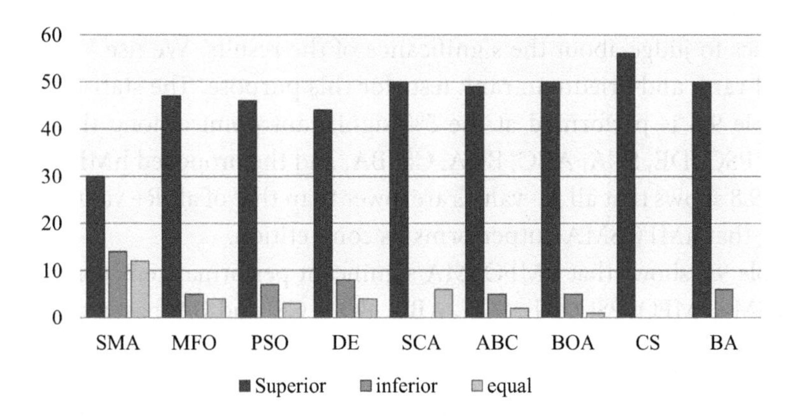

FIGURE 9.1 Schematic view of superior, inferior, and equal to hMFOSMA compared to other algorithms.

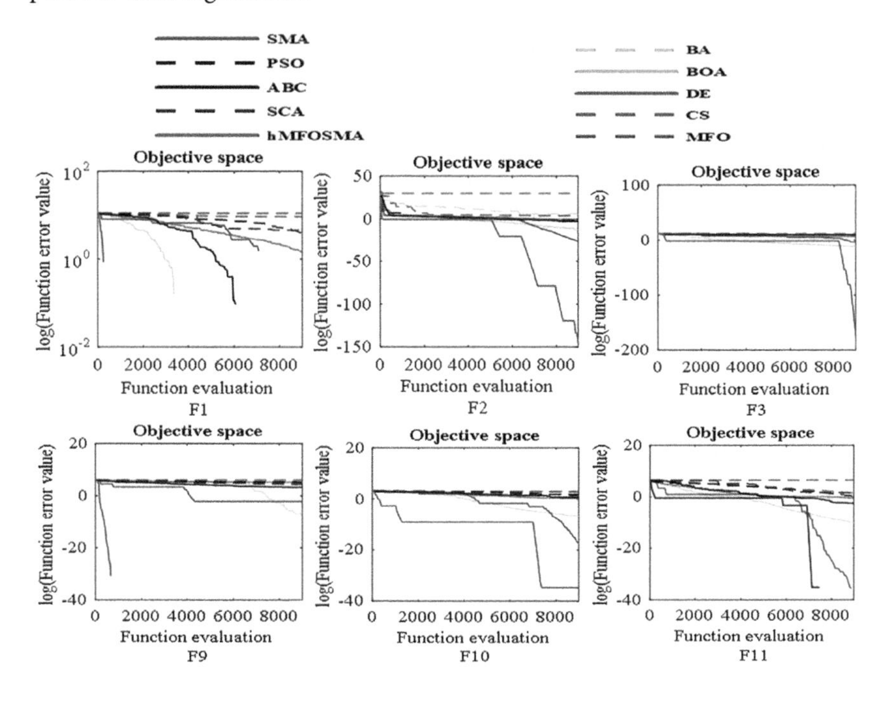

FIGURE 9.2 Convergence graph of some selected function.

9.4.3 Statistical Analysis

The statistical analysis of the collected results is critical for evaluating the substantial change. All the algorithms in the comparison (SMA, MFO, PSO, DE, SCA, ABC, BOA, CS, BA, and the proposed hMFOSMA) are stochastic; so in addition to using descriptive statistical measures (mean, median, std. dev., etc.) which provides an overall performance between algorithms over 30 runs, we need to use statistical test from inferential statistics to judge about the significance of the results. We use Wilcoxon signed-rank and Friedman rank tests for this purpose. The statistical test in Table 9.8 is performed at the 5% significant point among the SMA, MFO, PSO, DE, SCA, ABC, BOA, CS, BA, and the proposed hMFOSMA. Table 9.8 shows that all *R*− values are lower than that of all *R*+ values, indicating that hMFOSMA outperforms its competition.

Table 9.9 shows that hMFOSMA significant performance is compatible with SMA, MFO, PSO, DE, SCA, ABC, BOA, CS, and BA, based on a 95%

TABLE 9.8 Results of Wilcoxon Rank Sum Test

hMFOSMA vs Algorithm	*P*-value	*R*+	*R*−	Winner
SMA	1.4667E-02	464	166	hMFOSMA
MFO	3.2583E-09	1165	11	hMFOSMA
PSO	9.2649E-09	1148	28	hMFOSMA
DE	4.8364E-08	1040	41	hMFOSMA
SCA	1.6310E-09	1176	0	hMFOSMA
ABC	9.2140E-09	1190	35	hMFOSMA
BOA	7.0558E-08	1238	88	hMFOSMA
CS	7.5475E-11	1696	0	hMFOSMA
BA	1.2437E-09	1448	37	hMFOSMA

TABLE 9.9 Performance of Friedman Rank Test Results

Algorithm	Mean Rank	Rank
hMFOSMA	2.13	1
SMA	2.64	2
MFO	6.43	7
PSO	5.3	6
DE	4.37	3
SCA	6.51	8
ABC	5.2	5
BOA	5.08	4
CS	10	10
BA	7.35	9

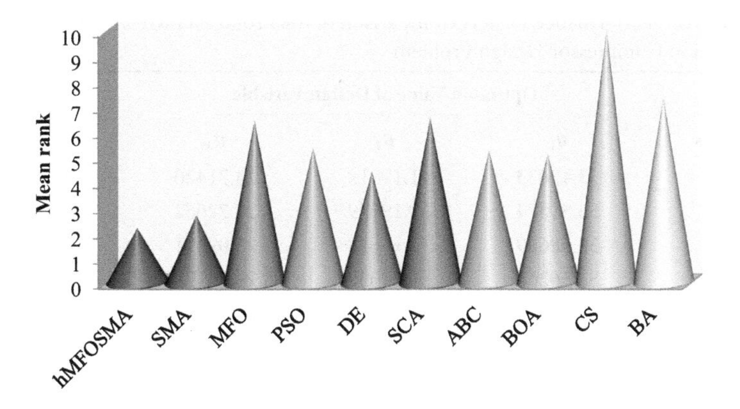

FIGURE 9.3 Schematic view of the results of Friedman rank test.

significance evaluation by Friedman rank test. Table 9.9 reveals that this algorithm's average value is lower than others, indicating that hMFOSMA is the ultimate rank. Figure 9.3 depicts a schematic representation of the results of Friedman rank test. As a consequence of the results on various benchmarks, it is clear that the proposed hMFOSMA has greatly improved numeric performance, upgrade domain exploitation, and convergence rate over SMA, MFO, PSO, DE, SCA, ABC, BOA, CS, and BA. From the discussion, it is concluded that the suggested hMFOSMA is statistically considerably better than basic SMA, MFO, PSO, DE, SCA, ABC, BOA, CS, and BA.

9.5 APPLICATION OF hMFOSMA ON ENGINEERING PROBLEMS

In this division, in two real-world optimization problems, the efficiency of the suggested hMFOSMA algorithm is studied. Unconstrained real-engineering optimization problems are (a) gas transmission compressor design problem and (b) optimal capacity of gas production facilities.

9.5.1 EP.1: Gas Transmission Compressor Design Problem [29]

The mathematical preparation of the gas transmission compressor design problem is given below:

$$\text{Minimize } f_1(\vec{u}) = 8.61*10^5 * u_1^{\frac{1}{2}} u_2 u_3^{\frac{-2}{3}} \left(u_2^2 - 1\right)^{-\frac{1}{2}} + 3.69*10^4 u_3 + 7.72*10^8 u_1^{-1} u_2^{0.219} - 765.43*10^6 u_1^{-1}$$

such that $10 \le u_1 \le 55$, $1.1 \le u_2 \le 2$, and $10 \le u_3 \le 40$.

TABLE 9.10 Performance Results Comparison of the Proposed hMFOSMA for the Gas Transmission Compressor Design Problem

	Optimum Value of Design Variable			
Algorithms	u_1	u_2	u_3	f_{min}
ABC	53.42633	1.19012	24.71420	2.96438E+06
BA	53.51374	1.19029	24.72662	2.96438E+06
BOA	52.58629	1.17679	25.09491	2.96556E+06
CS	53.56266	1.19030	24.74178	2.96438E+06
DE	53.44678	1.19010	24.71854	2.96438E+06
MFO	53.44646	1.19010	24.71862	2.96438E+06
PSO	53.44056	1.19009	24.71769	2.96438E+06
SCA	55.00000	1.19822	24.51343	2.96458E+06
SMA	53.44961	1.19011	24.71863	2.96438E+06
hMFOSMA	**53.44837**	**1.19011**	**24.71880**	**2.96438E+06**

In order to determine the solution, 30 independent runs are performed with 2,000 function evaluations and 30 population, and the performance metrics are recorded in Table 9.10. To compare the performance metrics of the hMFOSMA, the compared algorithms SMA, MFO, PSO, DE, SCA, ABC, BOA, CS, and BA are implemented with same parameter setting. The optimum values of design variable are 53.44837, 1.19011, and 24.71880, and the optimum value of gas transmission compressor design is 2.96438E+06. Table 9.11 also shows the statistical measurement results achieved by various algorithms. From Table 9.11, the statistic measurements of hMFOSMA in all statistics can be found to be higher than the other algorithms compared. The analysis of the results provided in tables validates that the suggested hMFOSMA algorithm has greater search proficiency than other algorithms in terms of different statistics.

9.5.2 EP.2: Optimal Capacity of Gas Production Facilities [29]

The mathematical formulation of the optimal capacity of gas production facilities design problem is given below:

$$\text{Minimize} \quad f_2(\vec{u}) = 61.8 + 5.72 * u_1 * 0.2623 * \left[(40 - u_1) \times \ln\left(\frac{u_2}{200}\right) \right]^{-0.85}$$

$$+ 0.087 * (40 - u_1) \times \ln\left(\frac{u_2}{200}\right) + 700.23 * u_2^{-0.75} \tag{9.8}$$

such that $17.5 \le u_1$ and $300 \le u_2 \le 600$.

TABLE 9.11 Statistical Measurement Comparison of the Proposed hMFOSMA for the Gas Transmission Compressor Design Problem

Algorithms	Minimum	Maximum	Median	Mean	Standard Deviation
ABC	2.96438E+06	2.96438E+06	2.96440E+06	2.96438E+06	7.56410E+00
BA	2.96438E+06	2.96477E+06	2.97228E+06	2.96569E+06	2.00505E+03
BOA	2.96556E+06	2.97335E+06	3.00465E+06	2.97905E+06	1.39814E+04
CS	2.96438E+06	2.96438E+06	2.96448E+06	2.96440E+06	2.66934E+01
DE	2.96438E+06	2.96438E+06	2.96438E+06	2.96438E+06	1.13984E-04
MFO	2.96438E+06	2.96438E+06	2.96447E+06	2.96439E+06	3.28481E+01
PSO	2.96438E+06	2.96438E+06	2.96438E+06	2.96438E+06	5.58271E-01
SCA	2.96458E+06	2.96582E+06	2.97517E+06	2.96679E+06	2.36225E+03
SMA	2.96438E+06	2.96438E+06	2.96438E+06	2.96438E+06	1.54834E-01
hMFOSMA	**2.96438E+06**	**2.96438E+06**	**2.96438E+06**	**2.96438E+06**	**7.29343E-03**

Table 9.12 displays the best solution obtained by hMFOSMA. Thirty independent experiments using 30 candidate solutions and 2,000 function evaluations are listed in the table. For comparison of hMFOSMA results, a comparable experimental model is used (i.e., same quantity of function evaluations) to compare algorithms SMA, MFO, PSO, DE, SCA, ABC, BOA, CS, and BA. The optimum values of parameters are 17.50150 and 599.99670 and the optimal capacity of gas production facilities design problem is 7.14451E+01. Table 9.13 also displays the effects of statistical measurement by the hMFOSMA for comparison of the algorithms. The table reveals that the algorithms SMA, MFO, PSO, DE, SCA, ABC, BOA, CS, and BA have the same effects as the hMFOSMA suggested. The cumulative analysis of results seen in the table indicates that the proposed hMFOSMA algorithm has a higher search performance with respect to different statistics relative to the other algorithms.

9.6 CONCLUSION AND FUTURE DIRECTIONS

A novel hybrid algorithm using the strengths of MFO and SMA is introduced in this research. The principal aim is to merge SMA's capabilities in execution with MFO's in investigation. In comparison with SMA, MFO, PSO, DE, SCA, ABC, BOA, CS, and BA, 56 benchmarking functions are utilized to verify hMFOSMA performance. The results demonstrate that hMFOSMA is superior in most minimal functions. The findings are also shown to be quicker than MFO and SMA at the convergence speed of hMFOSMA. The suggested method has been shown to be superior in

TABLE 9.12 Performance Results Comparison of the Proposed hMFOSMA for the Optimal Capacity of Gas Production Facilities

	Optimum Value of Design Variable		
Algorithms	u_1	u_2	f_{min}
ABC	17.5	600	7.14451E+01
BA	17.50005	599.99967	7.14451E+01
BOA	17.5	600	7.14451E+01
CS	17.50023	599.97276	7.14453E+01
DE	17.50000	599.99993	7.14451E+01
MFO	17.5	600	7.14451E+01
PSO	17.53402	599.86544	7.14482E+01
SCA	17.5	600	7.14451E+01
SMA	17.5	600	7.14451E+01
hMFOSMA	**17.50150**	**599.99670**	**7.14451E+01**

TABLE 9.13 Statistical Measurement Comparison of the Proposed hMFOSMA for the Optimal Capacity of Gas Production Facilities Problem

Algorithms	Minimum	Maximum	Median	Mean	Standard Deviation
ABC	7.14451E+01	7.14451E+01	7.14451E+01	7.14451E+01	2.89076E-14
BA	7.14451E+01	7.14802E+01	7.16202E+01	7.14949E+01	4.55632E-02
BOA	7.14451E+01	7.14452E+01	7.23038E+01	7.15877E+01	2.21725E-01
CS	7.14453E+01	7.14478E+01	7.14558E+01	7.14486E+01	2.58583E-03
DE	7.14451E+01	7.14451E+01	7.14451E+01	7.14451E+01	4.81896E-06
MFO	7.14451E+01	7.14451E+01	7.14451E+01	7.14451E+01	2.89076E-14
PSO	7.14482E+01	7.14580E+01	7.14684E+01	7.14581E+01	5.99855E-03
SCA	7.14451E+01	7.14451E+01	7.14451E+01	7.14451E+01	2.89076E-14
SMA	7.14451E+01	7.14451E+01	7.14451E+01	7.14451E+01	2.89076E-14
hMFOSMA	**7.14451E+01**	**7.14463E+01**	**7.14482E+01**	**7.14444E+01**	**7.46311E-04**

terms of improved convergence speed and modified local minimal escaping for the majority of benchmark functions.

This document also explores the use of hMFOSMA algorithm to solve two conventional real-life optimization problems. Comparative findings demonstrate the outstanding performance of the hMFOSMA method on hard unknown space of problem. The hMFOSMA attempts to make use of MFO and SMA to prevent optimally local results in this study. In combination, hMFOSMA may balance investigation with exploitation, solving complicated challenges and real-world engineering challenges in an effective manner.

In future studies, a range of actual optimization problem may be explored in the hybrid hMFOSMA algorithm, including vehicle routing, job shop planning, multi-objective problem, and workflow planning.

REFERENCES

1 J.H. Holland, Genetic algorithms, *Sci. Am.* 267 (1992) 66–72. doi: 10.1038/scientificamerican0792-66.
2. Y. Shi, R. Eberhart, Modified particle swarm optimizer. *In Proceedings of IEEE Conference on Evolutionary Computation ICEC*, IEEE (1998) pp. 69–73. doi: 10.1109/icec.1998.699146.
3. S. Li, H. Chen, M. Wang, A.A. Heidari, S. Mirjalili, Slime mould algorithm: A new method for stochastic optimization, *Futur. Gener. Comput. Syst.* 111 (2020) 300–323. doi: 10.1016/j.future.2020.03.055.
4. S. Mirjalili, Moth-flame optimization algorithm: A novel nature-inspired heuristic paradigm, *Knowl. Based Syst.* 89 (2015) 228–249. doi: 10.1016/j.knosys.2015.07.006.
5. R. Storn, K. Price, Differential evolution: A simple and efficient heuristic for global optimization over continuous spaces, *J. Glob. Optim.* 11 (1997) 341–359. doi: 10.1023/A:1008202821328.
6. S. Mirjalili, SCA: A sine cosine algorithm for solving optimization problems, *Knowl. Based Syst.* 96 (2016) 120–133. doi: 10.1016/j.knosys.2015.12.022.
7. D. Karaboga, B. Basturk, A powerful and efficient algorithm for numerical function optimization: Artificial bee colony (ABC) algorithm, *J. Glob. Optim.* 39 (2007) 459–471. doi: 10.1007/s10898-007-9149-x.
8. S. Arora, S. Singh, Butterfly optimization algorithm: A novel approach for global optimization, *Soft Comput.* 23 (2019) 715–734. doi: 10.1007/s00500-018-3102-4.
9. X.-S. Yang, S. Deb, Cuckoo search via Lévy flights, *In 2009 World Congress on Nature and Biologically Inspired Computing*, IEEE (2009) pp. 210–214. doi: 10.1109/NABIC.2009.5393690.
10. X.S. Yang, A new metaheuristic Bat-inspired Algorithm, In: Studies in Computational Intelligence, Springer, Berlin, Heidelberg (2010) pp. 65–74. doi: 10.1007/978-3-642-12538-6_6.
11. M. Crepinsek, S.H. Liu, M. Mernik, Exploration and exploitation in evolutionary algorithms: A survey, *ACM Comput. Surv.* 45 (2013). doi: 10.1145/2480741.2480752.
12. A.E. Eiben, C.A. Schippers, On evolutionary exploration and exploitation, *Fundam. Informaticae.* 35 (1998) 35–50. doi: 10.3233/fi-1998-35123403.
13. D.H. Wolpert, W.G. Macready, No free lunch theorems for optimization, *IEEE Trans. Evol. Comput.* 1 (1997) 67–82. doi: 10.1109/4235.585893.
14. M. Abd Elaziz, D. Yousri, S. Mirjalili, A hybrid Harris hawks-moth-flame optimization algorithm including fractional-order chaos maps and evolutionary population dynamics, *Adv. Eng. Softw.* 154 (2021) 102973. doi: 10.1016/j.advengsoft.2021.102973.

15. Z. Li, Y. Zhou, S. Zhang, J. Song, Lévy-flight moth-flame algorithm for function optimization and engineering design problems, *Math. Probl. Eng.* 2016 (2016). doi: 10.1155/2016/1423930.
16. A. Dabba, A. Tari, S. Meftali, Hybridization of moth flame optimization algorithm and quantum computing for gene selection in microarray data, *J. Ambient Intell. Humaniz. Comput.* 12 (2021) 2731–2750. doi: 10.1007/s12652-020-02434-9.
17. P. Singh, S.K. Bishnoi, Modified moth-flame optimization for strategic integration of fuel cell in renewable active distribution network, *Electr. Power Syst. Res.* 197 (2021) 107323. doi: 10.1016/j.epsr.2021.107323.
18. R. Ramachandran, J. Satheesh Kumar, B. Madasamy, V. Veerasamy, A hybrid MFO-GHNN tuned self-adaptive FOPID controller for ALFC of renewable energy integrated hybrid power system, *IET Renew. Power Gener.* 15 (2021) 1582–1595. doi: 10.1049/rpg2.12134.
19. S. Nama, A.K. Saha, A novel hybrid backtracking search optimization algorithm for continuous function optimization, *Decis. Sci. Lett.* (2019) 163–174. doi: 10.5267/j.dsl.2018.7.002.
20. S. Nama, A.K. Saha, A new hybrid differential evolution algorithm with self-adaptation for function optimization, *Appl. Intell.* 48 (2018) 1657–1671. doi: 10.1007/s10489-017-1016-y.
21. S. Nama, A.K. Saha, S. Ghosh, A new ensemble algorithm of differential evolution and backtracking s algorithm with adaptive control parameter for function optimization, *Int. J. Ind. Eng. Comput.* 7 (2016) 323–338. doi: 10.5267/j.ijiec.2015.9.003.
22. S. Nama, A.K. Saha, S. Sharma, Performance up-gradation of symbiotic organisms search by backtracking search algorithm, *J. Ambient Intell. Humaniz. Comput.* 1 (2021) 3. doi: 10.1007/s12652-021-03183-z.
23. S. Nama, A. Kumar Saha, S. Ghosh, A hybrid symbiosis organisms search algorithm and its application to real world problems, *Memetic Comput.* 9 (2017) 261–280. doi: 10.1007/s12293-016-0194-1.
24. A. Bala Krishna, S. Saxena, V.K. Kamboj, hSMA-PS: A novel memetic approach for numerical and engineering design challenges, *Eng. Comput.* (2021) 1–35. doi: 10.1007/s00366-021-01371-1.
25. R.M. Rizk-Allah, A.E. Hassanien, D. Song, Chaos-opposition-enhanced slime mould algorithm for minimizing the cost of energy for the wind turbines on high-altitude sites, *ISA Trans.* (2021). doi: 10.1016/j.isatra.2021.04.011.
26. M. Banaie-Dezfouli, M. H. Nadimi-Shahraki, Z. Beheshti, R-GWO: Representative-based grey wolf optimizer for solving engineering problems. *Appl. Soft Comput.* 106 (2021) 107328.
27. M. H. Nadimi-Shahraki, S. Taghian, S. Mirjalili, H. Faris, MTDE: An effective multi-trial vector-based differential evolution algorithm and its applications for engineering design problems. *Appl. Soft Comput.* 97 (2020) 106761.
28. M. H. Nadimi-Shahraki, S. Taghian, S. Mirjalili, An improved grey wolf optimizer for solving engineering problems. *Expert Syst Appl.*, 166 (2021) 113917.
29. R.A. Cuninghame-Green, C.S. Beightler, D.T. Phillips, Applied geometric programming, *Oper. Res. Q.* 28 (1977) 477. doi: 10.2307/3009002.

CHAPTER 10

Hybrid Aquila Optimizer with Moth-Flame Optimization Algorithm for Global Optimization

Laith Abualigah

Amman Arab University
Universiti Sains Malaysia

Seyedali Mirjalili

Torrens University Australia
Yonsei University

Mohamed Abd Elaziz

Zagazig University

Heming Jia

Sanming University

Canan Batur Şahin

Malatya Turgut Ozal University

Ala' Khalifeh

German Jordanian University

DOI: 10.1201/9781003205326-13

Amir H. Gandomi
University of Technology Sydney

CONTENTS

10.1 Introduction 178
10.2 The Proposed Hybrid Aquila Optimizer with Moth-Flame Optimization Algorithm (AOMFO) 180
10.2.1 Aquila Optimizer (AO) 180
10.2.1.1 Expanded Exploration 181
10.2.1.2 Narrowed Exploration 181
10.2.1.3 Expanded Exploitation 183
10.2.1.4 Narrowed Exploitation 183
10.2.2 Moth-Flame Optimization (MFO) Algorithm 184
10.2.2.1 Generate the Initial Population of MFO 184
10.2.2.2 Updating the Positions 185
10.2.2.3 Updating the Number of Flames 186
10.2.3 The Proposed AOMFO Method 187
10.2.3.1 Solutions Initialization 187
10.2.3.2 Structure of the Proposed AOMFO 187
10.3 Experiments and Results 188
10.3.1 Parameter Settings 188
10.3.2 Description of the Tested Benchmark Functions 188
10.3.3 Results and Discussions 189
10.4 Conclusion and Promising Potential Future Works 202
References 204

10.1 INTRODUCTION

There are various types of optimization problems (OPs) that really should be solved [4,5,19]. Their problem-solving approach can be defined as single-objective [14], big [6,25], multi-objective [9], multiple objectives [59], and fuzzy [33,50]. Metaheuristic algorithms (MAs) motivated by physiological and biological processes have emerged as a pioneer in solving easy and challenging OPs from the disciplines of engineering, energy, intelligent systems, and computing in the last two decades, which is the age of big data [17, 18, 23]. In optimization and machine learning domains, the principle of "trial and error" has already become a pivotal reference to discovering the best solution [31,49]. MAs, in general, have a set of operations, such as crossover and mutation. In subsequent implementations, these technicians produce new individuals, and the consistency is continuously increased [22, 32]. MAs have also been shown to outperform gradient-based algorithms in terms of performance [2,12,57]. However, there are already certain issues with MAs that really need to be addressed. For example, convergence is difficult and slow, and there was no way to find a model that could be applied to any challenge. For the reasons mentioned above, there is still a long way to go in terms of research and implementations and MAs [11,16,58].

Finding appropriate solutions to real-world problems pervades all aspects of life, catching researchers' attention in different disciplines, particularly for engineering and industrial applications, feature selection, clustering problems, multi-objective problems, machine learning and prediction, convolutional neural networks, text mining, parameter estimation, and so on [1,51,52]. Classic pure mathematical optimization algorithms and MAs and genetic algorithm are often used to solve OPs. Another one is focused on nature's established developmental processes. Conventional computational approaches, such as gradient descent, have stringent criteria for the type and scale of problems, as well as complicated constraints or implementation scope. When it comes to neural networks, they are remarkably inflexible. MAs, unlike conventional mathematical approaches [39], show new approaches for solving OPs, whether in engineering applications [47], economic considerations [48], green use estimation [35], air pollution [36], image processing [24,60], the energy sphere [61], classification problems [26,27], wireless sensor network [53], machine vision [30], text clustering [13], the internet of everything (IoE) [37,38], feature selection [7,15], multiple objective optimizations [40], multi-objective problems [46], or functional problems for which we seek feasible solutions [20,28]. For more related comprehensive papers refer to [3,8,10,29].

In the literature, several optimization algorithms have been proposed [34,44,45]. Aquila Optimizer (AO) [21] is a novel powerful search method introduced recently by Abualigah et al. in 2021 to solve OPs. After deep searching in the literature, we found that Moth-Flame Optimization (MFO) algorithm is also an exciting search method that can be used to improve the conventional AO's ability when combined together. It is proposed by Seyedali Mirjalili in 2015 [41].

Two leading critical causes may explain these phenomena. From this point of view, a minimal scan of the pending problems reveals several peaks and valleys, indicating that they are very complex processes. On the other hand, a flaw in the construction of the traditional AO existed from the start, making it impossible to guarantee that it did not crash into the local valley trough during the search. Furthermore, as the dimensions of these problems are raised in certain test functions, the AO's output suffers. The critical search technique of the traditional AO is updated in the current work to resolve all of these problems and to increase the efficiency of discovery and exploitation. The proposed hybrid Aquila Optimizer with Moth-Flame Optimization algorithm is referred to as AOMFO in the research. The goal of improving the efficiency of exploration and exploitation operators is to create a more excellent equilibrium between these two operators than is feasible with the traditional method. Exploring is the concept of visiting completely new areas of a search space while exploiting is the idea of visiting areas of a search space close to the previously visited sites.

The proposed AOMFO's validity is tested using a collection of 23 benchmark problems ranging in difficulty from unimodal to multimodal to fixed-dimensional multimodal. The research problems used in the original paper of AO are also used in this benchmark collection for a fair comparison. When comparing the proposed strategies to the traditional approach and other well-known comparative approaches, the convergence behavior of the elite solution during iterations of the algorithms indicates that the proposed hybrid AOMFO method has improved the initial convergence rate. A non-parametric Friedman rank test is used to draw concrete conclusions regarding the importance of variations in the results of the proposed AOMFO and other algorithms. The experimental findings, as compared to the standard AO, the MFO Algorithm, and different well-known algorithms, reveal that the AOMFO achieves superior efficiency and efficacy.

The rest of this research is designed as follows. Section 10.2 shows the proposed hybrid Aquila Optimizer with Moth-Flame Optimization algorithm. Section 10.3 presents experiments and decisions. Finally, the conclusion and interesting future works are given in Section 10.4.

10.2 THE PROPOSED HYBRID AQUILA OPTIMIZER WITH MOTH-FLAME OPTIMIZATION ALGORITHM (AOMFO)

In this section, the proposed hybrid Aquila Optimizer with Moth-Flame Optimization algorithm is presented, called AOMFO.

10.2.1 Aquila Optimizer (AO)

The AO simulates Aquila's hunting activity by presenting the behaviors taken at each point of the hunt [21].

The AO will move from exploration to exploitation using different behaviors based on this condition: if $t \leq \left(\frac{2}{3}\right) * T$, the exploration steps will be excited; otherwise e, the exploitation steps will be executed. The mathematical equations of the AO are given as below.

10.2.1.1 Expanded Exploration

The Aquila knows the prey area and chooses the right hunting area by high soar with the vertical stoop in the first approach (X_1). The AO uses high-altitude explorers to assess the area of the search space where the prey is located. The following equation is a mathematical expression of this action:

$$X_1(t+1) = X_{\text{best}}(t) \times \left(1 - \frac{t}{T}\right) + \left(X_M(t) - X_{\text{best}}(t) * \text{rand}\right), \qquad (10.1)$$

where $X_1(t+1)$ is the solution of the next iteration of t generated by the first search method (X_1). The best-obtained solution before the tth iteration is $X_{\text{best}}(t)$, which represents the estimated location of the prey. This equation $\left(\frac{1-t}{T}\right)$ is used to monitor the number of iterations in the extended quest (exploration). The position means the value of the current solutions associated with the tth iteration, which is determined using Equation (10.2) and is denoted by $X_M(t)$. A random value between 0 and 1 is represented by rand. The current iteration and the maximum number of iterations are expressed by t and T, respectively.

$$X_M(t) = \frac{1}{N} \sum_{i=1}^{N} X_i(t), \forall j = 1, 2, \ldots, \text{Dim} \qquad (10.2)$$

where Dim denotes the problem's dimension scale and N represents the number of candidate solutions (population size).

10.2.1.2 Narrowed Exploration

When the prey area is discovered from a high soar in the second approach (X_2), the Aquila circles over the target prey, prepares the ground, and then strikes. Contour flight with fast glide attack is the name assigned to this technique. In preparation for the attack, AO narrowly explores the chosen area of the target prey. The following equation is a mathematical expression of this action:

$$X_2(t+1)=X_{\text{best}}(t)\times \text{Levy}(D)+X_R(t)+(y-x)*\text{rand}, \tag{10.3}$$

The solution of the next iteration of t, which is provided by the second search method (X_2), is $X_2(t+1)$. The levy flight distribution function is determined using Equation (10.4). D is the dimension space and Levy(D) is the levy flight distribution function. At the ith iteration, $X_R(t)$ is a random answer taken in the context of $[1\ N]$.

$$\text{Levy}(D)=s\times\frac{u\times\sigma}{|\upsilon|^{\frac{1}{\beta}}} \tag{10.4}$$

where s is a fixed value of 0.01, u is a random number between 0 and 1, and υ is a random number between 0 and 1. The following equation is used to measure σ:

$$\sigma=\left(\frac{\Gamma(1+\beta)\times \text{sine}\left(\frac{\pi\beta}{2}\right)}{\Gamma\left(\frac{1+\beta}{2}\right)\times\beta\times 2^{\left(\frac{\beta-1}{2}\right)}}\right) \tag{10.5}$$

where β is a specified constant value of 1.5, y and x are used to present the spiral form in the quest in Equation (10.3), which are determined as follows:

$$y=r\times\cos(\theta) \tag{10.6}$$

$$x=r\times\sin(\theta) \tag{10.7}$$

where

$$r=r_1+U\times D_1 \tag{10.8}$$

$$\theta=-\omega\times D_1+\theta_1 \tag{10.9}$$

$$\theta_1=\frac{3\times\pi}{2} \tag{10.10}$$

For a fixed number of quest cycles, r_1 takes a value between 1 and 20, and U is a small value set to 0.00565. D_1 is an array of integer numbers ranging from 1 to the search space length (Dim) and ω is a small value set to 0.005.

10.2.1.3 Expanded Exploitation

The third approach (X_3) is employed once the prey location has been accurately determined and the Aquila has been able to land and strike. The Aquila makes a vertical descent for a practice attack to see how the prey reacts. This approach is known as low flying with a gradual fall assault. A mathematical depiction of this activity is as follows:

$$X_3(t+1)=\left(X_{\text{best}}(t)-X_M(t)\right)\times\alpha-\text{rand}+\left((UB-LB)\times\text{rand}+LB\right)\times\delta, \tag{10.11}$$

where $X_3(t+1)$ is the solution of the third search method (X_3), which is generated by the next iteration of t. $X_{\text{best}}(t)$ denotes the prey's estimated location before the ith iteration (the best-obtained solution) and $X_M(t)$ denotes the current solution's mean value at the tth iteration, which is determined using Equation (10.2). A random value between 0 and 1 is represented by rand. The exploitation modification parameters α and δ are set to a limited value in this chapter (0.1). The lower bound of the given problem is LB, while the upper bound is UB.

10.2.1.4 Narrowed Exploitation

When the Aquila reaches the prey in the fourth step (X_4), the Aquila uses stochastic movements to target the prey over land. This strategy is known as "walk and capture prey." Finally, in the final position, AO strikes the prey. The following equation illustrates this action mathematically:

$$\begin{aligned} X_4(t+1)=&\,QF\times X_{\text{best}}(t)-\left(G_1\times X(t)\times\text{rand}\right)\\ &-G_2\times\text{Levy}(D)+\text{rand}\times G_1, \end{aligned} \tag{10.12}$$

where $X_4(t+1)$ is the solution of the fourth search method (X_3), which is generated by the next iteration of t; the consistency function QF is used to match the search strategies and is determined using Equation (10.13). G_1 represents various AO motions used to monitor the prey during the slope, which are created using Equation (10.14). The flight slope of the AO is developed using Equation (10.15) and is represented by G_2, which has decreasing values from 2 to 0. At the tth iteration, $X(t)$ is the new solution.

$$\text{QF}(t)=t^{\left(\frac{2\times\text{rand}()-1}{(1-T)^2}\right)} \tag{10.13}$$

$$G_1=2\times\text{rand}()-1 \tag{10.14}$$

$$G_2 = 2 \times \left(1 - \frac{t}{T}\right) \tag{10.15}$$

rand is a random value between 0 and 1, and $QF(t)$ is the consistency function value at the tth iteration. The current iteration and the maximum number of iterations are described by t and T, respectively. The levy flight distribution function $Levy(D)$ is determined using Equation (10.4).

10.2.2 Moth-Flame Optimization (MFO) Algorithm

MFO is a MA that is population dependent [41]. It starts by creating moths at random in the global optimum followed by which the fitness values (i.e., location) are measured for each moth and the best spot with flame is indicated. After that, a spiral motion function is used to change the moths' locations and update the current best individual placements, resulting in better positions marked by a flame. Before the termination requirements are met, the preceding processes (updating the moth locations and establishing new ones) are repeated. In the MFO Algorithm, there are three critical steps [55,56]. These are the stages that are explained in the following paragraphs.

10.2.2.1 Generate the Initial Population of MFO

Each moth is thought to be capable of flying in 1-D, 2-D, 3-D, or hyperdimensional space. The following is how the group of moths can be expressed:

$$M = \begin{bmatrix} m_{1,1} & m_{1,2} & \cdots & \cdots & m_{1,d} \\ m_{2,1} & m_{2,2} & \cdots & \cdots & m_{2,d} \\ \vdots & \vdots & \vdots & \vdots & \vdots \\ m_{n,1} & m_{n,2} & \cdots & \cdots & m_{n,d} \end{bmatrix} \tag{10.16}$$

where d is the number of dimensions in the solution space and n is the number of moths. In addition, all of the moths' fitness values are stored in an array as follows:

$$OM = \begin{bmatrix} OM_1 \\ OM_2 \\ \vdots \\ OM_n \end{bmatrix} \tag{10.17}$$

Flames are the other elements of the MFO Algorithm. The following matrix depicts the flames in D-dimensional space, along with the fitness function vector that corresponds to them:

$$F = \begin{bmatrix} F_{1,1} & F_{1,2} & \cdots & \cdots & F_{1,d} \\ F_{2,1} & F_{2,2} & \cdots & \cdots & F_{2,d} \\ \vdots & \vdots & \vdots & \vdots & \vdots \\ F_{n,1} & F_{n,2} & \cdots & \cdots & F_{n,d} \end{bmatrix} \tag{10.18}$$

$$\text{OF} = \begin{bmatrix} \text{OF}_1 \\ \text{OF}_2 \\ \vdots \\ \text{OF}_n \end{bmatrix} \tag{10.19}$$

We should deduce from the above that both moths and fires are remedies. Each of them uses a particular approach to keep track of the best location (solution). Moths, for example, can be seen flying about in the quest room. However, without flames, which serve as flags for the moths in the quest room, they would not be able to reach the best spot. As a result, the moths' best solutions are never missed.

10.2.2.2 Updating the Positions

MFO converges the global-optimal combinatorial optimization using three functions. The following are the definitions for these functions:

$$\text{MFO} = (I, P, T) \tag{10.20}$$

where I denotes the initialization of the moths' random locations ($I:\phi \rightarrow \{M, \text{OM}\}$), P presents the movement of the moths in the solution space ($P: M \rightarrow M$) and T presents the termination of the search ($T: M \rightarrow$ true, false). In the I function, the following equation describes the implementation of any random distribution:

$$M(i,j) = (ub(i) - lb(j)) * \text{rand}() + lb(i) \tag{10.21}$$

where lb and ub indicate the lower and upper bounds of variables, respectively.

The moths fly in the quest space in a transverse direction, as already noted. When using a logarithmic spiral submitted, the following three conditions must be met:

- The initial point of the spiral should be the moth.
- The location of the flame will be the final point of the spiral.
- The range of the spiral does not fluctuate more than the search room.

As a result, the MFO Algorithm's logarithmic spiral can be described as follows:

$$S(M_i, F_j) = D_i \cdot e^{bt} \cdot \cos(2\pi t) + F_j \tag{10.22}$$

where D_i presents the distance between the i-th moth and the j-th flame $\left(\text{i.e., } D_i = \left|F_j - M_i\right|\right)$, b indicates a constant to define the shape of the logarithmic spiral, and t indicates a random number between [–1, 1].

The constant b is used to describe the form of the logarithmic spiral, while t is a random number in the range [–1, 1].

In MFO, the moth's spiral trip around the flame in quest space maintains a balance of manipulation and discovery. The best choices are kept in each iteration to avoid being stuck in local optima. The moths use the OF and OM matrices to fly around the flames (i.e., each moth flies around the nearest flame).

10.2.2.3 Updating the Number of Flames

This section focused on increasing the exploitation of the MFO method (for example, changing the moths' positions at n distinct locations in the search space might reduce the likelihood of exploiting the most promising solutions). As a consequence, as shown by the equation below, limiting the number of flames assists in the resolution of this issue:

$$\text{Flame no} = \text{Round}\left(N - 1 * \frac{N-1}{T}\right) \tag{10.23}$$

where N denotes the maximum number of fires, l denotes the current iteration number, and T indicates the maximum number of iterations.

10.2.3 The Proposed AOMFO Method

In this section, the procedure of the proposed AOMFO is presented as follows.

10.2.3.1 Solutions Initialization

The optimization rule in AO is focused on the population of candidate solutions (X), which is developed stochastically here between the upper bound (UB) and the lower bound (LB) of the specific issue, as shown in the following equation:

$$X = \begin{bmatrix} x_{1,1} & \cdots & x_{1,j} & x_{1,\text{Dim}-1} & x_{1,\text{Dim}} \\ x_{2,1} & \cdots & x_{2,j} & \cdots & x_{2,\text{Dim}} \\ \cdots & \cdots & x_{i,j} & \cdots & \cdots \\ \vdots & \vdots & \vdots & \vdots & \vdots \\ x_{N-1,1} & \cdots & x_{N-1,j} & \cdots & x_{N-1,\text{Dim}} \\ x_{N,1} & \cdots & x_{N,j} & x_{N,\text{Dim}-1} & x_{N,\text{Dim}} \end{bmatrix} \quad (10.24)$$

where X represents a collection of current candidate solutions created at random by Equation (10.25), X_i denotes the decision values (positions) of the ith solution, N denotes the total number of search agents (population), and Dim denotes the problem's dimension scale.

$$X_{ij} = rand \times \left(UB_j - LB_j\right) + LB_j,\ i = 1,2,\ldots\ldots,N\ j = 1,2,\ldots,Dim \quad (10.25)$$

where rand is a random integer, LB_j is the problem's j^{th} lower range, and UB_j is the problem's j^{th} upper range.

10.2.3.2 Structure of the Proposed AOMFO

In this section, the proposed AOMFO method is presented in Figure 10.1. The procedure of the AOMFO starts with (a) determining the parameters' values of the used algorithms; (b) generating the candidate solutions; (c) calculating the fitness functions; (d) selecting the best solution; (e) a condition being given: if the condition is true, then the AO will be executed to update the solutions; otherwise, the MFO will be executed to update the solutions; (f) then, another condition being used to decide whether to continue with the optimization process or stop; and finally, (g) returning to the best-obtained solution. The flowchart for the proposed AOMFO is seen in Figure 10.1.

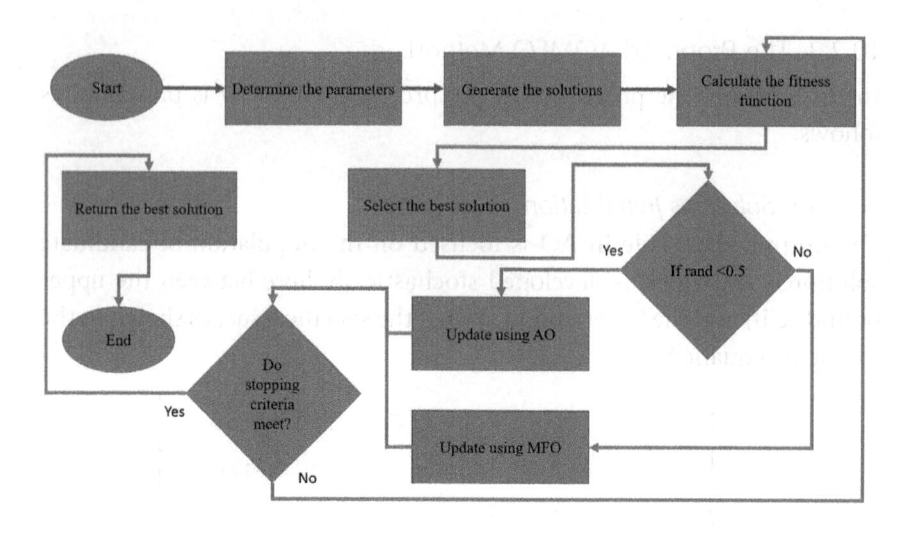

FIGURE 10.1 The flowchart of the proposed AOMFO.

10.3 EXPERIMENTS AND RESULTS

In this section, the experimental setting, details, experiments, and discussions are given as follows.

10.3.1 Parameter Settings

In this section, the comparative methods' main parameter settings are given as shown in Table 10.1. The comparative methods are AO [21], Arithmetic Optimization Algorithm (AOA) [14], Dragonfly Algorithm (DA) [42], Sine Cosine Algorithm (SCA) [45], Gray Wolf Optimizer (GWO), Grasshopper Optimization Algorithm (GOA) [54], and MFO algorithm [41]. The number of solutions and the maximum number of iterations are fixed at 30 and 1000, respectively. Thirty independent runs, for each tested method, are conducted to collect the results.

Note that MATLAB® 2015b is used to run the tests and studies on a CPU Core i7 running on Windows 10 64-bit with 16 GB RAM.

10.3.2 Description of the Tested Benchmark Functions

The details of the tested benchmark functions (F1–F23) are given in Table 10.2. Several types of functions are given including unimodal, multimodal, and fixed-dimension multimodal benchmark functions.

TABLE 10.1 Parameters Values of the Proposed HHMV Method and other Selected Comparative Methods

Algorithm	Parameter	Value
AO	α	0.1
	δ	0.1
AOA	α	5
	μ	0.5
DA	w	0.2-0.9
	s, a, and c	0.1
	f and e	1
SCA	α	0.05
	r_1–r_4	Random numbers
GWO	Convergence parameter (a)	Linear reduction from 2 to 0
GOA	l	1.5
	f	0.5
PSO	Topology	Fully connected
	Cognitive and social constant	(C_1, C_2) 2, 2
	Inertia weight	Linear reduction from 0.9 to 0.1
	Velocity limit	10% of dimension range
MFO	Convergence constant a	[−2, −1]
	Spiral factor b.	1

10.3.3 Results and Discussions

The results of the tested techniques are shown in this part to demonstrate the capabilities of the suggested hybrid AOMFO approach compared to other well-known methods previously discussed. The fitness function values' average, best, worst, and standard deviation (STD) are among the metrics used to assess the generated outcomes.

The results of the proposed AOMFO and other comparison approaches utilizing F1–F13 benchmark functions, where the dimension size is 10, are presented in Table 10.3. In general, as compared to the previous approaches, the suggested AOMFO produced many superior outcomes in virtually all of the examples studied. In F1, F2, F3, F4, F6, F7, F8, F9, F10, F11, F12, and F13, AOMFO had the most outstanding performance. Using the Friedman ranking test, AOMFO has ranked the best method (first ranking), followed by AO, AOA, GWO, MFO, GOA, DA, and GOA. This has proved that the performance of the AOMFO is promising, and it can find new best solutions for the tested problems. According to the SDV, the spread of the achieved results over many runs clearly shows that the obtained results are

TABLE 10.2 Unimodal, Multimodal, and Fixed-Dimension Multimodal Benchmark Functions

Function	Description	Dimensions	Range	$f_{\min}$
F1	$f(x)=\sum_{i=1}^{n} x_i^2$	10,100	[100,100]	0
F2	$f(x)=\sum_{i=0}^{n} \lvert x_i\rvert+\Pi_{i=0}^{n}\lvert x_i\rvert$	10,100	[10,10]	0
F3	$f(x)=\sum_{i=1}^{d}\left(\sum_{j=1}^{i} x_j\right)^2$	10,100	[100,100]	0
F4	$f(x)=\max_i\left[\lvert x_i\rvert,1\le i\le n\right]$	10,100	[100,100]	0
F5	$f(x)=\sum_{i=1}^{n-1}\left[100\left(x_i^2-x_{i+1}\right)^2+(1-x_i)^2\right]$	10,100	[30,30]	0
F6	$f(x)=\sum_{i=1}^{n}\left([x_i+0.5]\right)^2$	10,100	[−100,100]	0
F7	$f(x)=\sum_{i=0}^{n} ix_i^4+\text{random}[0,1)$	10,100	[−128, −28]	0
F8	$f(x)=\sum_{i=1}^{n}\left(-x_i\sin\left(\sqrt{\lvert x_i\rvert}\right)\right)$	10,100	[−500,500]	$-418.9829\times n$
F9	$f(x)=\sum_{i=1}^{n}\left[x_i^2-10\cos(2\pi x_i)+10\right]$	10,100	[−5,12,5.12]	0
F10	$f(x)=-20\exp\left(-0.2\sqrt{\frac{1}{n}\sum_{i=1}^{n} x_i^2}\right)-\exp\left(\frac{1}{n}\sum_{i=1}^{n}\cos(2\pi x_i)\right)+20+e$	10,100	[−32,32]	0
F11	$f(x)=1+\frac{1}{4000}\sum_{i=1}^{n} x_i^1-\prod_{i=1}^{n}\cos\left(\frac{x_i}{\sqrt{i}}\right)$	10,100	[−600,600]	0
F12	$f(x)=\frac{\pi}{n}\left\{10\sin(\pi y_1)\right\}+\sum_{i=1}^{n}(y_i-1)^2\left[1+10\sin^2(\pi y_i+1)+\sum_{i=1}^{n} u(x_i,10,100,4)\right]$	10,100	[−50,50]	0
	where $y_i=1+\frac{x_i+1}{4},u(x_i,a,k,m)\begin{cases}K(x_i-a)^m & \text{if } x_i>a\\ 0 & -a\le x_i\ge a\\ K(-x_i-a)^m & -a\le x_i\end{cases}$			

(Continued)

TABLE 10.2 (*Continued*) Unimodal, Multimodal, and Fixed-Dimension Multimodal Benchmark Functions

Function	Description	Dimensions	Range	f_{min}
F13	$f(x)=0.1\left(\sin^2(3\pi x_1)+\sum_{i=1}^{n}(x_i-1)^2\left[1+\sin^2(3\pi x_1+1)\right]+(x_n-1)^2 1+\sin^2(3\pi x_1+1)\right)$ $+\sum_{i=1}^{n}u(x_i,5,100,4)$	10,100	[−50,50]	0
F14	$f(x)=\left(\frac{1}{500}+\sum_{j=1}^{25}\frac{1}{j+\sum_{i=1}^{2}}(x_i-a_{ij})\right)^{-1}$	2	[−65,65]	1
F15	$f(x)=\sum_{i=1}^{11}\left[a_i-\frac{x_1(b_i^2+b_i x_2)}{b_i^2+b_i x_3+x_4}\right]^2$	4	[−5,5]	0.00030
F16	$f(x)=4x_1^2-2.1x_1^4+\frac{1}{3}x_1^6+x_1x_2-4x_2^2+4x_2^4$	2	[−5,5]	−1.0316
F17	$f(x)=\left(x_2-\frac{5.1}{4\pi}x_1^2+\frac{5}{\pi}x_1-6\right)+10\left(1-\frac{1}{8\pi}\right)\cos x_1+10$	2	[−5,5]	0.398
F18	$f(x)=\left[1+(x_1+x_2+1)(19-14x_1+3x_1^2-14x_2+6x_1x_2+3x_2^3)\right]$ $\times\left[30+(2x_1-3x_2)^2\times(18-32x_i+12x_1^2+48x_2-36x_1x_2+27x_2^2)\right]$	2	[−2,2]	3
F19	$f(x)=-\sum_{i=1}^{4}c_i\exp\left(\sum_{i=1}^{3}a_{ij}(x_j-p_{ij})^2\right)$	3	[−1,2]	−3.86
F20	$f(x)=-\sum_{i=1}^{4}c_i\exp\left(\sum_{i=1}^{6}a_{ij}(x_j-p_{ij})^2\right)$	6	[0,1]	−0.32
F21	$f(x)=-\sum_{i=1}^{5}\left[(X-a_i)(X-a_i)^T+c_i\right]^{-1}$	4	[0,1]	−10.1532
F22	$f(x)=-\sum_{i=1}^{7}\left[(X-a_i)(X-a_i)^T+c_i\right]^{-1}$	4	[0,1]	−10.4028
F23	$f(x)=-\sum_{i=1}^{10}\left[(X-a_i)(X-a_i)^T+c_i\right]^{-1}$	4	[0,1]	−10.5363

TABLE 10.3 The Results of the Proposed AOMFO on Various Optimization Problems F1–F13, Dim=10

Function	Measure	Comparative Algorithms							
		AO	AOA	DA	SCA	GWO	GOA	MFO	AOMFO
	Average	2.82953E−161	4.24692E−98	7.44385E+01	2.64188E−10	5.40982E−41	1.08306E+00	3.79085E−06	1.77382E−170
	Best	4.52182E−178	0.00000E+00	3.40246E+01	1.89933E−17	5.04463E−44	5.67452E−01	1.15461E−09	2.01840E−231
	STD	3.92617E−161	9.49641E−98	5.57841E+01	5.90695E−10	1.16798E−40	6.72901E−01	4.95255E−06	0.00000E+00
	Rank	2	3	8	5	4	7	6	1
F2	Worst	5.04916E−82	0.00000E+00	1.98289E+01	1.19666E−11	6.60119E−23	4.36555E+01	3.79739E−02	0.00000E+00
	Average	1.05816E−82	0.00000E+00	8.58224E+00	4.98259E−12	1.46952E−23	3.76908E+01	8.92738E−03	0.00000E+00
	Best	1.05480E−84	0.00000E+00	2.64841E+00	5.00877E−13	6.05036E−25	2.63391E+01	1.09320E−04	0.00000E+00
	STD	2.23174E−82	0.00000E+00	7.36609E+00	5.59147E−12	2.87440E−23	7.47300E+00	1.63345E−02	0.00000E+00
	Rank	3	1	7	5	4	8	6	1
F3	Worst	2.57551E−47	3.43741E−150	7.36724E+03	2.11402E−03	2.50061E−15	1.70396E+02	1.82310E−01	1.26352E−165
	Average	5.15103E−48	6.87483E−151	4.72080E+03	4.61044E−04	5.23423E−16	6.88072E+01	8.67823E−02	2.63176E−166
	Best	0.00000E+00	5.73140E−230	3.94747E+02	2.07495E−07	2.39782E−19	9.45528E+00	1.63747E−02	1.22880E−190
	STD	1.15181E−47	1.53726E−150	3.11844E+03	9.25075E−04	1.10590E−15	6.82051E+01	6.06641E−02	0.00000E+00
	Rank	3	2	8	5	4	7	6	1
F4	Worst	4.99498E−20	4.69120E−75	1.63378E+01	1.04323E+00	4.74425E−13	2.65781E+01	3.09737E−01	6.90446E−82
	Average	1.01353E−20	9.38239E−76	1.22863E+01	2.08735E−01	1.61155E−13	1.58103E+01	1.56781E−01	1.49656E−82
	Best	0.00000E+00	4.62330E−140	3.67692E+00	4.22112E−05	1.05711E−14	1.81514E+00	7.76823E−03	1.72372E−88
	STD	2.22577E−20	2.09797E−75	5.16463E+00	4.66497E−01	2.01851E−13	1.15225E+01	1.21546E−01	3.03301E−82
	Rank	3	2	7	6	4	8	5	1
F5	Worst	6.59083E−02	8.06047E+00	9.04132E+04	8.73381E+00	8.08852E+00	7.02808E+04	6.96426E+01	6.701540518
	Average	2.36645E−02	7.56282E+00	2.32398E+04	7.85755E+00	7.54324E+00	1.79326E+04	1.99270E+01	3.971465862
	Best	8.82759E−04	7.03838E+00	5.09197E+02	7.36366E+00	7.16343E+00	2.19451E+03	6.66146E+00	2.13884E−06
	STD	2.66707E−02	4.90476E−01	3.80481E+04	5.61275E−01	4.85378E−01	2.93314E+04	2.77999E+01	3.619184254

(Continued)

TABLE 10.3 (*Continued*) The Results of the Proposed AOMFO on Various Optimization Problems F1–F13, Dim=10

Function	Measure	Comparative Algorithms							
		AO	AOA	DA	SCA	GWO	GOA	MFO	AOMFO
	Rank	1	4	8	5	3	7	6	2
F6	Worst	2.05603E−02	3.36037E−01	8.08215E+02	8.28485E−01	2.51527E−01	1.98381E+00	1.83982E−06	1.7921E−06
	Average	4.29647E−03	2.04990E−01	2.81738E+02	6.45558E−01	2.00127E−01	1.06063E+00	4.42865E−07	9.15162E−07
	Best	1.58007E−06	9.22885E−02	4.68204E+01	4.72130E−01	6.52160E−06	7.32138E−01	7.32463E−10	5.40886E−08
	STD	9.09723E−03	9.57149E−02	3.05281E+02	1.39874E−01	1.11874E−01	5.21983E−01	7.86679E−07	6.71207E−07
	Rank	3	5	8	6	4	7	1	2
F7	Worst	4.33475E−04	7.61212E−04	1.96907E−01	3.83059E−02	2.02837E−03	8.59154E+00	7.64214E−02	2.20959E−04
	Average	2.99641E−04	3.25835E−04	8.85537E−02	1.75943E−02	9.37430E−04	4.63799E+00	2.64595E−02	1.12225E−04
	Best	9.44665E−05	1.22383E−04	8.44657E−03	2.47528E−03	1.92698E−04	2.26741E+00	7.85038E−03	4.58962E−06
	STD	1.28506E−04	2.57307E−04	8.21347E−02	1.38420E−02	6.84922E−04	2.78674E+00	2.85211E−02	8.21644E−05
	Rank	2	3	7	5	4	8	6	1
F8	Worst	−1.80017E+03	−2.60560E+03	−1.86386E+03	−1.88193E+03	−2.06441E+03	−1.68238E+03	−1.22467E+03	−3.47920E+03
	Average	−2.33453E+03	−2.65915E+03	−2.21450E+03	−1.98480E+03	−2.29909E+03	−2.28508E+03	−1.73267E+03	−4.04708E+03
	Best	−3.00401E+03	−2.76958E+03	−2.76341E+03	−2.12499E+03	−2.56675E+03	−2.64940E+03	−2.18852E+03	−4.18983E+03
	STD	4.60962E+02	6.64340E+01	3.55931E+02	1.12342E+02	2.45967E+02	3.80920E+02	4.15556E+02	3.17460E+02
	Rank	3	2	6	7	4	5	8	1
F9	Worst	9.22500E−03	0.00000E+00	6.81751E+01	2.83928E+00	0.00000E+00	8.61488E+01	1.60037E+01	0.00000E+00
	Average	1.86606E−03	0.00000E+00	4.77176E+01	5.67933E−01	0.00000E+00	6.21862E+01	1.19872E+01	0.00000E+00
	Best	0.00000E+00	0.00000E+00	3.29083E+01	0.00000E+00	0.00000E+00	3.74334E+01	5.97125E+00	0.00000E+00
	STD	4.11403E−03	0.00000E+00	1.46299E+01	1.26972E+00	0.00000E+00	2.12344E+01	3.77931E+00	0.00000E+00
	Rank	4	1	7	5	1	8	6	1
F10	Worst	8.88178E−16	8.88178E−16	1.56976E+01	1.99582E+01	2.93099E−14	1.99434E+01	1.65123E+00	8.88178E−16
	Average	8.88178E−16	8.88178E−16	9.13633E+00	3.99164E+00	1.86517E−14	1.90695E+01	3.38304E−01	8.88178E−16

(*Continued*)

TABLE 10.3 (*Continued*) The Results of the Proposed AOMFO on Various Optimization Problems F1–F13, Dim=10

Function	Measure	Comparative Algorithms							
		AO	AOA	DA	SCA	GWO	GOA	MFO	AOMFO
	Best	8.88178E−16	8.88178E−16	4.23840E+00	3.94917E−10	1.15463E−14	1.85736E+01	9.38117E−05	8.88178E−16
	STD	0.00000E+00	0.00000E+00	5.08132E+00	8.92558E+00	7.10543E−15	5.14825E−01	7.33995E−01	0.00000E+00
	Rank	1	1	7	6	4	8	5	1
F11	Worst	0.00000E+00	1.22417E−10	5.76667E+00	2.31518E−01	3.75561E−02	5.91055E−01	5.86310E−01	0.00000E+00
	Average	0.00000E+00	2.44833E−11	2.56340E+00	4.63143E−02	1.64322E−02	4.94344E−01	3.16886E−01	0.00000E+00
	Best	0.00000E+00	0.00000E+00	1.08170E+00	3.33067E−16	0.00000E+00	3.26512E−01	6.64195E−02	0.00000E+00
	STD	0.00000E+00	5.47463E−11	2.05610E+00	1.03532E−01	1.39822E−02	1.22623E−01	2.38715E−01	0.00000E+00
	Rank	1	3	8	5	4	7	6	1
F12	Worst	2.46083E−04	1.82610E−01	9.33624E+00	2.27897E−01	1.56828E−01	7.16197E+01	2.99489E−07	1.59680E−07
	Average	1.00103E−04	1.14478E−01	5.86401E+00	1.60514E−01	5.50266E−02	2.23576E+01	6.58271E−08	5.23678E−08
	Best	1.59757E−06	8.22194E−02	3.50106E+00	1.28675E−01	8.32089E−06	3.77516E+00	5.37240E−12	9.63532E−10
	STD	9.97735E−05	4.04230E−02	2.22941E+00	3.88899E−02	6.09737E−02	2.79386E+01	1.31211E−07	6.62039E−08
	Rank	3	5	7	6	4	8	2	1
F13	Worst	2.67018E−04	9.95928E−01	8.61696E+03	5.67113E−01	4.97382E−01	3.05206E+01	9.74214E−02	4.83827E−05
	Average	1.36080E−04	9.50804E−01	1.73161E+03	4.56137E−01	2.20372E−01	1.98000E+01	1.94844E−02	1.34568E−05
	Best	9.62683E−06	8.84665E−01	3.04348E+00	2.69735E−01	9.71032E−02	1.57051E+01	2.53970E−09	6.84804E−10
	STD	9.91453E−05	4.64248E−02	3.84904E+03	1.18552E−01	1.77576E−01	6.23322E+00	4.35681E−02	2.06585E−05
	Rank	2	6	8	5	4	7	3	1
	Summation	31	38	96	71	48	95	66	15
	Mean Ranking	2.38	2.92	7.38	5.46	3.69	7.31	5.08	1.15
	Final Ranking	2	3	8	6	4	7	5	1

stable, and the proposed AOMFO got its results with small distribution values compared to other methods. Consequently, the obtained results in the tested 13 benchmark functions, where the dimension size is fixed to 10, proved the proposed hybrid method's ability. This means that integrating the Aquila Optimizer with Moth-Flame Optimization algorithm generates a robust search method that reached an excellent performance compared to other methods.

In Table 10.4, the results of the proposed AOMFO and other comparative methods using F1–F13 benchmark functions are given, where the dimension size is 100. In general, as compared to other approaches, the proposed AOMFO generated significantly better results in almost all of the cases studied. In F1, F2, F3, F4, F5, F6, F8, F9, F10, F11, F12, and F13, AOMFO had the highest results. According to the Friedman ranking test, AOMFO is the strongest form (first ranking), followed by AO (second ranking), AOA (third ranking), GWO (fourth ranking), MFO (fifth ranking), GOA (sixth ranking), DA (seventh ranking), and GOA (eighth ranking). This shows that the AOMFO's success is promising and that it is capable of finding new best solutions to the tested problems. The spread of the achieved results over several runs, according to the SDV, clearly shows that the obtained results are robust, and the proposed AOMFO obtained its results with limited distribution values compared to other approaches. As a result, the results obtained in the tested 13 benchmark functions, where the dimension size is set to 10, demonstrated the proposed hybrid process's capability. This means that combining the Aquila Optimizer with the Moth-Flame Optimization algorithm results in a rigorous search process that outperforms other approaches.

In Table 10.5, the tests of the suggested AOMFO and other state-of-the-art methods using F14–F23 benchmark functions are given. In all of the cases tested, the suggested AOMFO yielded significantly better results than other approaches. F14–F23 yielded the best results by AOMFO. Using the Friedman ranking test, AOMFO is the strongest method, which got the best ranking over all the comparative methods (first ranking), followed by GWO (second ranking), MFO (third ranking), AO (fourth ranking), DA (fifth ranking), SCA (sixth ranking), AOA (seventh ranking), and GOA (eighth ranking). This demonstrates that the AOMFO's success is promising and that it is capable of discovering new best solutions to the issues that have been discussed. The spread of the achieved results over several runs, according to the SDV, clearly shows that the obtained

TABLE 10.4 The Results of the Proposed AOMFO on Various Optimization Problems F1–F13, Dim=100

		Comparative algorithms							
Function	**Measure**	**AO**	**AOA**	**DA**	**SCA**	**GWO**	**GOA**	**MFO**	**AOMFO**
F1	Worst	1.16841E−166	2.16565E−15	8.84274E+03	7.67489E+01	1.68618E−22	1.06789E+03	3.01001E+00	0.00000E+00
	Average	2.35243E−167	4.33131E−16	3.59078E+03	3.07420E+01	7.34617E−23	5.57269E+02	1.55756E+00	0.00000E+00
	Best	2.10823E−189	7.38711E−69	2.31638E+02	1.38678E−02	6.56830E−24	3.54969E+02	5.53553E−01	0.00000E+00
	STD	0.00000E+00	9.68510E−16	3.85778E+03	4.12567E+01	7.76581E−23	2.91790E+02	1.01284E+00	0.00000E+00
	Rank	2	4	8	6	3	7	5	1
F2	Worst	4.61320E−82	5.35601E−150	1.38458E+02	9.01009E−03	1.15607E−13	6.98357E+05	6.72614E+00	0.00000E+00
	Average	9.29381E−83	1.07120E−150	5.82026E+01	2.47427E−03	4.60581E−14	1.43596E+05	3.86827E+00	0.00000E+00
	Best	8.22484E−95	2.14883E−185	2.51228E+01	6.69678E−05	1.62020E−14	1.67818E+02	1.76478E+00	0.00000E+00
	STD	2.05935E−82	2.39528E−150	4.56265E+01	3.72631E−03	4.11578E−14	3.10227E+05	2.20067E+00	0.00000E+00
	Rank	3	2	7	5	4	8	6	1
F3	Worst	4.24867E−18	5.01352E−02	9.29319E+04	2.44155E+04	9.15147E−02	1.57317E+04	1.62672E+03	7.43878E−157
	Average	8.49734E−19	1.69628E−02	5.27949E+04	1.37294E+04	4.17823E−02	1.14676E+04	9.23113E+02	1.52069E−157
	Best	0.00000E+00	5.45212E−35	2.40759E+04	4.75928E+03	4.16740E−05	6.66769E+03	5.23462E+02	8.60633E−178
	STD	1.90006E−18	2.19193E−02	2.50993E+04	8.37484E+03	3.86104E−02	3.39721E+03	4.19356E+02	3.30908E−157
	Rank	2	3	8	7	4	6	5	1
F4	Worst	2.46075E−19	6.48642E−02	4.06437E+01	5.91666E+01	1.21278E−04	4.96990E+01	1.16047E+01	3.02821E−80
	Average	7.18591E−20	4.95033E−02	3.37412E+01	4.76181E+01	3.92115E−05	4.38471E+01	6.93134E+00	7.23754E−81
	Best	0.00000E+00	4.09787E−02	1.28691E+01	3.25153E+01	7.05207E−06	4.00427E+01	3.46138E+00	9.22976E−88
	STD	1.01216E−19	9.76113E−03	1.18568E+01	1.22265E+01	4.71104E−05	4.00103E+00	3.14100E+00	1.31283E−80
	Rank	2	4	6	8	3	7	5	1
F5	Worst	6.52552E+00	8.23581E+00	7.89167E+03	8.73986E+00	8.72049E+00	1.33842E+04	2.76186E+02	3.10181E−02
	Average	6.39565E+00	7.74715E+00	3.53935E+03	7.94752E+00	7.54337E+00	1.05495E+04	6.47129E+01	1.18384E−02
	Best	6.21792E+00	7.21658E+00	6.93895E+02	7.42374E+00	6.24771E+00	4.65694E+03	8.47381E+00	1.97084E−05

(*Continued*)

TABLE 10.4 (*Continued*) The Results of the Proposed AOMFO on Various Optimization Problems F1–F13, Dim = 100

Function	Measure	Comparative algorithms							
		AO	AOA	DA	SCA	GWO	GOA	MFO	AOMFO
	STD	1.44291E−01	4.10624E−01	2.75949E+03	5.00565E−01	1.06431E+00	3.48621E+03	1.18253E+02	1.39042E−02
	Rank	2	4	7	5	3	8	6	1
F6	Worst	1.92488E+00	4.10925E+00	1.31469E+04	1.55507E+02	3.49393E+00	7.55381E+02	1.10394E+00	3.94700E−02
	Average	7.36630E−01	3.88441E+00	5.24718E+03	3.63080E+01	2.84749E+00	4.94853E+02	6.26979E−01	8.70823E−03
	Best	2.21997E−04	3.68628E+00	1.96660E+02	5.18505E+00	1.76024E+00	3.50004E+02	1.66999E−01	1.84674E−04
	STD	1.00962E+00	1.54562E−01	5.09818E+03	6.66428E+01	6.69248E−01	1.58951E+02	3.88619E−01	1.72112E−02
	Rank	3	5	8	6	4	7	2	1
F7	Worst	5.72491E−04	8.18234E−05	3.45703E+00	7.62005E−01	8.28272E−03	1.13402E+02	2.88041E+00	8.60481E−04
	Average	3.66463E−04	5.55268E−05	1.33178E+00	2.01376E−01	5.03905E−03	7.99803E+01	1.55155E+00	2.95761E−04
	Best	5.89126E−05	2.76747E−05	2.78745E−01	2.35977E−02	2.16283E−03	5.43056E+01	6.57112E−01	9.00365E−05
	STD	2.62702E−04	2.20671E−05	1.25851E+00	3.15388E−01	2.94426E−03	2.30666E+01	1.05009E+00	3.28239E−04
	Rank	3	1	6	5	4	8	7	2
F8	Worst	−3.37926E+03	−4.06009E+03	−4.01618E+03	−2.91873E+03	−3.00817E+03	−6.22403E+03	−1.50930E+03	−9.00167E+03
	Average	−8.93416E+03	−4.95474E+03	−4.80989E+03	−3.26854E+03	−5.15940E+03	−6.91056E+03	−2.31145E+03	−1.17754E+04
	Best	−1.24999E+04	−5.94106E+03	−5.61637E+03	−3.67324E+03	−5.93015E+03	−7.87326E+03	−3.64267E+03	−1.25526E+04
	STD	4.85497E+03	8.33438E+02	7.35345E+02	3.39315E+02	1.22686E+03	6.84294E+02	8.80051E+02	1.55274E+03
	Rank	2	5	6	7	4	3	8	1
F9	Worst	0.00000E+00	0.00000E+00	2.92447E+02	7.93448E+01	3.95319E+00	3.36301E+02	1.36052E+02	0.00000E+00
	Average	0.00000E+00	0.00000E+00	2.27038E+02	3.24561E+01	2.05597E+00	2.93613E+02	1.11702E+02	0.00000E+00
	Best	0.00000E+00	0.00000E+00	1.51753E+02	1.49164E−01	5.68434E−14	2.57799E+02	8.72691E+01	0.00000E+00
	STD	0.00000E+00	0.00000E+00	5.83506E+01	2.97665E+01	1.96341E+00	2.93178E+01	1.86423E+01	0.00000E+00
	Rank	1	1	7	5	4	8	6	1
F10	Worst	4.44089E−15	8.88178E−16	1.18109E+01	2.03298E+01	2.69385E−12	2.02724E+01	3.52703E+00	8.88178E−16

(*Continued*)

TABLE 10.4 (*Continued*) The Results of the Proposed AOMFO on Various Optimization Problems F1–F13, Dim=100

		Comparative algorithms							
Function	**Measure**	**AO**	**AOA**	**DA**	**SCA**	**GWO**	**GOA**	**MFO**	**AOMFO**
	Average	2.30926E−15	8.88178E−16	1.07743E+01	1.53680E+01	1.48876E−12	1.99221E+01	2.74841E+00	8.88178E−16
	Best	8.88178E−16	8.88178E−16	9.71560E+00	2.34257E+00	6.79456E−13	1.95582E+01	2.00897E+00	8.88178E−16
	STD	1.94590E−15	0.00000E+00	8.00017E−01	7.70354E+00	8.88070E−13	2.89335E−01	6.12683E−01	0.00000E+00
	Rank	3	1	6	7	4	8	5	1
F11	Worst	0.00000E+00	4.77120E−01	1.17450E+02	1.20535E+00	3.68471E−02	1.47631E+00	8.07081E−01	0.00000E+00
	Average	0.00000E+00	3.40620E−01	6.09170E+01	8.97368E−01	1.19066E−02	1.37772E+00	4.39007E−01	0.00000E+00
	Best	0.00000E+00	2.50356E−01	1.01890E+01	4.00807E−01	0.00000E+00	1.22803E+00	1.04155E−01	0.00000E+00
	STD	0.00000E+00	9.30230E−02	4.12185E+01	3.06213E−01	1.70552E−02	9.26341E−02	3.02871E−01	0.00000E+00
	Rank	1	4	8	6	3	7	5	1
F12	Worst	5.39743E−04	7.49800E−01	2.79695E+07	1.25643E+06	6.62604E−01	9.31542E+05	1.46794E+00	3.71376E−04
	Average	1.28543E−04	7.08555E−01	9.52683E+06	3.11617E+05	3.00498E−01	2.67125E+05	7.32374E−01	7.82867E−05
	Best	3.06005E−06	6.66785E−01	5.69145E+00	1.45348E+00	1.01957E−01	2.09849E+04	3.55314E−03	4.34111E−07
	STD	2.31091E−04	3.12351E−02	1.33688E+07	5.43870E+05	2.26332E−01	3.75412E+05	6.45543E−01	1.63982E−04
	Rank	2	4	8	7	3	6	5	1
F13	Worst	9.02437E−04	9.39561E−01	4.65250E+03	6.32053E−01	3.99925E−01	6.17698E+01	1.19417E−04	1.20724E−01
	Average	2.83757E−04	8.84514E−01	1.70950E+03	4.97872E−01	2.23114E−01	2.49566E+01	2.53471E−05	2.41452E−02
	Best	6.68305E−06	8.28888E−01	7.40943E−01	3.78852E−01	2.12653E−05	5.43278E+00	5.61814E−11	3.54475E−08
	STD	3.65925E−04	5.47190E−02	2.35026E+03	9.99745E−02	1.49454E−01	2.22566E+01	5.26795E−05	5.39889E−02
	Rank	2	6	8	5	4	7	1	
	Summation	28	44	93	79	47	90	66	13
	Mean Ranking	2.15	3.38	7.15	6.08	3.62	6.92	5.08	1.08
	Final Ranking	2	3	8	6	4	7	5	1

TABLE 10.5 The Results of the Proposed AOMFO on Various Optimization Problems F14–F23

Function	Measure	Comparative algorithms							
		AO	AOA	DA	SCA	GWO	GOA	MFO	AOMFO
F14	Worst	1.26705E+01	1.26705E+01	2.38094E+01	1.26705E+01	1.26705E+01	2.29006E+01	2.19884E+01	1.07632E+01
	Average	6.26263E+00	1.07328E+01	7.54046E+00	7.62004E+00	8.03222E+00	1.57588E+01	1.14028E+01	3.35213E+00
	Best	9.98004E−01	2.98211E+00	1.99203E+00	9.98004E−01	2.98211E+00	5.92885E+00	9.98004E−01	9.98004E−01
	STD	5.89147E+00	4.33278E+00	9.16122E+00	6.09501E+00	4.67540E+00	8.56006E+00	8.74471E+00	4.23043E+00
	Rank	2	6	3	4	5	8	7	1
F15	Worst	1.62229E−03	1.02391E−01	2.25533E−02	1.89303E−03	2.03634E−02	2.40520E−02	2.49610E−03	1.18744E−03
	Average	7.79546E−04	2.39968E−02	1.11497E−02	1.26475E−03	4.69292E−03	1.41126E−02	1.29801E−03	6.01071E−04
	Best	4.40268E−04	4.40642E−04	1.88122E−03	5.73113E−04	3.66799E−04	1.21421E−03	9.70333E−04	3.10808E−04
	STD	4.82118E−04	4.42548E−02	8.20429E−03	5.58330E−04	8.77361E−03	1.16120E−02	6.69991E−04	3.79383E−04
	Rank	2	8	6	3	5	7	4	1
F16	Worst	−1.02761E+00	−1.03163E+00	−1.03124E+00	−1.03120E+00	−1.03163E+00	−2.15464E−01	−1.03163E+00	−1.03163E+00
	Average	−1.03057E+00	−1.03163E+00	−1.03153E+00	−1.03144E+00	−1.03163E+00	−7.05163E−01	−1.03163E+00	−1.03163E+00
	Best	−1.03154E+00	−1.03163E+00	−1.03163E+00	−1.03158E+00	−1.03163E+00	−1.03163E+00	−1.03163E+00	−1.03163E+00
	STD	1.66659E−03	3.54939E−07	1.63645E−04	1.53119E−04	1.03293E−07	4.47032E−01	1.57009E−16	0.00000E+00
	Rank	7	4	5	6	3	8	1	1
F17	Worst	3.98527E−01	4.47728E−01	3.98340E−01	5.04011E+00	3.97892E−01	3.97892E−01	3.97887E−01	3.97887E−01
	Average	3.98168E−01	4.24521E−01	3.98003E−01	2.26378E+00	3.97890E−01	3.97890E−01	3.97887E−01	3.97887E−01
	Best	3.97904E−01	4.05800E−01	3.97889E−01	3.98487E−01	3.97888E−01	3.97888E−01	3.97887E−01	3.97887E−01
	STD	2.32228E−04	1.54897E−02	1.90492E−04	2.53446E+00	1.46876E−06	1.46876E−06	2.19004E−14	0.00000E+00
	Rank	6	7	5	8	3	3	2	1

(Continued)

TABLE 10.5 (*Continued*) The Results of the Proposed AOMFO on Various Optimization Problems F14–F23

Function	Measure	Comparative algorithms							
		AO	AOA	DA	SCA	GWO	GOA	MFO	AOMFO
F18	Worst	3.44382E+00	1.92661E+02	8.40024E+01	3.00198E+00	3.00095E+00	8.40000E+02	8.40006E+01	3.00000E+00
	Average	3.18068E+00	6.63207E+01	1.92008E+01	3.00071E+00	3.00040E+00	1.94674E+02	1.92001E+01	3.00000E+00
	Best	3.00651E+00	3.00000E+00	3.00000E+00	3.00001E+00	3.00000E+00	3.00000E+00	3.00000E+00	3.00000E+00
	STD	1.66155E–01	7.39085E+01	3.62252E+01	8.02709E–04	4.96555E–04	3.62805E+02	3.62245E+01	1.77636E–15
	Rank	4	7	6	3	2	8	5	1
F19	Worst	–3.79405E+00	–3.84330E+00	–3.67340E+00	–3.84725E+00	–3.85580E+00	–1.33756E+00	–3.85470E+00	–3.86278E+00
	Average	–3.84011E+00	–3.84991E+00	–3.78933E+00	–3.84993E+00	–3.86020E+00	–2.96403E+00	–3.86072E+00	–3.86278E+00
	Best	–3.85614E+00	–3.85311E+00	–3.85766E+00	–3.85319E+00	–3.86277E+00	–3.81097E+00	–3.86278E+00	–3.86278E+00
	STD	2.61540E–02	3.90023E–03	9.25764E–02	2.99935E–03	3.37720E–03	9.58350E–01	3.50185E–03	4.44089E–16
	Rank	6	5	7	4	3	8	2	1
F20	Worst	–5.08756E+00	–2.63206E+00	–2.53662E+00	–3.72438E–01	–5.08763E+00	–1.22350E+00	–2.76590E+00	–1.04029E+01
	Average	–9.29620E+00	–3.44876E+00	–4.80441E+00	–2.66708E+00	–9.33857E+00	–4.17040E+00	–7.53980E+00	–1.04029E+01
	Best	–1.04024E+01	–4.61965E+00	–1.03266E+01	–4.65842E+00	–1.04016E+01	–1.04003E+01	–1.04029E+01	–1.04029E+01
	STD	2.35341E+00	8.01468E–01	3.24543E+00	2.09347E+00	2.37635E+00	3.59580E+00	3.93513E+00	3.28357E–08
	Rank	3	7	5	8	2	6	4	1
F21	Worst	–7.76472E–01	–2.82374E+00	–2.57681E+00	–3.51363E–01	–2.63023E+00	–8.57665E–01	–2.68286E+00	–9.73545E+00
	Average	–6.23865E+00	–5.06299E+00	–4.96059E+00	–1.39789E+00	–6.60841E+00	–3.71200E+00	–6.63411E+00	–9.96931E+00
	Best	–1.01532E+01	–8.79486E+00	–9.40082E+00	–4.90974E+00	–1.01518E+01	–1.01491E+01	–1.01532E+01	–1.01529E+01
	STD	3.97755E+00	2.33483E+00	2.76657E+00	1.97504E+00	3.38184E+00	3.67326E+00	3.35188E+00	1.94344E–01
	Rank	4	5	6	8	3	7	2	1

(*Continued*)

TABLE 10.5 (*Continued*) The Results of the Proposed AOMFO on Various Optimization Problems F14–F23

		Comparative algorithms							
Function	**Measure**	**AO**	**AOA**	**DA**	**SCA**	**GWO**	**GOA**	**MFO**	**AOMFO**
F22	Worst	−5.08756E+00	−2.63206E+00	−2.53662E+00	−3.72438E−01	−5.08763E+00	−1.22350E+00	−2.76590E+00	−1.04029E+01
	Average	−9.29620E+00	−3.44876E+00	−4.80441E+00	−2.66708E+00	−9.33857E+00	−4.17040E+00	−7.53980E+00	−1.04029E+01
	Best	−1.04024E+01	−4.61965E+00	−1.03266E+01	−4.65842E+00	−1.04016E+01	−1.04003E+01	−1.04029E+01	−1.04029E+01
	STD	2.35341E+00	8.01468E−01	3.24543E+00	2.09347E+00	2.37635E+00	3.59580E+00	3.93513E+00	3.28357E−08
	Rank	3	7	5	8	2	6	4	1
F23	Worst	−1.03613E+01	−1.59505E+00	−2.30796E+00	−4.05694E−01	−2.42157E+00	−1.85948E+00	−2.42734E+00	−1.05364E+01
	Average	−1.04442E+01	−2.70212E+00	−4.28706E+00	−7.28819E+00	−2.95716E+00	−4.87100E+00	−1.05364E+01	−9.37771E−01
	Best	−1.05362E+01	−5.02222E+00	−9.79525E+00	−2.22927E+00	−1.05358E+01	−5.12034E+00	−1.05364E+01	−1.05364E+01
	STD	8.05190E−02	1.42588E+00	3.12390E+00	7.49261E−01	4.44258E+00	1.27268E+00	3.36075E+00	5.04775E−09
	Rank	2	7	5	8	3	6	4	1
	Summation	39	63	53	60	31	67	35	10
	Mean Ranking	3.90	6.30	5.30	6.00	3.10	6.70	3.50	1.00
	Final Ranking	4	7	5	6	2	8	3	1

results are robust, and the proposed AOMFO got its results with limited distribution values compared to other approaches. As a result, the results obtained in the tested ten benchmark functions (F14–F23) demonstrated the proposed hybrid process's capability. This means that combining the Aquila Optimizer with the Moth-Flame Optimization algorithm results in a rigorous search process that outperforms other approaches. Figure 10.2 shows the convergence behavior of the tested methods. This figure also proved the ability of the proposed AOMFO in solving the OPs compared to other methods. The proposed AOMFO is clearly superior to other algorithms in terms of convergence rate, as shown by the curves in Figure 10.2. One of the most significant features of AOMFO is that it also manages demographic diversity.

10.4 CONCLUSION AND PROMISING POTENTIAL FUTURE WORKS

AO is a recent metaheuristic method proposed by Abualigah et al. in 2021 for solving different OPs. MFO algorithm wins a wide reputation over several search optimization methods published in the literature. In this research, a novel variant of AO is proposed to solve optimization problems, called AOMFO. The proposed AOMFO method is based on hybridizing the AO and MFO algorithm to gather various search strategies together. The main advantages of the hybrid method focus on the AO exploration search and the exploitation of the MFO algorithm. This is to employ the survival-of-the-fittest theory appropriately, and thus, active convergence can be accomplished.

Several benchmark functions are used to highlight the ability of the proposed AOMFO in solving the OPs and ensure that the proposed method reached better results in comparison with other well-known methods published in the literature, including AO, AOA, DA, SCA, GWO, GOA, PSO, and MFO algorithm. Finally, a qualified evaluation of nine well-known search methods with two different dimensions (i.e., 10 and 100) is carried out. The proposed AOMFO obtained impressive results for several benchmark functions. Interestingly, AOMFO can achieve the best results. In conclusion, the researchers interested in employing HAOMFO for their problems are recommended to apply Hybrid Aquila Optimizer and Moth-Flame Optimization (HAOMFO) algorithm for their implementations.

For future works, the proposed method can be tested in solving different problems, such as text clustering, data clustering, feature selection,

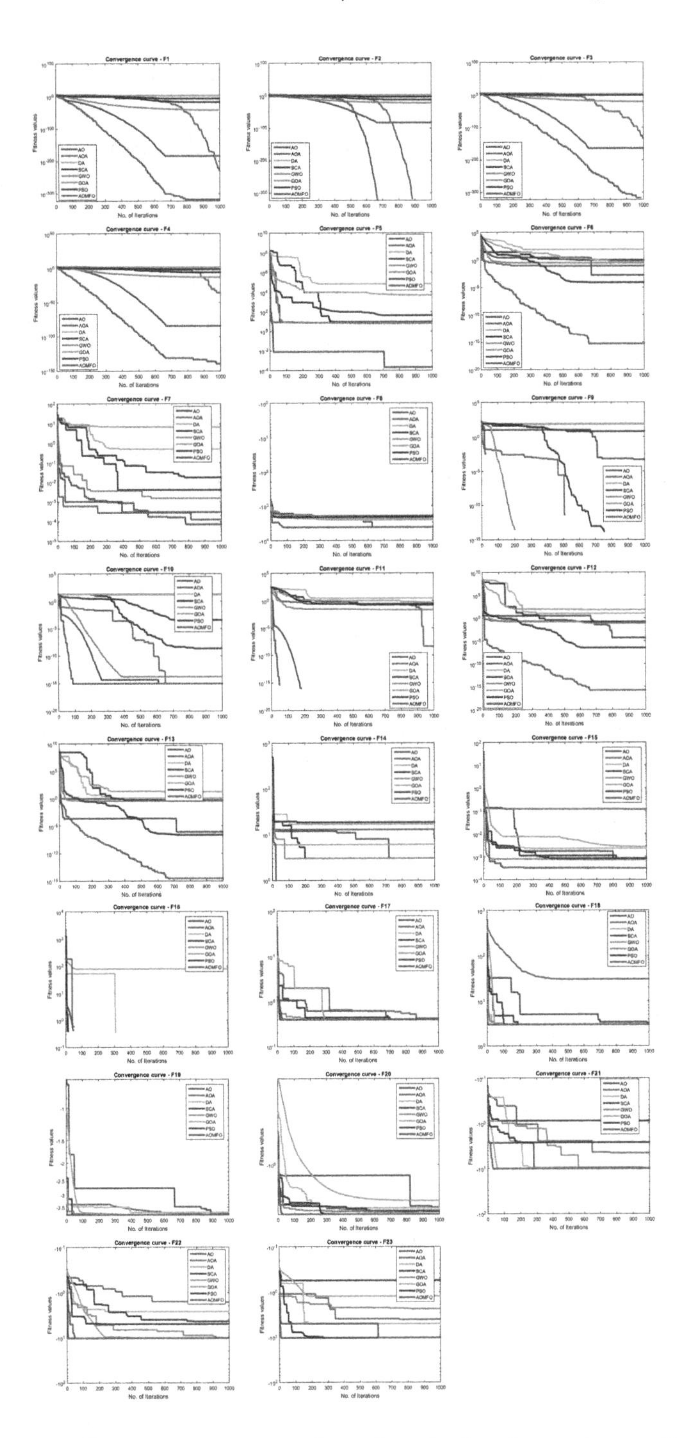

FIGURE 10.2 Convergence curves of tested methods using various optimization problems (F_1–F_{23}).

cloud-based task scheduling, wireless sensor networks, big data analysis, foresting process, image segmentation problem, engineering design problems, industrial OP, and parameters estimation. Also, other operators can be tested to show the effectiveness of them over the proposed method.

REFERENCES

1. Abdelkader Abbassi, Rached Ben Mehrez, Bilel Touaiti, Laith Abualigah, and Ezzeddine Touti. Parameterization of photovoltaic solar cell double-diode model based on improved arithmetic optimization algorithm. *Optik*, 253: 168600, 2022.
2. Mohamed Abd Elaziz, Ahmed A Ewees, Dalia Yousri, Laith Abualigah, and Mohammed AA Al-qaness. Modified marine predators algorithm for feature selection: case study metabolomics. *Knowledge and Information Systems*, 64: 1–27, 2022.
3. Laith Abualigah. Group search optimizer: a nature-inspired meta-heuristic optimization algorithm with its results, variants, and applications. *Neural Computing and Applications*, 33: 1–24, 2020.4. Laith Abualigah, Mohamed Abd Elaziz, Abdelazim G Hussien, Bisan Alsalibi, Seyed Mohammad Jafar Jalali, and Amir H Gandomi. Lightning search algorithm: a comprehensive survey. *Applied Intelligence*, 51: 1–24, 2020.
5. Laith Abualigah, Mohamed Abd Elaziz, Putra Sumari, Zong Woo Geem, and Amir H Gandomi. Reptile search algorithm (rsa): A nature-inspired meta-heuristic optimizer. *Expert Systems with Applications*, 191:116158, 2022.
6. Laith Abualigah and Bahaa Al Masri. Advances in mapreduce big data processing: Platform, tools, and algorithms. *Artificial Intelligence and IoT: Smart Convergence for Eco-Friendly Topography*, 85: 105–128, 2021.
7. Laith Abualigah, Bisan Alsalibi, Mohammad Shehab, Mohammad Alshinwan, Ahmad M Khasawneh, and Hamzeh Alabool. A parallel hybrid krill herd algorithm for feature selection. *International Journal of Machine Learning and Cybernetics*, 12(3):783–806, 2021.
8. Laith Abualigah and Ali Diabat. A comprehensive survey of the grasshopper optimization algorithm: results, variants, and applications. *Neural Computing and Applications*, 32: 1–24, 2020.
9. Laith Abualigah and Ali Diabat. A novel hybrid antlion optimization algorithm for multi-objective task scheduling problems in cloud computing environments. *Cluster Computing*, 24: 1–19, 2020.
10. Laith Abualigah and Ali Diabat. Advances in sine cosine algorithm: a comprehensive survey. *Artificial Intelligence Review*, 54: 1–42, 2021.
11. Laith Abualigah and Ali Diabat. Improved multi-core arithmetic optimization algorithm-based ensemble mutation for multidisciplinary applications. *Journal of Intelligent Manufacturing*, 1–42, 2022, in press.
12. Laith Abualigah, Ali Diabat, Maryam Altalhi, and Mohamed Abd Elaziz. Improved gradual change-based Harris hawks optimization for real-world engineering design problems. *Engineering with Computers*, 1–41, 2022, in press.

13. Laith Abualigah, Ali Diabat, and Zong Woo Geem. A comprehensive survey of the harmony search algorithm in clustering applications. *Applied Sciences*, 10(11):3827, 2020.
14. Laith Abualigah, Ali Diabat, Seyedali Mirjalili, Mohamed Abd Elaziz, and Amir H. Gandomi. The arithmetic optimization algorithm. *Computer Methods in Applied Mechanics and Engineering*, 376:113609, 2021.
15. Laith Abualigah and Akram Jamal Dulaimi. A novel feature selection method for data mining tasks using hybrid sine cosine algorithm and genetic algorithm. *Cluster Computing*, 24: 1–16, 2021.
16. Laith Abualigah, Mohamed Abd Elaziz, Ahmad M Khasawneh, Mohammad Alshinwan, Rehab Ali Ibrahim, Mohammed AA Al-qaness, Seyedali Mirjalili, Putra Sumari, and Amir H Gandomi. Meta-heuristic optimization algorithms for solving real-world mechanical engineering design problems: a comprehensive survey, applications, comparative analysis, and results. *Neural Computing and Applications*, 34: 1–30, 2022.
17. Laith Abualigah, Amir H. Gandomi, Mohamed Abd Elaziz, Husam Al Hamad, Mahmoud Omari, Mohammad Alshinwan, and Ahmad M Khasawneh. Advances in meta-heuristic optimization algorithms in big data text clustering. *Electronics*, 10(2):101, 2021.
18. Laith Abualigah, Amir H Gandomi, Mohamed Abd Elaziz, Abdelazim G Hussien, Ahmad M Khasawneh, Mohammad Alshinwan, and Essam H Houssein. Nature-inspired optimization algorithms for text document clustering: a comprehensive analysis. *Algorithms*, 13(12):345, 2020.
19. Laith Abualigah, Mohammad Shehab, Mohammad Alshinwan, and Hamzeh Alabool. Salp swarm algorithm: a comprehensive survey. *Neural Computing and Applications*, 32: 1–21, 2019.
20. Laith Abualigah, Mohammad Shehab, Ali Diabat, and Ajith Abraham. Selection scheme sensitivity for a hybrid salp swarm algorithm: analysis and applications. *Engineering with Computers*, 38: 1–27, 2020.
21. Laith Abualigah, Dalia Yousri, Mohamed Abd Elaziz, Ahmed A Ewees, Mohammed AA Al-qaness, and Amir H Gandomi. Aquila optimizer: a novel meta-heuristic optimization algorithm. *Computers & Industrial Engineering*, 157: 107250, 2021.
22. Laith Mohammad Qasim Abualigah. *Feature Selection and Enhanced Krill Herd Algorithm for Text Document Clustering*. Springer, Berlin, Germany, 2019.
23. Jeffrey O Agushaka, Absalom E Ezugwu, and Laith Abualigah. Dwarf mongoose optimization algorithm. *Computer Methods in Applied Mechanics and Engineering*, 391:114570, 2022.
24. Mohammed AA Al-Qaness, Ahmed A Ewees, Hong Fan, Laith Abualigah, and Mohamed Abd Elaziz. Marine predators algorithm for forecasting confirmed cases of covid-19 in Italy, USA, Iran and Korea. *International Journal of Environmental Research and Public Health*, 17(10):3520, 2020.
25. Zaher Ali Al-Sai and Laith Mohammad Abualigah. Big data and e-government: a review. *In* 2017 8*th International Conference on Information Technology (ICIT)*, 580–587. IEEE, Jordan, 2017.

26. Osama Ahmad Alomari, Ahamad Tajudin Khader, M Azmi Al-Betar, and Laith Mohammad Abualigah. MRMR BA: a hybrid gene selection algorithm for cancer classification. *Journal of Theoretical and Applied Information Technology*, 95(12):2610–2618, 2017.
27. Osama Ahmad Alomari, Ahamad Tajudin Khader, Mohammed Azmi Al-Betar, and Laith Mohammad Abualigah. Gene selection for cancer classification by combining minimum redundancy maximum relevancy and bat-inspired algorithm. *International Journal of Data Mining and Bioinformatics*, 19(1):32–51, 2017.
28. Bisan Alsalibi, Laith Abualigah, and Ahamad Tajudin Khader. A novel bat algorithm with dynamic membrane structure for optimization problems. *Applied Intelligence*, 51: 1–26, 2020.
29. Mohammad Alshinwan, Laith Abualigah, Mohammad Shehab, Mohamed Abd Elaziz, Ahmad M Khasawneh, Hamzeh Alabool, and Husam Al Hamad. Dragonfly algorithm: a comprehensive survey of its results, variants, and applications. *Multimedia Tools and Applications*, 80: 1–38, 2021.
30. Zaid Abdi Alkareem Alyasseri, Ahamad Tajudin Khader, Mohammed Azmi Al-Betar, and Laith Mohammad Abualigah. ECG signal denoising using β-hill climbing algorithm and wavelet transform. In *2017 8th International Conference on Information Technology (ICIT)*, 96–101. IEEE, Jordan, 2017.
31. Asaju La'aro Bolaji, Mohammed Azmi Al-Betar, Mohammed A Awadallah, Ahamad Tajudin Khader, and Laith Mohammad Abualigah. A comprehensive review: Krill herd algorithm (kh) and its applications. *Applied Soft Computing*, 49:437–446, 2016.
32. Ali Danandeh Mehr, Amir Rikhtehgar Ghiasi, Zaher Mundher Yaseen, Ali Unal Sorman, and Laith Abualigah. A novel intelligent deep learning predictive model for meteorological drought forecasting. *Journal of Ambient Intelligence and Humanized Computing*, 1–15, 2022, in press.
33. Amir H. Gandomi, Fang Chen, and Laith Abualigah. Machine learning technologies for big data analytics. *Electronics*, 11(3), 2022.
34. Amir Hossein Gandomi and Amir Hossein Alavi. Krill herd: A new bio-inspired optimization algorithm. *Communications in Nonlinear Science and Numerical Simulation*, 17(12):4831–4845, 2012.
35. Ankit Garg, Jinhui Li, Jinjun Hou, Christian Berretta, and Akhil Garg. A new computational approach for estimation of wilting point for green infrastructure. *Measurement*, 111:351–358, 2017.
36. Hessam Golmohamadi, Reza Keypour, and Pouya Mirzazade. Multi-objective co-optimization of power and heat in urban areas considering local air pollution. *Engineering Science and Technology, an International Journal*, 24(2):372–383, 2021.
37. Ahmad M Khasawneh, Laith Abualigah, and Mohammad Al Shinwan. Void aware routing protocols in underwater wireless sensor networks: Variants and challenges. *In Journal of Physics: Conference Series*, 1550:032145, 2020.
38. Ahmad M Khasawneh, Omprakash Kaiwartya, Laith M Abualigah, Jaime Lloret, et al. Green computing in underwater wireless sensor networks pressure centric energy modeling. *IEEE Systems Journal*, 14(4):4735–4745, 2020.

39. Yongbo Li, Hamed Soleimani, and Mostafa Zohal. An improved ant colony optimization algorithm for the multi-depot green vehicle routing problem with multiple objectives. *Journal of Cleaner Production*, 227:1161–1172, 2019.
40. Duc-Long Luong, Duc-Hoc Tran, and Phong Thanh Nguyen. Optimizing multi-mode time-cost-quality trade-off of construction project using opposition multiple objective difference evolution. *International Journal of Construction Management*, 21(3):271–283, 2021.
41. Seyedali Mirjalili. Moth-flame optimization algorithm: A novel nature-inspired heuristic paradigm. *Knowledge-Based Systems*, 89:228–249, 2015.
42. Seyedali Mirjalili. Dragonfly algorithm: a new meta-heuristic optimization technique for solving single-objective, discrete, and multi-objective problems. *Neural Computing and Applications*, 27(4):1053–1073, 2016.
43. Seyedali Mirjalili. SCA: a sine cosine algorithm for solving optimization problems. *Knowledge-Based Systems*, 96:120–133, 2016.
44. Seyedali Mirjalili and Andrew Lewis. The whale optimization algorithm. *Advances in Engineering Software*, 95:51–67, 2016.
45. Seyedali Mirjalili, Seyed Mohammad Mirjalili, and Andrew Lewis. Grey wolf optimizer. *Advances in Engineering Software*, 69:46–61, 2014.
46. Seyedeh Zahra Mirjalili, Seyedali Mirjalili, Shahrzad Saremi, Hossam Faris, and Ibrahim Aljarah. Grasshopper optimization algorithm for multi-objective optimization problems. *Applied Intelligence*, 48(4):805–820, 2018.
47. Mohammad H Nadimi-Shahraki, Shokooh Taghian, and Seyedali Mirjalili. An improved grey wolf optimizer for solving engineering problems. *Expert Systems with Applications*, 166:113917, 2021.
48. Javad Olamaei, Mohammad Esmaeil Nazari, and Sepideh Bahravar. Economic environmental unit commitment for integrated CCHP-thermal-heat only system with considerations for valve-point effect based on a heuristic optimization algorithm. *Energy*, 159:737–750, 2018.
49. Mohammed Otair, Osama Talab Ibrahim, Laith Abualigah, Maryam Altalhi, and Putra Sumari. An enhanced grey wolf optimizer based particle swarm optimizer for intrusion detection system in wireless sensor networks. *Wireless Networks*, 28: 1–24, 2022.
50. Zhouhua Peng, Jun Wang, and Dan Wang. Distributed maneuvering of autonomous surface vehicles based on neurodynamic optimization and fuzzy approximation. *IEEE Transactions on Control Systems Technology*, 26(3):1083–1090, 2017.
51. Hasan Rashaideh, Ahmad Sawaie, Mohammed Azmi Al-Betar, Laith Mohammad Abualigah, Mohammed M Al-Laham, M Ra'ed, and Malik Braik. A grey wolf optimizer for text document clustering. *Journal of Intelligent Systems*, 29(1):814–830, 2020.
52. Hussein Mohammed Ridha, Hashim Hizam, Seyedali Mirjalili, Mohammad Lutfi Othman, Mohammad Effendy Ya'acob, and Laith Abualigah. A novel theoretical and practical methodology for extracting the parameters of the single and double diode photovoltaic models. *IEEE Access*, 10: 11110–11137, 2022.

53. Mukaram Safaldin, Mohammed Otair, and Laith Abualigah. Improved binary gray wolf optimizer and SVM for intrusion detection system in wireless sensor networks. *Journal of Ambient Intelligence and Humanized Computing*, 12(2):1559–1576, 2021.
54. Shahrzad Saremi, Seyedali Mirjalili, and Andrew Lewis. Grasshopper optimisation algorithm: theory and application. *Advances in Engineering Software*, 105:30–47, 2017.
55. Mohammad Shehab, Laith Abualigah, Husam Al Hamad, Hamzeh Alabool, Mohammad Alshinwan, and Ahmad M Khasawneh. Moth–flame optimization algorithm: variants and applications. *Neural Computing and Applications*, 32(14):9859–9884, 2020.
56. Mohammad Shehab, Hanadi Alshawabkah, Laith Abualigah, and AL-Madi Nagham. Enhanced a hybrid moth-flame optimization algorithm using new selection schemes. *Engineering with Computers*, 37: 1–26, 2020.
57. Mohammad Shehab, Mohammad Sh Daoud, Hani Mahmouad AlMimi, Laith Mohammad Abualigah, and Ahamad Tajudin Khader. Hybridising cuckoo search algorithm for extracting the ODF maxima in spherical harmonic representation. *International Journal of Bio-Inspired Computation*, 14(3):190–199, 2019.
58. Mohammad Shehab, Ahamad Tajudin Khader, Mohammed Azmi Al-Betar, and Laith Mohammad Abualigah. Hybridizing cuckoo search algorithm with hill climbing for numerical optimization problems. In *2017 8th International Conference on Information Technology (ICIT)*, 36–43. IEEE, Jordan, 2017.
59. Sarah E Shukri, Rizik Al-Sayyed, Amjad Hudaib, and Seyedali Mirjalili. Enhanced multi-verse optimizer for task scheduling in cloud computing environments. *Expert Systems with Applications*, 168:114230, 2021.
60. Dalia Yousri, Mohamed Abd Elaziz, Laith Abualigah, Diego Oliva, Mohammed AA Al-Qaness, and Ahmed A Ewees. Covid-19 X-ray images classification based on enhanced fractional-order cuckoo search optimizer using heavy-tailed distributions. *Applied Soft Computing*, 101:107052, 2021.
61. Dalia Yousri, Mohamed Abd Elaziz, Diego Oliva, Laith Abualigah, Mohammed AA Al-qaness, and Ahmed A Ewees. Reliable applied objective for identifying simple and detailed photovoltaic models using modern metaheuristics: Comparative study. *Energy Conversion and Management*, 223:113279, 2020.

CHAPTER 11

Boosting Moth-Flame Optimization Algorithm by Arithmetic Optimization Algorithm for Data Clustering

Laith Abualigah

Amman Arab University
Universiti Sains Malaysia

Seyedali Mirjalili

Torrens University Australia
Yonsei University

Mohammed Otair

Amman Arab University

Putra Sumari

Universiti Sains Malaysia

Mohamed Abd Elaziz

Zagazig University

DOI: 10.1201/9781003205326-14

Heming Jia
Sanming University

Amir H. Gandomi
University of Technology Sydney

CONTENTS

11.1 Introduction 210
11.2 DC Application 212
11.3 The Proposed Clustering Algorithm (MFOAOA) 213
11.3.1 MFO Algorithm 213
11.3.1.1 *Generate the Initial Population of MFO* 213
11.3.1.2 *Updating the Positions* 214
11.3.1.3 *Updating the Number of Flames* 215
11.3.2 Arithmetic Optimization Algorithm (AOA) 216
11.3.3 Exploration Phase 216
11.3.4 Exploitation Phase 217
11.3.5 The Proposed Hybrid Moth-Flame Optimization with Arithmetic Optimization Algorithm (MFOAOA) 217
11.3.5.1 Solutions Initialization 217
11.3.5.2 Structure of the Proposed MFOAOA 218
11.4 Experiments and Discussion 219
11.4.1 Experiments Settings 219
11.4.2 Datasets Description 219
11.4.3 Results 220
11.5 Conclusion and Future Works 226
References 233

11.1 INTRODUCTION

Pattern detection systems may provide invaluable knowledge to help decision-making processes in real-world issues affecting large datasets [11,54–56]. Clus ter analysis, also known as data clustering (DC), is an unsupervised machine learning task that searches out the most close clusters of objects based on a predefined similarity metric [20, 41,57]. Cluster analysis has practical uses in a variety of fields, including information processing [20], text clustering [16], big data [15,32], mathematical problems [35], engineering problems [1,36], energy problems [19], image analysis

[31,64], text classification [37], wireless sensor networks [45,46], and bioinformatics [33,34,40].

Clustering is the process for extracting natural groupings from data and converting them into readable information [12,39]. The clusters created by the resulting groups collect related artifacts based on their characteristics. Clustering methods have been applied to a variety of new fields, including network analysis, industry, marketing, education, data science, and medical diagnosis [4,8].

DC is a technique for grouping related objects together like items clustered together and distinct items grouped together in different classes [3]. It is an unsupervised learning method in which artifacts are clustered into undefined preset clusters. In contrast to conceptualization, a classification is a form of supervised learning in which subjects are assigned to predetermined classes (clusters). DC is utilized in a wide variety of applications, including data processing, computational data analysis, machine learning, pattern recognition, image analysis, and information retrieval [43]. This is due to different clustering techniques, which can be classified as partitional, hierarchical, density-based, grid-based, and model-based methods [59].

Several optimization methods proved their ability in solving the clustering problems [25,42] such as Particle Swarm Optimization [21,27], Krill Herd Algorithm [29], Harmony Search Algorithm [23,24], Genetic Algorithm [22], and many other hybrid search methods [26]. Also, many other optimization methods can be used to solve the DC problem such as Sine Cosine Algorithm (SCA) [7,49], Dragonfly Algorithm [38,48], Gray Wolf Optimizer (GWO) [53,60], Lightning Search Algorithm [5], Whale Optimization Algorithm (WOA) [51], Grasshopper Optimization Algorithm [9,61], Salp Swarm Algorithm [17,50], Reptile Search Algorithm (RSA) [6],

Dwarf Mongoose Optimization Algorithm (DMOA) [30], Arithmetic Optimization Algorithm [13], Aquila Optimizer (AO) [18], and others [2,10,14,58].

In this research, the hybrid Moth-Flame Optimization (MFO) algorithm with Arithmetic Optimization Algorithm (AOA) is proposed for deciding the best centroid for performing optimal DC, called MFOAOA. In a dynamic data system, optimum clustering yields more usable and important knowledge needed for undertaking the decision-making process. The MFOAOA employs fitness metrics that are defined by three fitness constraints: inter-cluster size, intra-cluster distance, and cluster density. The centroid referring to the minimum fitness value is calculated, with the minimum inter-cluster size, maximum intra-cluster distance, and maximum cluster density contributing to this minimum fitness value. The data are clustered using the optimal centroid calculated using the MFOAOA algorithm, and the data are clustered depending on the minimal distance between the centroid and the data point. The data points are clustered together in a single category based on this minimum distance, and all data points are grouped separately. The proposed MFOAOA is tested using several benchmark datasets and compared with several well-known methods. The results illustrated that the proposed MFOAOA is a promising clustering method, and it got better results than other tested clustering-based approaches.

The rest of the research is given as follows. The DC definition and description are given in Section 11.2. Section 11.3 presents the proposed hybrid clustering method (MFOAOA). Experimental results and discussion are presented in Section 11.4. Section 11.5 gives the conclusion and possible future work directions.

11.2 DC APPLICATION

DC is a common data mining technique used to partition set of objects (D) into a predefined subset of clusters (K) based on the distance or similarity measure. Assume that $D = d_1, d_2, \ldots, d_i, \ldots, d_n$ is a set of n objects to be partitioned on K clusters and the object number d_i is given as vector $d_i = d_{i1}, d_{i2}, \ldots, d_{ij}, \ldots, d_{it}$, where $i = 1$ to n, d_{ij} is the j_{th} value of the object number i.

The main purpose of the clustering algorithm is to determine a collection of K clusters. Each cluster has a centroid, where $C = C = C_1, C_2, \ldots, C_k, C_K$ and $C_K \neq \varnothing$. The fitness function is determined as follows:

$$\mathrm{FF} = \sum_{i=1}^{K} \sum_{j=1}^{K} \min\left(Ds\left(d_i, c_j\right)\right), \tag{11.1}$$

where c_j is the j_{th} center, d_j is the j_{th} object, and Ds is the distance value between the object d_i and the center c_j. Euclidean distance is defined as follows:

$$D\left(d_i, c_j\right) = \sqrt{\sum_{j=1}^{t} \left(d_{1j}, c_{2j}\right)^2}, \tag{11.2}$$

where $D\left(d_i, c_j\right)$ is the distance value of the object i and the cluster j, d_{1j} is the instance j in object 1, c_{2j} is the instance j in cluster 2, and c_1 is the cluster center of the cluster number 1, calculated as follows:

$$c_j = \frac{1}{n_j} \sum_{d_i \in c_j} d_i, \tag{11.3}$$

where d_i is the objects i, c_j is the centroid of cluster j, and n_j presents the total number of objects in cluster j.

11.3 THE PROPOSED CLUSTERING ALGORITHM (MFOAOA)

11.3.1 MFO Algorithm

MFO is a population-dependent metaheuristic algorithm [47]. It starts by creating moths at random in the global optimum followed by which the fitness values (i.e., location) for each moth are measured and the best spot with flame is indicated. After that, a spiral motion feature is used to change the moths' placements and update the current best individual positions, resulting in improved positions marked by a flame. Before the termination requirements are met, the preceding processes (updating the moth locations and establishing new ones) are repeated. In the MFO Algorithm, there are three important steps [62,63]. These are the stages that are explained in the following paragraphs.

11.3.1.1 Generate the Initial Population of MFO

Each moth is thought to be capable of flying in 1D, 2D, 3D, or hyperdimensional space. The following is how the group of moths can be expressed:

$$M = \begin{bmatrix} m_{1,1} & m_{1,2} & \cdots & \cdots & m_{1,d} \\ m_{2,1} & m_{2,2} & \cdots & \cdots & m_{2,d} \\ \vdots & \vdots & \vdots & \vdots & \vdots \\ m_{n,1} & m_{n,2} & \cdots & \cdots & m_{n,d} \end{bmatrix} \tag{11.4}$$

where d is the number of dimensions in the solution space and n is the number of moths. In addition, all of the moths' fitness values are stored in an array as follows:

$$\text{OM} = \begin{bmatrix} \text{OM}_1 \\ \text{OM}_2 \\ \vdots \\ \text{OM}_n \end{bmatrix} \tag{11.5}$$

The MFO Algorithm's other components are flames. The flames in D-dimensional space, as well as the fitness function matrix that correlates to them, are shown in the following matrix:

$$F = \begin{bmatrix} F_{1,1} & F_{1,2} & \cdots & \cdots & F_{1,d} \\ F_{2,1} & F_{2,2} & \cdots & \cdots & F_{2,d} \\ \vdots & \vdots & \vdots & \vdots & \vdots \\ F_{n,1} & F_{n,2} & \cdots & \cdots & F_{n,d} \end{bmatrix} \tag{11.6}$$

$$\text{OF} = \begin{bmatrix} \text{OF}_1 \\ \text{OF}_2 \\ \vdots \\ \text{OF}_n \end{bmatrix} \tag{11.7}$$

We should deduce from the above that both moths and fires are remedies. Each of them uses a particular approach to keep track of the best location (solution). Moths, for example, can be seen flying about in the quest room. However, without flames, which serve as flags for the moths in the quest room, they would not be able to reach the best spot. As a result, the moths' best solutions are never missed.

11.3.1.2 Updating the Positions

MFO converges the global-optimal combinatorial optimization using three functions. The following are the definitions for these functions:

$$\text{MFO} = (I, P, T) \tag{11.8}$$

where I refers to the initialization of the random positions of the moths ($I : \phi \rightarrow \{M, \text{OM}\}$), P presents the movement of moths in the solution space ($P : M \rightarrow M$), and T presents the termination of the search ($T : M \rightarrow \text{true}, \text{false}$). In the I function, the following equation describes the implementation of any random distribution:

$$M(i, j) = (ub(i) - lb(j)) * \text{rand}() + lb(i) \tag{11.9}$$

where the lower and upper ranges of variables are denoted by lb and ub, respectively.

The moths fly in the quest space in a transverse direction, as already noted. When using a logarithmic spiral submitted, the following three conditions must be met:

- The initial point of the spiral should be the moth.
- The location of the flame will be the final point of the spiral.
- The range of the spiral does not fluctuate more than the search room.

As a result, the MFO algorithm's logarithmic spiral can be described as follows:

$$S(M_i, F_j) = D_i \cdot e^{bt} \cdot \cos(2\pi t) + F_j \quad (11.10)$$

where D_i presents the distance between the *i-th* moth and the *j-th* flame (i.e., $D_i = |F_j - M_i|$), b indicates a constant to define the shape of the logarithmic spiral, and t indicates a random number between [–1, 1].

The constant b is used to describe the form of the logarithmic spiral, while t is a random number in the range [–1, 1].

The spiral travel of the moth around the flame in the quest space ensures a balance between manipulation and discovery in MFO. To prevent getting lost in local optima, the best options are held in each iteration, and the moths fly around the flames (i.e., each moth flies around the closest flame) using the OF and OM matrices.

11.3.1.3 Updating the Number of Flames

This section focused on increasing the exploitation of the MFO method (for example, changing the moths' positions at n distinct locations in the search space might reduce the likelihood of exploiting the most promising solutions). As a consequence, as shown by the equation below, limiting the number of flames assists in the resolution of this issue:

$$\text{Flame no} = \text{Round}\left(N - l * \frac{N-l}{T}\right) \quad (11.11)$$

where N denotes the maximum number of fires, l denotes the current iteration number, and T indicates the maximum number of iterations.

11.3.2 Arithmetic Optimization Algorithm (AOA)

In AOA, the optimization step begins with selecting possible solutions (X). Before beginning its job, the AOA should decide on a mission methodology (i.e., exploration or exploitation). The Math Optimizer Accelerated (MOA) function is used in the following search steps as a coefficient specified by the following equation:

$$\text{MOA}(C\_\text{Iter}) = \text{Min} + C\_\text{Iter} \times \left(\frac{\text{Max} - \text{Min}}{M\_\text{Iter}} \right) \tag{11.12}$$

where $\text{MOA}(C\_\text{Iter})$ is the function value, which is determined by using Equation (11.12). C_Iter is the current iteration which is in range 1 and M_Iter is the maximum iterations. Min and Max are the minimum and maximum values given for the MOA function, respectively.

11.3.3 Exploration Phase

The AOA discovery operators randomly search multiple regions of the search field in order to find a better solution using two primary search strategies (Division (D) and Multiplication (M)), which are modeled in Equation (11.13). The MOA function (see Equation (11.12)) conditions this step of searching (exploration search by executing D or M) for the condition of r_1 >MOA. The first operator (D) is conditioned by r_2<0.5 in this step (the first rule in Equation (11.13)) and the other operator (M) will be ignored before this operator completes its current mission. Otherwise, instead of D, the second operator (M) will be used to complete the current assignment (r_1 and r_2 are random numbers).

$$x_{i,j}(C\_\text{Iter}+1) = \begin{cases} \text{best}(x_j) \div \text{MOP} \times ((\text{UB}_j - \text{LB}_j) \times \mu + \text{LB}_j), & r_2 < 0.5 \\ \text{best}(x_j) \times \text{MOP} \times ((\text{UB}_j - \text{LB}_j) \times \mu + \text{LB}_j), & \text{otherwise} \end{cases} \tag{11.13}$$

where $x_{i,j}(C\_\text{Iter}+1)$ is the i_{th} solution with the current iteration +1, $x_{i,j}(C\_\text{Iter})$ is the j_{th} position of the i_{th} solution, and $best(x_j)$ is the j_{th} position of the best solution. UB_j and LB_j are the upper bound and the lower bound of the j_{th} position, respectively. μ is a control parameter fixed equal to 0.5 according to Equation (11.13).

$$\text{MOP}(C\_\text{Iter}) = 1 - \frac{C\_\text{Iter}^{1/\alpha}}{M\_\text{Iter}^{1/\alpha}} \tag{11.14}$$

where (MOP) is a coefficient and $\text{MOP}(C\_\text{Iter})$ is the function value. α is a sensitive parameter fixed equal to 5 according to Equation (11.14).

11.3.4 Exploitation Phase

The MOA function for the condition of r_1 is not higher than the existing $\text{MOA}(C\_\text{Iter})$ value (see Equation (11.12)). The manipulation operators of AOA (Subtraction (*S*) and Addition (A)) investigate the quest field extensively on many dense regions and find a better answer focused on two key search techniques (Subtraction (*S*) and Addition (A)), which are modeled in the following equation:

$$x_{i,j}(C\_\text{Iter}+1) = \begin{cases} \text{best}(x_j) - \text{MOP} \times \left((\text{UB}_j - \text{LB}_j) \times \mu + \text{LB}_j\right), & r_3 < 0.5 \\ \text{best}(x_j) + \text{MOP} \times \left((\text{UB}_j - \text{LB}_j) \times \mu + \text{LB}_j\right), & \text{otherwise} \end{cases} \tag{11.15}$$

The first operator (*S*) is conditioned by $r_3 < 0.5$ in this step (first law in Equation (11.15)) and the other operator (A) would be ignored before this operator completes its current mission. Otherwise, instead of *S*, the second operator (A) would complete the current task. This phase's processes are identical to the previous phase's partitions. Exploitation search operators (*S* and A), on the other hand, often want to avoid being trapped in the local search.

11.3.5 The Proposed Hybrid Moth-Flame Optimization with Arithmetic Optimization Algorithm (MFOAOA)

In this section, the procedure of the proposed MFOAOA is presented as follows.

11.3.5.1 Solutions Initialization

The optimization rule in the proposed MFOAOA is focused on the population of candidate solutions (X), which is developed stochastically here between the upper bound (UB) and the lower bound (LB) of the specific issue, as shown in the following equation:

$$X = \begin{bmatrix} x_{1,1} & \cdots & x_{1,j} & x_{1,\text{Dim}-1} & x_{1,\text{Dim}} \\ x_{2,1} & \cdots & x_{2,j} & \cdots & x_{2,\text{Dim}} \\ \cdots & \cdots & x_{i,j} & \cdots & \cdots \\ \vdots & \vdots & \vdots & \vdots & \vdots \\ x_{N-1,1} & \cdots & x_{N-1,j} & \cdots & x_{N-1,\text{Dim}} \\ x_{N,1} & \cdots & x_{N,j} & x_{N,\text{Dim}-1} & x_{N,\text{Dim}} \end{bmatrix} \quad (11.16)$$

where X represents a collection of current candidate solutions created at random by Equation (11.17), X_i denotes the decision values (positions) of the ith solution, N denotes the total number of search agents (population), and Dim denotes the problem's dimension scale.

$$X_{ij} = \text{rand} \times \left(\text{UB}_j - \text{LB}_j\right) + \text{LB}_j,\ i = 1,2,\ldots,N\ j = 1,2,\ldots,\text{Dim} \quad (11.17)$$

where rand is a random number, LB_j denotes the jth lower bound, and UB_j denotes the jth upper bound of the given problem.

11.3.5.2 Structure of the Proposed MFOAOA

In this section, the proposed MFOAOA method is presented in Figure 11.1. The procedure of the MFOAOA starts with (a) determining the parameters' values of the used algorithms; (b) generating the candidate solutions; (c)

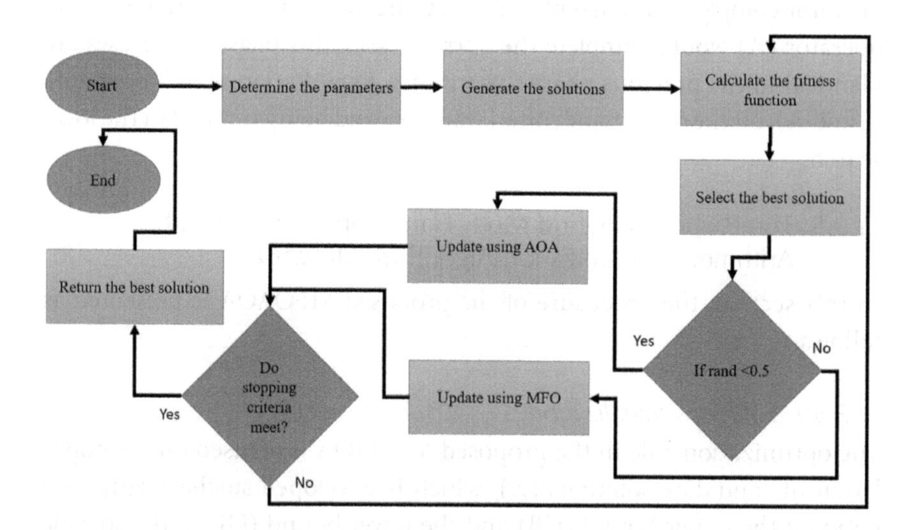

FIGURE 11.1 The flowchart of the proposed MFOAOA.

calculating the fitness functions; (d) selecting the best solution; (e) a condition being given: if the condition is true, then the AOA will be executed to update the solutions; otherwise, the MFO will be executed to update the solutions; (f) then, another condition being used to decide whether to continue with the optimization process or stop; and finally, (g) returning to the best-obtained solution. The flowchart for the proposed MFOAOA is seen in Figure 11.1.

11.4 EXPERIMENTS AND DISCUSSION

11.4.1 Experiments Settings

In this section, the effectiveness of the proposed MFOAOA is evaluated using several benchmark datasets and the obtained results are compared with other several well-known methods (i.e., AOA, Particle Swarm Optimizer (PSO), GWO, SCA, AO, WOA, and MFO algorithm). The number of solutions and the maximum number of iterations are fixed at 30 and 1,000, respectively. Thirty independent runs, for each tested method, are conducted to collect the results. Four measures are utilized, these measures are worst, best, average, and standard deviation (STD) of the fitness values. Besides, the Friedman ranking and Wilcoxon rank tests are used to test the significant differences between the proposed MFOAOA and other tested clustering-based approaches or not at p-value <0.05.

11.4.2 Datasets Description

Eight UCI datasets are used in this experiment, namely Cancer, Heart, CMC, Iris, Seeds, Vowels, Wine, and Glass. The details of given datasets are shown in Table 11.1.

TABLE 11.1 UCI Benchmark Datasets

Dataset	Features No.	Instances No.	Classes No.
Cancer	9	683	2
Heart	13	270	2
CMC	10	1473	3
Iris	4	150	3
Seeds	7	210	3
Vowels	6	871	3
Wine	13	178	3
Glass	9	214	7

11.4.3 Results

The findings of the tested clustering-based techniques are shown in this section for all of the evaluated datasets. The findings of the clustering-based techniques that were evaluated using the Cancer dataset are shown in Table 11.2. It is evident from this table that the suggested MFOAOA produced the best results in this dataset. Furthermore, according to the Wilcoxon rank test, it outperformed all other comparable algorithms. According to the Friedman ranking test, the suggested MFOAOA is ranked first, followed by PSO, which is ranked second, GWO, which is ranked third, SCA, which is ranked fourth, AO, which is ranked fifth, MFO, which is ranked sixth, AOA, which is ranked seventh, and WOA, which is ranked eighth. The acquired findings demonstrated the suggested MFOAOA's capacity to solve the given problem.

The findings of the clustering-based techniques evaluated using the CMC dataset are shown in Table 11.3. It is also apparent from this table that the suggested MFOAOA gave the best outcomes in this dataset. Furthermore, according to the Wilcoxon rank test, it outperformed all other comparable algorithms. According to the Friedman ranking test, the suggested MFOAOA is ranked first, followed by PSO, which is ranked second, GWO, which is ranked third, AOA, which is ranked fourth, AO, which is ranked fifth, SCA, which is ranked sixth, MFO, which is ranked seventh, and WOA, which is ranked eighth. These findings reveal that the proposed MFOAOA outperforms all existing clustering-based techniques that have been examined.

The evaluated clustering-based techniques utilizing the Glass dataset are shown in Table 11.4. It is evident from this table that the suggested MFOAOA produced the best results in this dataset. Furthermore, according to the Wilcoxon rank test, it outperformed all other comparable algorithms. According to the Friedman ranking test, the suggested MFOAOA is ranked top, followed by PSO, which is ranked second, GWO, which is ranked third, WOA, which is ranked fourth, AOA, which is ranked fifth, SCA, which is ranked sixth, MFO, which is ranked seventh, and AO, which is ranked eighth. The acquired findings demonstrated the suggested MFOAOA's capacity to solve the given problem.

Table 11.5 shows the results of the comparative techniques using the Iris dataset. As shown in the table, the suggested MFOAOA gave the best results in this dataset. Furthermore, it beats all other comparable algorithms in the Wilcoxon rank test. The suggested MFOAOA comes in #1 in the Friedman ranking test, followed by PSO, GWO, AOA, SCA, AO, WOA,

TABLE 11.2 The Results of the Tested Clustering-Based Approaches Using Cancer Dataset

	Tested Clustering-Based Approaches							
Metric	**AOA**	**PSO**	**GWO**	**SCA**	**AO**	**WOA**	**MFO**	**MFOAOA**
Worst	3.3469E+03	1.8584E+03	2.8951E+03	3.4737E+03	3.5005E+03	3.4022E+03	3.4602E+03	3.8543E+02
Average	3.2928E+03	1.0652E+03	2.5564E+03	3.1143E+03	3.1685E+03	3.3298E+03	3.2358E+03	3.1812E+02
Best	3.2180E+03	5.8341E+02	2.4480E+03	2.8031E+03	2.8740E+03	3.2234E+03	3.0300E+03	1.9942E+02
STD	5.0029E+01	5.2072E+02	1.8991E+02	2.5629E+02	2.8849E+02	7.7708E+01	2.0101E+02	7.0499E+01P-
Value	9.0724E-13	1.3018E-02	7.6970E-09	1.1349E-08	2.3387E-08	3.8619E-12	1.4031E-09	1.0000E+00
h	1	1	1	1	1	1	1	NAN
Rank	7	2	3	4	5	8	6	1

TABLE 11.3 The Results of the Tested Clustering-Based Approaches Using CMC Dataset

	Tested Clustering-Based Approaches							
Metric	**AOA**	**PSO**	**GWO**	**SCA**	**AO**	**WOA**	**MFO**	**MFOAOA**
Worst	3.3348E+02	1.3965E+02	3.1825E+02	3.3476E+02	3.3343E+02	3.3451E+02	3.3429E+02	7.9054E+01
Average	3.3285E+02	1.3598E+02	3.1149E+02	3.3388E+02	3.3328E+02	3.3408E+02	3.3393E+02	7.6740E+01
Best	3.3227E+02	1.3372E+02	3.0744E+02	3.3241E+02	3.3312E+02	3.3370E+02	3.3368E+02	7.4486E+01
STD	6.0466E-01	3.2100E+00	5.8925E+00	1.2794E+00	1.5917E-01	4.0813E-01	3.2036E-01	2.2842E+00P-
Value	4.8288E-09	1.2916E-05	3.4958E-07	7.1628E-09	4.2297E-09	4.4054E-09	4.3116E-09	1.0000E+00
h	1	1	1	1	1	1	1	NAN
Rank	4	2	3	6	5	8	7	1

TABLE 11.4 The Results of the Tested Clustering-Based Approaches Using Glass Dataset

	Tested Clustering-Based Approaches							
Metric	**AOA**	**PSO**	**GWO**	**SCA**	**AO**	**WOA**	**MFO**	**MFOAOA**
Worst	3.4819E+01	1.0789E+01	3.0683E+01	3.4892E+01	3.5156E+01	3.4312E+01	3.5108E+01	1.2339E+00
Average	3.4156E+01	6.3974E+00	2.8928E+01	3.4374E+01	3.4871E+01	3.3668E+01	3.4420E+01	7.6719E-01
Best	3.3643E+01	0.0000E+00	2.7384E+01	3.3750E+01	3.4396E+01	3.2272E+01	3.3672E+01	0.0000E+00
STD	4.6500E-01	4.2028E+00	1.4023E+00	4.1559E-01	3.3531E-01	8.7444E-01	6.6635E-01	4.6153E-01P-
Value	3.9293E-14	1.7665E-02	1.0063E-10	2.4331E-14	1.0967E-14	1.1861E-12	2.0235E-13	1.0000E+00
H	1	1	1	1	1	1	1	NAN
Rank	5	2	3	6	8	4	7	1

TABLE 11.5 The Results of the Tested Clustering-Based Approaches Using Iris Dataset

	Tested Clustering-Based Approaches							
Metric	**AOA**	**PSO**	**GWO**	**SCA**	**AO**	**WOA**	**MFO**	**MFOAOA**
Worst	2.3879E+01	6.1854E+00	1.6620E+01	2.4327E+01	2.4743E+01	2.4724E+01	2.4769E+01	2.1568E+00
Average	2.3669E+01	4.4847E+00	1.5420E+01	2.3704E+01	2.3725E+01	2.4030E+01	2.4389E+01	1.5982E+00
Best	2.3336E+01	6.1644E-01	1.4206E+01	2.2933E+01	2.2675E+01	2.3344E+01	2.3910E+01	9.0277E-01
STD	2.6766E-01	2.2264E+00	1.0745E+00	5.4736E-01	8.9844E-01	5.3093E-01	3.3891E-01	5.3802E-01P-
Value	5.3884E-13	2.2566E-02	5.6017E-09	3.7581E-12	4.4539E-11	2.9593E-12	6.5502E-13	1.0000E+00
h	1	1	1	1	1	1	1	NAN
Rank	4	2	3	5	6	7	8	1

and MFO. The acquired findings indicated the capacity of the suggested MFOAOA to address the problem.

Table 11.6 displays the effects of the tested clustering-based approaches using the Seeds dataset. The proposed MFOAOA provided the best results in this dataset, as shown in the table. Furthermore, it outperformed all other similar algorithms in the Wilcoxon rank test. The suggested MFOAOA comes first in the Friedman ranking test, followed by PSO, GWO, AOA, SCA, AO, WOA, and MFO. The obtained results demonstrated the potential of the proposed MFOAOA to solve the problem.

Table 11.7 displays the effects of the tested clustering-based approaches using the Statlog dataset. The proposed MFOAOA provided the best results in this dataset, as shown in the table. Furthermore, it outperformed all other similar algorithms in the Wilcoxon rank test. The suggested MFOAOA comes first in the Friedman ranking test, followed by AOA, AO, SCA, MFO, WOA, GWO, and PSO. The achieved results proved the effectiveness of the proposed MFOAOA to solve this problem.

The results of the comparative approaches using the Vowels dataset are seen in Table 11.8. As seen in the table, the proposed MFOAOA generated the best results in this dataset. In the Wilcoxon rank test, it also outperformed all other related algorithms. In the Friedman ranking test, the PSO comes first, followed by MFOAOA, GWO, WOA, AOA, MFO, AO, and SCA. The obtained results demonstrated that the planned MFOAOA has the strength to address the problem.

Table 11.9 shows the results of the comparative techniques using the Cancer dataset. As shown in the table, the suggested MFOAOA gave the best results in this dataset. Furthermore, it beats all other comparable algorithms in the Wilcoxon rank test. The suggested MFOAOA comes in #1 in the Friedman ranking test, followed by PSO, GWO, AOA, AO, MFO, WOA, and SCA. The acquired findings indicated the capacity of the suggested MFOAOA to address the problem. The final ranking of the evaluated algorithms across all datasets is shown in Figure 11.2. PSO, GWO, AOA, AO, SCA, WOA, and MFO came in top place among all examined techniques across all datasets, followed by PSO, GWO, AOA, AO, SCA, WOA, and MFO. The best cluster centroids for all the tested datasets are shown in Tables 11.10–11.17.

The results of the clustering analysis are shown in Figure 11.3 as the coloring of the multiplication signs (objects) into *k* clusters (cycle sign). Figure 11.3 shows the object distributions over the given clusters and it shows how the proposed MFOAOA grouped the objects for the tested

TABLE 11.6 The Results of the Tested Clustering-Based Approaches Using Seeds Dataset

	Tested Clustering-Based Approaches							
Metric	**AOA**	**PSO**	**GWO**	**SCA**	**AO**	**WOA**	**MFO**	**MFOAOA**
Worst	2.3879E+01	6.1854E+00	1.6620E+01	2.4327E+01	2.4743E+01	2.4724E+01	2.4769E+01	2.1568E+00
Average	2.3669E+01	4.4847E+00	1.5420E+01	2.3704E+01	2.3725E+01	2.4030E+01	2.4389E+01	1.5982E+00
Best	2.3336E+01	6.1644E-01	1.4206E+01	2.2933E+01	2.2675E+01	2.3344E+01	2.3910E+01	9.0277E-01
STD	2.6766E-01	2.2264E+00	1.0745E+00	5.4736E-01	8.9844E-01	5.3093E-01	3.3891E-01	5.3802E-01P-
Value	5.3884E-13	2.2566E-02	5.6017E-09	3.7581E-12	4.4539E-11	2.9593E-12	6.5502E-13	1.0000E+00
h	1	1	1	1	1	1	1	NAN
Rank	4	2	3	5	6	7	8	1

TABLE 11.7 The Results of the Tested Clustering-Based Approaches Using Statlog (Heart) Dataset

	Tested Clustering-based Approaches							
Metric	**AOA**	**PSO**	**GWO**	**SCA**	**AO**	**WOA**	**MFO**	**MFOAOA**
Worst	1.5659E+03	2.6630E+02	8.2140E+02	1.6256E+03	1.6034E+03	1.5648E+03	1.5794E+03	1.5415E+03
Average	2.3387E+01	2.1217E+02	1.1204E+02	4.6375E+01	3.8454E+01	8.4028E+01	6.6896E+01	2.2257E+01
Best	9.0933E-08	2.2757E-02	1.5338E-04	5.5784E-07	3.2365E-07	5.1891E-06	2.3602E-06	1.5200E-05
STD	5.4029E-01	4.5165E+00	3.0689E+00	2.1601E-01	5.3280E-01	9.7262E-01	6.9292E-01	1.2411E+00P-
Value	2.6756E-10	9.1231E-03	3.1508E-08	1.0044E-10	3.4637E-13	8.4803E-12	1.2834E-10	1.0000E+00
h	1	1	1	1	1	1	1	NAN
Rank	2	8	7	4	3	6	5	1

TABLE 11.8 The Results of the Tested Clustering-Based Approaches Using Vowels Dataset

	Tested Clustering-Based Approaches							
Metric	**AOA**	**PSO**	**GWO**	**SCA**	**AO**	**WOA**	**MFO**	**MFOAOA**
Worst	1.5277E+02	1.2075E+01	1.3927E+02	1.5341E+02	1.5335E+02	1.5277E+02	1.5340E+02	2.1256E+01
Average	1.5211E+02	7.0645E+00	1.3493E+02	1.5316E+02	1.5287E+02	1.5158E+02	1.5264E+02	1.9585E+01
Best	1.5137E+02	0.0000E+00	1.3192E+02	1.5296E+02	1.5199E+02	1.5018E+02	1.5186E+02	1.7762E+01
STD	5.4029E-01	4.5165E+00	3.0689E+00	2.1601E-01	5.3280E-01	9.7262E-01	6.9292E-01	1.2411E+00P-
Value	2.1218E-16	3.3182E-04	8.2105E-13	1.1214E-16	1.9914E-16	7.4254E-16	3.0387E-16	1.0000E+00
h	1	1	1	1	1	1	1	NAN
Rank	5	1	3	8	7	4	6	2

TABLE 11.9 The Results of the Tested Clustering-Based Approaches Using Wine Dataset

	Tested Clustering-Based Approaches							
Metric	**AOA**	**PSO**	**GWO**	**SCA**	**AO**	**WOA**	**MFO**	**MFOAOA**
Worst	3.8210E+03	1.44	68E+03	2.8652E+03	3.9875E+03	3.9153E+03	3.9306E+03	3.9513E+03 4.6831E+02
Average	3.7753E+03	1.1205E+03	2.5006E+03	3.9211E+03	3.8236E+03	3.8795E+03	3.8408E+03	3.1662E+02
Best	3.7283E+03	6.8156E+02	2.1515E+03	3.7207E+03	3.5891E+03	3.8349E+03	3.5500E+03	2.5110E+02
STD	3.7688E+01	2.8814E+02	2.9013E+02	1.1287E+02	1.3317E+02	4.5488E+01	1.6666E+02	9.0559E+01 P-
Value	7.4641E-13	3.4147E-04	2.2584E-07	1.1982E-11	3.5000E-11	7.6412E-13	1.2407E-10	1.0000E+00
h	1	1	1	1	1	1	1	NAN
Rank	4	2	3	8	5	7	6	1

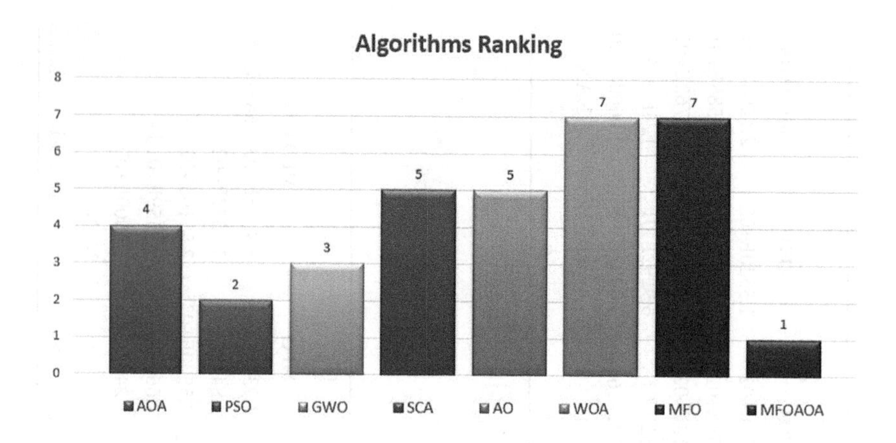

FIGURE 11.2 Ranking of all the tested clustering-based approaches using all the tested datasets.

eight datasets. For each of the eight datasets, Figure 11.4 shows graphical comparisons of the proposed MFOAOA and the other methods using the convergence rates. MFOAOA strongly converges faster than all other methods (i.e., AOA, PSO, GWO, SCA, AO, WOA, and MFO) in all test cases. Whereas the curves on the tested clustering-based approaches (i.e., AOA, SCA, AO, WOA, and MFO) are not smoothly expressed, the curves on the MFOAOA are more substantial, that is, tighter. Identically, PSO converged better than other tested clustering-based approaches, but its convergence continues to be weak.

11.5 CONCLUSION AND FUTURE WORKS

DC is a typical data mining problem in which a collection of objects is separated into a number of clusters. Each cluster contains objects that are identical, whereas other clusters contain objects that are dissimilar. A novel hybrid clustering approach is adopted in this research to solve a number of DC problems. The proposed approach, dubbed MFOAOA, merged the MFO algorithm and the AOA. The proposed MFOAOA's key purpose is to merge the benefits of the MFO algorithm's exploration search ability and the AOA's exploitation search. Experiments are conducted using eight standard benchmark datasets in the DC domain. The suggested MFOAOA is compared to a number of well-known algorithms, including the AOA, PSO, GWO, SCA, AO, WOA, and Moth-Flame Optimizer (MFO). The findings revealed that the planned MFOAOA outperformed all other

TABLE 11.10 Determining the Centroids of the Clusters for the Cancer Dataset

	Calculated Centroids								
Centroids	**ATR.1**	**ATR.2**	**ATR.3**	**ATR.4**	**ATR.5**	**ATR.6**	**ATR.7**	**ATR.8**	**ATR.9**
Centroid1	57.2087	27.51274	97.50435	10.02828	2.695522	26.69112	10.16593	14.81758	537.5709
Centroid2	53.07817	53.07817	53.07817	53.07817	53.07817	53.07817	53.07817	53.07817	53.07817

TABLE 11.11 Determining the Centroids of the Clusters for the CMC Dataset

	Calculated Centroids								
Centroids	**ATR.1**	**ATR.2**	**ATR.3**	**ATR.4**	**ATR.5**	**ATR.6**	**ATR.7**	**ATR.8**	**ATR.9**
Centroid1	32.53233	2.958475	3.430225	3.257999	0.850238	0.749489	2.137509	3.133424	0.0742
Centroid2	38	2.5	2.5	5.5	1	0.5	3	3	0
Centroid3	31.5	3.5	4	3.5	1	1	1.5	3.5	0

TABLE 11.12 Determining the Centroids of the Clusters for the Glass Dataset

	Calculated Centroids								
Centroids	**ATR.1**	**ATR.2**	**ATR.3**	**ATR.4**	**ATR.5**	**ATR.6**	**ATR.7**	**ATR.8**	**ATR.9**
Centroid1	1.517827	13.54182	2.385227	1.493409	72.77068	0.412273	8.986136	0.202045	0.065
Centroid2	1.518549	13.48515	2.399091	1.619697	72.63697	0.396364	9.032424	0.270909	0.077879
Centroid3	1.519066	13.341	2.744	1.361333	72.56167	0.578333	8.97	0.282	0.068667
Centroid4	1.518261	13.33267	2.944444	1.423778	72.58711	0.565556	8.851333	0.092667	0.050444
Centroid5	1.518558	13.468	2.2175	1.443	72.59	0.6765	9.256	0.1495	0.0735
Centroid6	1.517898	13.26222	3.278519	1.338889	72.7663	0.491481	8.63037	0.058889	0.024074
Centroid7	1.519036	13.386	2.845333	1.342	72.574	0.37	9.185333	0.161333	0.021333

TABLE 11.13 Determining the Centroids of the Clusters for the Iris Dataset

	Calculated Centroids			
Centroids	**ATR.1**	**ATR.2**	**ATR.3**	**ATR.4**
Centroid1	5.91746	3.038095	3.92381	1.273016
Centroid2	5.770588	3.083824	3.539706	1.113235
Centroid3	5.857895	3	3.994737	1.257895

TABLE 11.14 Determining the Centroids of the Clusters for the Seeds Dataset

	Calculated Centroids						
Centroids	**ATR.1**	**ATR.2**	**ATR.3**	**ATR.4**	**ATR.5**	**ATR.6**	**ATR.7**
Centroid1	14.8493	14.5620	0.8707	5.6321	3.2573	3.7140	5.4123
Centroid2	14.9067	14.4467	0.8906	5.4343	3.3717	3.8337	5.2543
Centroid3	14.5800	14.4550	0.8773	5.5570	3.2220	2.0840	5.2055

comparison approaches. Furthermore, the Friedman ranking test and Wilcoxon rank test are used to demonstrate the importance of the received development. The suggested MFOAOA achieves considerably better performance than other comparable approaches. The suggested approach can be utilized to address other optimization issues in the future, such as text clustering, feature selection, image segmenting, parameters estimation, engineering problems, industry problems, and advanced benchmark function problems. Also, the proposed method can be improved using other search methods.

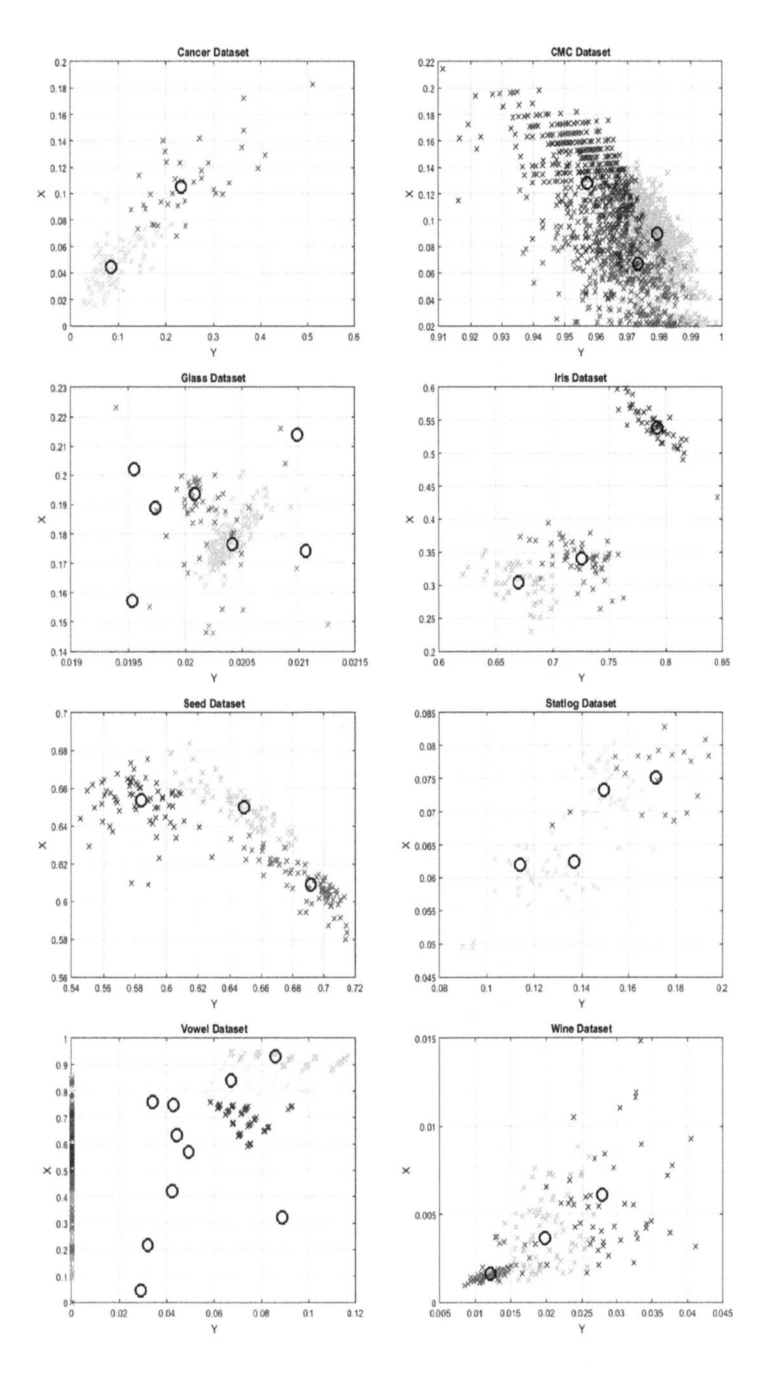

FIGURE 11.3 The results of the clustering analysis are shown as the coloring of the multiplication signs (objects) into k clusters (cycle sign).

TABLE 11.15 Determining the Centroids of the Clusters for the Statlog (Heart) Dataset

	Calculated Centroids								
Centroids	**ATR.1**	**ATR.2**	**ATR.3**	**ATR.4**	**ATR.5**	**ATR.6**	**ATR.7**	**ATR.8**	**ATR.9**
Centroid1	92.5914	44.76344	81.33333	168.4301	62.66667	8.924731	167.0968	41.31183	20.46237
Centroid2	0	0	0	0	0	0	0	0	0
Centroid3	93.5	93.5	93.5	93.5	93.5	93.5	93.5	93.5	93.5
Centroid4	0	0	0	0	0	0	0	0	0
	ATR.10	**ATR.11**	**ATR.12**	**ATR.13**	**ATR.14**	**ATR.15**	**ATR.16**	**ATR.17**	**ATR.18**
Centroid1	147.7419	188.6774	432.6452	176.1183	73.98925	5.376344	10.17204	188.172	194.7204
Centroid2	0	0	0	0	0	0	0	0	0
Centroid3	93.5	93.5	93.5	93.5	93.5	93.5	93.5	93.5	93.5
Centroid4	0	0	0	0	0	0	0	0	0

TABLE 11.16 Determining the Centroids of the Clusters for the Vowel Dataset

	Calculated Centroids						
Centroids	**ATR.1**	**ATR.2**	**ATR.3**	**ATR.4**	**ATR.5**	**ATR.6**	**ATR.7**
Centroid1	0.460215	6.955914	0.469892	−3.18264	1.854737	−0.50698	0.519796
Centroid2	0.580645	8.064516	0.483871	−3.54716	2.219548	−0.40661	0.522677
Centroid3	0.7	9	0.4	−4.104	2.6443	−0.528	0.7221
Centroid4	0.4	6	0.4	−2.9254	2.3728	−0.8216	−0.1558
Centroid5	0.8	9	0.4	−3.9726	2.6852	−0.1584	0.2502
Centroid6	0.25	3.5	0	−2.62375	1.68875	−1.1525	0.6455
Centroid7	0	0	0	0	0	0	0
Centroid8	0	0	0	0	0	0	0
Centroid9	0.621769	0.621769	0.621769	0.621769	0.621769	0.621769	0.621769
Centroid10	0.25	5.75	0.5	−3.0385	2.3425	−0.80575	−0.247
	ATR.8	**ATR.9**	**ATR.10**	**ATR.11**	**ATR.12**	**ATR.13**	
Centroid1	−0.30378	0.626337	−0.00731	0.332153	−0.29958	−0.06744	
Centroid2	−0.25439	0.659	0.020968	0.412806	−0.50487	−0.04548	
Centroid3	−0.3458	0.521	0.0445	0.3492	−0.2428	−0.0246	
Centroid4	−0.2412	0.6728	0.0616	0.2302	−0.2482	−0.59	
Centroid5	−0.7172	1.0394	0.1072	0.4844	0.1862	−0.1134	
Centroid6	−0.83375	1.01325	0.0605	0.61975	−0.15925	−0.25725	
Centroid7	0	0	0	0	0	0	
Centroid8	0	0	0	0	0	0	
Centroid9	0.621769	0.621769	0.621769	0.621769	0.621769	0.621769	
Centroid10	−0.1445	0.5335	0.105	0.404	−0.30425	−0.2845	

TABLE 11.17 Determining the Centroids of the Clusters for the Wine Dataset

	Calculated Centroids						
Centroids	**ATR.1**	**ATR.2**	**ATR.3**	**ATR.4**	**ATR.5**	**ATR.6**	**ATR.7**
Centroid1	13.01175	2.341017	2.367458	19.48362	99.82486	2.294181	2.028475
Centroid 2	41.68308	41.68308	41.68308	41.68308	41.68308	41.68308	41.68308
Centroid 3	0	0	0	0	0	0	0
	ATR.8	**ATR.9**	**ATR.10**	**ATR.11**	**ATR.12**	**ATR.13**	
Centroid 1	0.36096	1.588531	5.075932	0.953198	2.610226	748.8136	
Centroid 2	41.68308	41.68308	41.68308	41.68308	41.68308	41.68308	
Centroid 3	0	0	0	0	0	0	

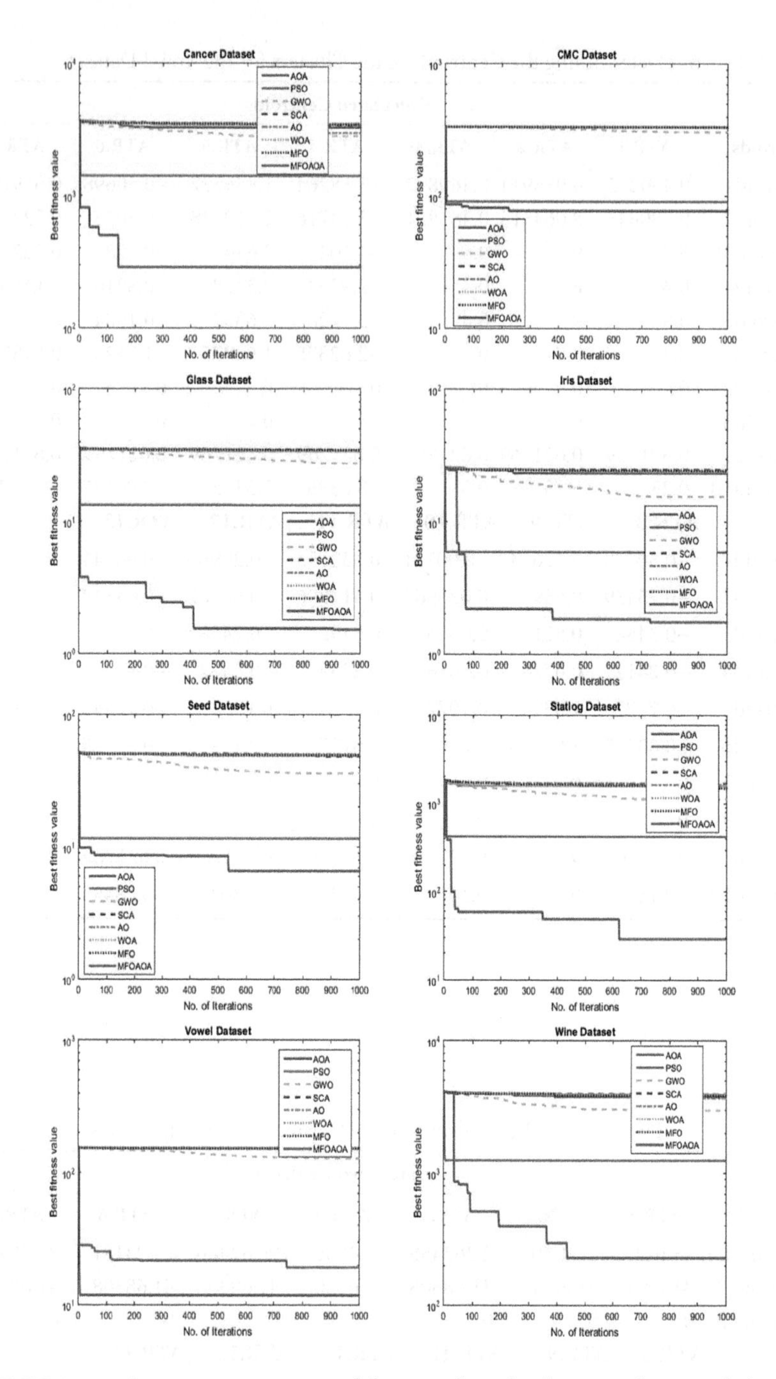

FIGURE 11.4 Convergence curves of the tested methods using the tested clustering datasets.

REFERENCES

1. Abdelkader Abbassi, Rached Ben Mehrez, Bilel Touaiti, Laith Abualigah, and Ezzeddine Touti. Parameterization of photovoltaic solar cell double-diode model based on improved arithmetic optimization algorithm, *Optik*, 253:168600, 2022.
2. Mohamed Abd Elaziz, Ahmed A Ewees, Dalia Yousri, Laith Abualigah, and Mohammed AA Al-qanes. Modified marine predators algorithm for feature selection: case study metabolomics. *Knowledge and Information Systems*, 64:1–27, 2022.
3. Haneen A Abdulwahab, A Noraziah, Abdulrahman A Alsewari, and Sinan Q Salih. An enhanced version of black hole algorithm via levy flight for optimization and data clustering problems, *IEEE Access* 7:142085–142096, 2019.
4. Laith Abualigah. Group search optimizer: a nature-inspired meta-heuristic optimization algorithm with its results, variants, and applications. *Neural Computing and Applications*, 33:1–24, 2020.
5. Laith Abualigah, Mohamed Abd Elaziz, Abdelazim G Hussien, Bisan Alsalibi, Seyed Mohammad Jafar Jalali, and Amir H Gandomi. Lightning search algorithm: a comprehensive survey. *Applied Intelligence*, 16:1–24, 2020.
6. Laith Abualigah, Mohamed Abd Elaziz, Putra Sumari, Zong Woo Geem, and Amir H Gandomi. Reptile search algorithm (RSA): A nature-inspired meta-heuristic optimizer. *Expert Systems with Applications*, 191:116158, 2022.
7. Laith Abualigah and Ali Diabat. Advances in sine cosine algorithm: a comprehensive survey. *Artificial Intelligence Review*,54:1–42.
8. Laith Abualigah and Ali Diabat. A comprehensive survey of the grasshopper optimization algorithm: results, variants, and applications. *Neural Computing and Applications*, 32:1–24, 2020.
9. Laith Abualigah and Ali Diabat. A comprehensive survey of the grasshopper optimization algorithm: results, variants, and applications. *Neural Computing and Applications*, 32:1–24, 2020.
10. Laith Abualigah and Ali Diabat. Improved multi-core arithmetic optimization algorithm-based ensemble mutation for multidisciplinary applications. *Journal of Intelligent Manufacturing*, 1–42, 2022.
11. Laith Abualigah, Ali Diabat, Maryam Altalhi, and Mohamed Abd Elaziz. Improved gradual change-based Harris hawks optimization for real-world engineering design problems. *Engineering with Computers*, 1–41, 2022, in press.
12. Laith Abualigah, Ali Diabat, and Zong Woo Geem. A comprehensive survey of the harmony search algorithm in clustering applications. *Applied Sciences*, 10(11):3827, 2020.
13. Laith Abualigah, Ali Diabat, Seyedali Mirjalili, Mohamed Abd Elaziz, and Amir H Gandomi. The arithmetic optimization algorithm. *Computer Methods in Applied Mechanics and Engineering*, 376:113609, 2021.

14. Laith Abualigah, Mohamed Abd Elaziz, Ahmad M Khasawneh, Mohammad Alshinwan, Rehab Ali Ibrahim, Mohammed AA Al-qaness, Seyedali Mirjalili, Putra Sumari, and Amir H Gandomi. Meta-heuristic optimization algorithms for solving real-world mechanical engineering design problems: a comprehensive survey, applications, comparative analysis, and results. *Neural Computing and Applications*, 1–30, 2022, in press.
15. Laith Abualigah, Amir H Gandomi, Mohamed Abd Elaziz, Husam Al Hamad, Mahmoud Omari, Mohammad Alshinwan, and Ahmad M Khasawneh. Advances in meta-heuristic optimization algorithms in big data text clustering. *Electronics*, 10(2):101, 2021.
16. Laith Abualigah, Amir H Gandomi, Mohamed Abd Elaziz, Abdelazim G Hussien, Ahmad M Khasawneh, Mohammad Alshinwan, and Essam H Houssein. Nature-inspired optimization algorithms for text document clustering: a comprehensive analysis. *Algorithms*, 13(12):345, 2020.
17. Laith Abualigah, Mohammad Shehab, Mohammad Alshinwan, and Hamzeh Alabool. Salp swarm algorithm: a comprehensive survey. *Neural Computing and Applications*, 32:1–21, 2019.
18. Laith Abualigah, Dalia Yousri, Mohamed Abd Elaziz, Ahmed A Ewees, Mohammed AA Al-qaness, and Amir H Gandomi. Aquila optimizer: a novel meta-heuristic optimization algorithm. *Computers & Industrial Engineering*, 157:107250, 2021.
19. Laith Abualigah, Raed Abu Zitar, Khaled H Almotairi, Ahmad MohdAziz Hussein, Mohamed Abd Elaziz, Mohammad Reza Nikoo, and Amir H Gandomi. Wind, solar, and photovoltaic renewable energy systems with and without energy storage optimization: a survey of advanced machine learning and deep learning techniques. *Energies*, 15(2):578, 2022.
20. Laith M Abualigah, Essam Said Hanandeh, Ahamad Tajudin Khader, Mohammed Abdallh Otair, and Shishir Kumar Shandilya. An improved b-hill climbing optimization technique for solving the text documents clustering problem. *Current Medical Imaging*, 16(4):296–306, 2020.
21. Laith Mohammad Abualigah and Ahamad Tajudin Khader. Unsupervised text feature selection technique based on hybrid particle swarm optimization algorithm with genetic operators for the text clustering. *The Journal of Supercomputing*, 73(11):4773–4795, 2017.
22. Laith Mohammad Abualigah, Ahamad Tajudin Khader, and Mohammed Azmi Al-Betar. Unsupervised feature selection technique based on genetic algorithm for improving the text clustering. In *2016 7th International Conference on Computer Science and Information Technology (CSIT)*, 1–6. IEEE, Jordan, 2016.
23. Laith Mohammad Abualigah, Ahamad Tajudin Khader, Mohammed Azmi Al-Betar, and Osama Ahmad Alomari. Text feature selection with a robust weight scheme and dynamic dimension reduction to text document clustering. *Expert Systems with Applications*, 84:24–36, 2017.
24. Laith Mohammad Abualigah, Ahamad Tajudin Khader, Mohammed Azmi AlBetar, and Essam Said Hanandeh. A new hybridization strategy for krill herd algorithm and harmony search algorithm applied to improve the data

clustering. In *1st EAI International Conference on Computer Science and Engineering, 54*. European Alliance for Innovation (EAI), Jordan, 2016.

25. Laith Mohammad Abualigah, Ahamad Tajudin Khader, and Essam Said Hanandeh. A combination of objective functions and hybrid krill herd algorithm for text document clustering analysis. *Engineering Applications of Artificial Intelligence*, 73:111–125, 2018.
26. Laith Mohammad Abualigah, Ahamad Tajudin Khader, and Essam Said Hanandeh. Hybrid clustering analysis using improved krill herd algorithm. *Applied Intelligence*, 48(11):4047–4071, 2018.
27. Laith Mohammad Abualigah, Ahamad Tajudin Khader, and Essam Said Hanandeh. A new feature selection method to improve the document clustering using particle swarm optimization algorithm. *Journal of Computational Science*, 25:456–466, 2018.
28. Laith Mohammad Abualigah, Ahamad Tajudin Khader, Essam Said Hanandeh, and Amir H Gandomi. A novel hybridization strategy for krill herd algorithm applied to clustering techniques. *Applied Soft Computing*, 60:423–435, 2017.
29. Laith Mohammad Qasim Abualigah. *Feature Selection and Enhanced Krill Herd Algorithm for Text Document Clustering*. Springer, Berlin, Germany, 2019.
30. Jeffrey O Agushaka, Absalom E Ezugwu, and Laith Abualigah. Dwarf mongoose optimization algorithm. *Computer Methods in Applied Mechanics and Engineering*, 391:114570, 2022.
31. Mohammed AA Al-Qaness, Ahmed A Ewees, Hong Fan, Laith Abualigah, and Mohamed Abd Elaziz. Marine predators algorithm for forecasting confirmed cases of covid-19 in Italy, USA, Iran and Korea. *International Journal of Environmental Research and Public Health*, 17(10):3520, 2020.
32. Zaher Ali Al-Sai and Laith Mohammad Abualigah. Big data and e-government: A review. In *2017 8th International Conference on Information Technology (ICIT)*, 580–587. IEEE, 2017.
33. Osama Ahmad Alomari, Ahamad Tajudin Khader, M Azmi Al-Betar, and Laith Mohammad Abualigah. MRMR BA: a hybrid gene selection algorithm for cancer classification. *Journal of Theoretical and Applied Information Technology*, 95(12):2610–2618, 2017.
34. Osama Ahmad Alomari, Ahamad Tajudin Khader, Mohammed Azmi Al-Betar, and Laith Mohammad Abualigah. Gene selection for cancer classification by combining minimum redundancy maximum relevancy and bat-inspired algorithm. *International Journal of Data Mining and Bioinformatics*, 19(1):32–51, 2017.
35. Bisan Alsalibi, Laith Abualigah, and Ahamad Tajudin Khader. A novel bat algorithm with dynamic membrane structure for optimization problems. *Applied Intelligence*, 51:1–26, 2020.
36. Bisan Alsalibi, Seyedali Mirjalili, Laith Abualigah, Amir H Gandomi, et al. A comprehensive survey on the recent variants and applications of membrane-inspired evolutionary algorithms. *Archives of Computational Methods in Engineering*, 1–17, 2022, in press.

37. Hadeel N Alshaer, Mohammed A Otair, Laith Abualigah, Mohammad Alshinwan, and Ahmad M Khasawneh. Feature selection method using improved chi square on Arabic text classifiers: analysis and application. *Multimedia Tools and Applications*, 80:1–18, 2020.
38. Mohammad Alshinwan, Laith Abualigah, Mohammad Shehab, Mohamed Abd Elaziz, Ahmad M Khasawneh, Hamzeh Alabool, and Husam Al Hamad. Dragonfly algorithm: a comprehensive survey of its results, variants, and applications. *Multimedia Tools and Applications*, 80:1–38, 2021.
39. Mohammed Alswaitti, Mohanad Albughdadi, and Nor Ashidi Mat Isa. Density-based particle swarm optimization algorithm for data clustering. *Expert Systems with Applications*, 91:170–186, 2018.
40. Ali Danandeh Mehr, Amir Rikhtehgar Ghiasi, Zaher Mundher Yaseen, Ali Unal Sorman, and Laith Abualigah. A novel intelligent deep learning predictive model for meteorological drought forecasting. *Journal of Ambient Intelligence and Humanized Computing*, 1–15, 2022, in press.
41. Amir H. Gandomi, Fang Chen, and Laith Abualigah. Machine learning technologies for big data analytics. *Electronics*, 11(3), 2022.
42. Amolkumar Narayan Jadhav and N Gomathi. WGC: Hybridization of exponential grey wolf optimizer with whale optimization for data clustering. *Alexandria Engineering Journal*, 57(3):1569–1584, 2018.
43. Anil K Jain, M Narasimha Murty, and Patrick J Flynn. Data clustering: a review. *ACM Computing Surveys (CSUR)*, 31(3):264–323, 1999.
44. James Kennedy and Russell Eberhart. Particle swarm optimization. In *Proceedings of ICNN'95-International Conference on Neural Networks*, vol. 4, 1942–1948. IEEE, 1995.
45. Ahmad M Khasawneh, Laith Abualigah, and Mohammad Al Shinwan. Void aware routing protocols in underwater wireless sensor networks: Variants and challenges. In *Journal of Physics: Conference Series*, 1550:032145. IOP Publishing, 2020.
46. Ahmad M Khasawneh, Omprakash Kaiwartya, Laith M Abualigah, Jaime Lloret, et al. Green computing in underwater wireless sensor networks pressure centric energy modeling. *IEEE Systems Journal*, 14(4):4735–4745, 2020.
47. Seyedali Mirjalili. Moth-flame optimization algorithm: a novel nature-inspired heuristic paradigm. *Knowledge-Based Systems*, 89:228–249, 2015.
48. Seyedali Mirjalili. Dragonfly algorithm: a new meta-heuristic optimization technique for solving single-objective, discrete, and multi-objective problems. *Neural Computing and Applications*, 27(4):1053–1073, 2016.
49. Seyedali Mirjalili. SCA: a sine cosine algorithm for solving optimization problems. *Knowledge-Based Systems*, 96:120–133, 2016.
50. Seyedali Mirjalili, Amir H Gandomi, Seyedeh Zahra Mirjalili, Shahrzad Saremi, Hossam Faris, and Seyed Mohammad Mirjalili. Salp swarm algorithm: A bio-inspired optimizer for engineering design problems. *Advances in Engineering Software*, 114:163–191, 2017.
51. Seyedali Mirjalili and Andrew Lewis. The whale optimization algorithm. *Advances in Engineering Software*, 95:51–67, 2016.

52. Seyedali Mirjalili and Andrew Lewis. The whale optimization algorithm. *Advances in Engineering Software*, 95:51–67, 2016.
53. Seyedali Mirjalili, Seyed Mohammad Mirjalili, and Andrew Lewis. Grey wolf optimizer. *Advances in Engineering Software*, 69:46–61, 2014.
54. Mohammed Otair, Osama Talab Ibrahim, Laith Abualigah, Maryam Altalhi, and Putra Sumari. An enhanced grey wolf optimizer based particle swarm optimizer for intrusion detection system in wireless sensor networks. *Wireless Networks*, 28:1–24, 2022.
55. Eduardo Queiroga, Anand Subramanian, and F Cabral Lucídio dos Anjos. Continuous greedy randomized adaptive search procedure for data clustering. *Applied Soft Computing*, 72:43–55, 2018.
56. Sandeep Rana, Sanjay Jasola, and Rajesh Kumar. A review on particle swarm optimization algorithms and their applications to data clustering. *Artificial Intelligence Review*, 35(3):211–222, 2011.
57. Hasan Rashaideh, Ahmad Sawaie, Mohammed Azmi Al-Betar, Laith Mohammad Abualigah, Mohammed M Al-Laham, M Ra'ed, and Malik Braik. A grey wolf optimizer for text document clustering. *Journal of Intelligent Systems*, 29(1):814–830, 2018.
58. Hussein Mohammed Ridha, Hashim Hizam, Seyedali Mirjalili, Mohammad Lutfi Othman, Mohammad Effendy Ya'acob, and Laith Abualigah. A novel theoretical and practical methodology for extracting the parameters of the single and double diode photovoltaic models (December 2021). *IEEE Access*, 10:11110–11137, 2022.
59. Lior Rokach and Oded Maimon. Clustering methods. In: *Data Mining and Knowledge Discovery Handbook*, pp. 321–352. Springer, Berlin, Germany, 2005.
60. Mukaram Safaldin, Mohammed Otair, and Laith Abualigah. Improved binary gray wolf optimizer and SVM for intrusion detection system in wireless sensor networks. *Journal of Ambient Intelligence and Humanized Computing*, 12:1–18, 2020.
61. Shahrzad Saremi, Seyedali Mirjalili, and Andrew Lewis. Grasshopper optimisation algorithm: theory and application. *Advances in Engineering Software*, 105:30–47, 2017.
62. Mohammad Shehab, Laith Abualigah, Husam Al Hamad, Hamzeh Alabool, Mohammad Alshinwan, and Ahmad M Khasawneh. Moth–flame optimization algorithm: variants and applications. *Neural Computing and Applications*, 32(14):9859–9884, 2020.
63. Mohammad Shehab, Hanadi Alshawabkah, Laith Abualigah, and AL-Madi Nagham. Enhanced a hybrid moth-flame optimization algorithm using new selection schemes. *Engineering with Computers*, 37:1–26, 2020.
64. Dalia Yousri, Mohamed Abd Elaziz, Laith Abualigah, Diego Oliva, Mohammed AA Al-Qaness, and Ahmed A Ewees. Covid-19 x-ray images classification based on enhanced fractional-order cuckoo search optimizer using heavy-tailed distributions. *Applied Soft Computing*, 101:107052, 2021.

IV

Applications of Moth-Flame Optimization Algorithm

CHAPTER 12

Moth-Flame Optimization Algorithm, Arithmetic Optimization Algorithm, Aquila Optimizer, Gray Wolf Optimizer, and Sine Cosine Algorithm

A Comparative Analysis Using Multilevel Thresholding Image Segmentation Problems

Laith Abualigah

Amman Arab University
Universiti Sains Malaysia

Nada Khalil Al-Okbi

University of Baghdad

DOI: 10.1201/9781003205326-16

Seyedali Mirjalili
Torrens University Australia
Yonsei University

Mohammad Alshinwan, Husam Al Hamad, and Ahmad M. Khasawneh
Amman Arab University

Waheeb Abu-Ulbeh
Al-Istiqlal University

Mohamed Abd Elaziz
Zagazig University

Heming Jia
Sanming University

Amir H. Gandomi
University of Technology Sydney

CONTENTS

12.1 Introduction 242
12.2 Problem Definitions 245
12.3 Experiments and Discussion 247
12.3.1 Experiments Details 247
12.3.2 Results 247
12.4 Conclusion and Future Works 258
References 259

12.1 INTRODUCTION

Image segmentation has an essential and primary role in many areas, including image processing and pattern recognition [25,35]. The main goal of the segmentation process is to separate the objects within an image from its background [39,58]. A large number of proposed methods have been presented to segment the images by many researchers, including clustering-based methods, region-growing methods, edge detection, graph partitioning-based methods, thresholding, and many other methods [22,23,43].

One of the most critical segmentation methods is the threshold method. The threshold technique is one of the most common methods used by many researchers due to its ease, simplicity, and efficiency. The threshold depends on two main factors, which are the histogram and the number of thresholds for the image [57]. The threshold can be classified into two classes: bi-level threshold and multilevel threshold. A bi-level divides an image into two parts by specifying a single threshold value. At the same time, the second type determines the number of objects in the image by selecting the number of thresholds. Whereas if the color images are divided into a number of areas depending on the colors by specifying the values of a number of thresholds, it is called color image multilevel thresholding. Still, there are some difficulties in determining the threshold for color images because it includes a 3D histogram, unlike grayscale images with a 1D histogram. The color images have their components that depend on the colors and their intensity [2].

A number of proposed methods are used with a threshold segmentation. The Otsu method [49] was proposed in 1979. The Otsu method searches for an optimal threshold value that maximizes the variance between-class. Kapoor's method [41], proposed in 1985, uses entropy to the histogram. The Tsallis entropy method [26] is based on the principle of preserving the moment. It is widely used for image thresholding segmentation. Various methods of determining the multilevel threshold for image segmentation are used in many applications, for example, medical image processing, satellite images, synthetic aperture radar images, and infrared image segmentation. Despite the various proposed methods for determining the threshold values, the problem of the computational cost of the algorithm used increases with the number of levels [30,33]. This puts many obstacles in front of the applications [36].

In recent decades, due to the difficulty and complexity of problems [9,29,37], the need for highly reliable techniques has emerged to improve algorithms' functioning, especially metaheuristic optimization algorithms [1,15,24,31]. Metaheuristic techniques are characterized by randomness and estimation of the optimal solution to various optimization problems and their ability to avoid local solutions, replacing traditional optimization algorithms [28,56]. These algorithms find the optimal decision variables for a specific problem by minimizing or maximizing the objective function [52,55,59]. These algorithms are characterized by a complex and high computational time and large search areas in addition to non-linear problems, and this makes them more difficult [2,8,51].

Metaheuristic optimization algorithms are based on two basic tasks [54]: (a) exploration, which represents the ability of the algorithm to explore non-local (global) search areas [16,19] and (b) exploitation, which represents the ability of the algorithm to explore high-quality optimization solutions within those global solutions [11,12,20]. There must be an appropriate balance between these strategies (exploration and exploitation); all optimization algorithms based on the population principle use these two strategies [17,18]. Still, they differ according to each algorithm's mechanisms and operating factors [7,13,34,53].

Among the most important classifications on which the metaheuristics are based are swarm intelligence algorithms, physics-based or human-based methods, in addition to evolutionary algorithms [3,10]. Where evolutionary algorithms are based on natural evolution [32]an use evolutionary mechanisms such as mutation [5]. One of the most widely used evolutionary algorithms is the genetic algorithm (GA), whose work is based on the Darwinian evolutionary theory [40].

Swarm algorithms are another class that many metaheuristics algorithms adopt. Most of these algorithms simulate the behavior of animal movement [27,38,48]. The most important feature of this group is the sharing of object movement information within the swarm to all swarm members during the optimization cycle. Among the algorithms in this group are the Salp Swarm, Krill Herd, and others [42]. The physics-based approach is another type adopted by metaheuristic algorithms. The work of these algorithms simulates physical laws within the realities of everyday life. The search solutions are based on the rules for controlling the physical methods as in the Gravitational Search Algorithm and the Charged System Search and others. The last method is the human-based method, which is based on human behavior within societies. One of the most important algorithms used is the Imperialism Competitive Algorithm and the Teaching–Learning-Based Optimization Algorithm [50].

Most of the published studies are modifying a specific algorithm, hybridizing different algorithms, or proposing a new algorithm. The failure to use a specific algorithm is due to an algorithm's inability to solve all optimization problems for segmentation according to the theory of "there is no free lunch" [6]. In this research, a comparative analysis of the most recent and used algorithms is presented and tested to solve multilevel thresholding image segmentation problems. The comparative methods are Moth-Flame Optimization (MFO) algorithm [44], Arithmetic Optimization Algorithm (AOA) [14], Aquila Optimizer (AO) [21], Gray Wolf Optimizer (GWO) [46], and Sine Cosine Algorithm (SCA) [45]. Four common gray and color images are used to evaluate the comparative methods to show their performance in solving the given problems. Two standard elevation measures (i.e., Peak signal-to-noise ratio (PSNR) and structural similarity index measure (SSIM)) are used to assist the comparative methods. The results are taken for several thresholds values (i.e., 2, 3, 4, 5, and 6). The results show that the comparative methods almost have the same ability in solving the given problems with different results according to the test cases.

The rest of the research is given as follows. Section 12.2 presents the definitions of the multilevel thresholding image segmentation problem. Experimental results and discussion are presented in Section 12.3. Section 12.4 gives the conclusion and possible future work directions.

12.2 PROBLEM DEFINITIONS

In this section, we concisely describe the problem of multilevel thresholding. If we possess I gray or color image, $K+1$ classes are produced. For segmenting the given image (I) into $K+1$ classes, the k threshold values are needed to proceed further; $\{t_k, k=1,\ldots,K\}$, and this formulation is expressed as follows:

$$\begin{aligned} C_0 &= \{I_{i,j} \mid 0 \le I_{i,j} \le t_1 - 1\}, \\ C_1 &= \{I_{i,j} \mid t_1 \le I_{i,j} \le t_2 - 1\}, \\ &\vdots \\ C_K &= \{I_{i,j} \mid t_K \le I_{i,j} \le L - 1\} \end{aligned} \tag{12.1}$$

where L denotes the maximum gray levels and C_K denotes the I's kth class. The parameter t_k is the k-th threshold, with $I_{i,j}$ being the gray level at the (i,j)-th pixel.

Furthermore, in Equation 12.2, a multilevel thresholding problem is known as a maximization problem that is used to find the optimal threshold. Here,K is the multilevel threshold values.

$$t_1^*, t_2^*, \ldots, t_K^* = \arg\max_{t_1,\ldots,t_K} \mathrm{Fit}(t_1,\ldots,t_K) \tag{12.2}$$

The objective function, fuzzy entropy, is defined as Fit. Various segmentation methods have used fuzzy entropy [4,47], which can be expressed as follows:

Fitness function:

$$\mathrm{Fit}(t_1,\ldots,t_K) = \sum_{k=1}^{K} H_i \tag{12.3}$$

$$H_k = -\sum_{i=0}^{L-1} \frac{p_i \times \mu_k(i)}{P_k} \times \ln\left(\frac{p_i \times \mu_k(i)}{P_k}\right), \tag{12.4}$$

$$P_k = \sum_{i=0}^{L-1} p_i \times \mu_k(i) \tag{12.5}$$

$$\mu_1(l) = \begin{cases} 1 & l \le a_1 \\ \dfrac{l - c_1}{a_1 - c_1} & a_1 \le l \le c_1 \\ 0 & l > c_1 \end{cases}$$

$$\mu_K(l) = \begin{cases} 1 & l \le a_{K-1} \\ \dfrac{l - a_K}{c_K - a_K} & a_{K-1} < l \le c_{K-1} \\ 0 & l > c_{K-1} \end{cases} \tag{12.6}$$

where p_i is the probability distribution, $h(i)$ is the number of pixels for the used gray level L, and N_p is the total numbers of pixels of the image I. p_i presents the probability value for the distribution, determined as $p_i = h(i)/N_p\,(0 < i < L-1)$. $h(i)$ and N_p are the number of pixels for the used gray level L and the total pixel of the image I. $a_1, c_1, \ldots, a_{k-1}, c_{k-1}$ are the used fuzzy parameters, and $0 \le a_1 \le c_1 \le \ldots \le a_{K-1} \le c_{K-1}$.

Then, $t_1 = \dfrac{a_1 + c_1}{2}, t_2 = \dfrac{a_2 + c_2}{2}, \ldots, t_{K-1} = \dfrac{a_{K-1} + c_{K-1}}{2}$. The best fitness function obtained is the highest value.

12.3 EXPERIMENTS AND DISCUSSION

12.3.1 Experiments Details

In this section, the effectiveness of the comparative methods is evaluated using several benchmark images. The comparative methods are MFO algorithm [44], AOA [14], AO [21], GWO [46], and SCA [45].

Experiments are conducted on four benchmark images to evaluate the comparative methods shown in Figure 12.1. The number of solutions and the maximum number of iterations are fixed at 30 and 1,000, respectively, for all methods. Twenty independent runs for each tested method are conducted to collect the results of the tested methods. Two standard elevation criteria, PSNR and SSIM, are employed to assess the comparative methods' performance. For each criterion, four metrics are utilized; worst, best, average, and standard deviation (STD). The results are taken for several thresholds values (i.e., 2, 3, 4, 5, and 6).

12.3.2 Results

In this section, the obtained results of different optimization methods are given. The bold font refer to the best value.

FIGURE 12.1 The benchmark images: Test 1, Test 2, Test 3, and Test 4.

TABLE 12.1 The PSNR Results of Test Case 1

		Comparative Methods				
Threshold	**Metric**	**MFO**	**AO**	**AOA**	**SCA**	**GWO**
2	MAX	16.297305	16.237918	16.307652	15.808943	16.293373
	MEAN	15.421774	**15.449475**	15.384342	13.873557	15.153816
	MIN	14.397054	14.834068	14.368238	10.527087	13.733983
	STD	0.800594	0.618138	0.877979	2.504772	1.180674
3	MAX	18.642747	17.081048	18.884399	18.760855	19.010879
	MEAN	**17.767499**	16.045926	17.386048	17.002439	17.283605
	MIN	15.065411	15.010152	16.675439	14.934842	15.245124
	STD	2.072525	0.846314	0.861472	1.471806	1.608113
4	MAX	20.214656	20.541846	20.105756	19.713000	19.325093
	MEAN	17.629577	18.134093	18.067505	**18.154539**	18.091335
	MIN	15.327319	15.930218	16.731453	16.696394	16.974990
	STD	2.047700	1.654368	1.379592	1.223796	0.952520
5	MAX	20.426919	21.246904	21.154259	20.281427	21.091155
	MEAN	18.812203	19.353525	**19.512460**	19.176371	19.204648
	MIN	16.780316	17.321631	18.277225	17.074232	17.506763
	STD	1.720927	1.512193	1.065610	1.306451	1.399939
6	MAX	20.536102	22.088999	20.318775	21.612630	21.357295
	MEAN	19.905910	20.142970	19.994213	**20.541370**	19.279130
	MIN	19.298422	18.519194	19.275081	19.706329	17.397531
	STD	0.555035	1.539479	0.422765	0.848872	1.730398

Note: Bold indicates the best results.

Table 12.1 shows the results of the comparative methods in terms of PNSR for the first image, where the studied threshold values are 2, 3, 4, 5, and 6. AO method got the best results when the threshold is 2, followed by MFO getting the second-best result, AOA getting the third-best result, GWO getting the fourth-best results, and SCA getting the worst outcome. MFO method got the best results when the threshold is 3, followed by AOA, GWO, SCA, and finally, AO got the worst result. SCA method got the best results when the threshold is 4, followed by AOA, GWO, AO, and finally, MFO got the worst outcome. AOA method got the best results when the threshold is 5, followed by AO, GWO, SCA, and finally, MFO got the worst result. SCA method got the best results when the threshold is 6, followed by AO, AOA, MFO, and finally, GWO got the worst outcome. From these results, the comparative method got almost the same results in which each technique got the best results in one case.

TABLE 12.2 The SSIM Results of Test Case 1

Threshold	Metric	Comparative Methods				
		MFO	**AO**	**AOA**	**SCA**	**GWO**
2	MAX	0.618157	0.617182	0.616877	0.608041	0.594538
	MEAN	0.572572	0.560885	0.572655	**0.577415**	0.502983
	MIN	0.526230	0.492952	0.532610	0.549951	0.338779
	STD	0.040032	0.061668	0.036698	0.027347	0.122014
3	MAX	0.705912	0.638260	0.710383	0.708534	0.700482
	MEAN	0.632398	0.610280	0.649286	0.641998	**0.653744**
	MIN	0.534424	0.582726	0.587206	0.563616	0.617822
	STD	0.075901	0.026636	0.052899	0.057680	0.033944
4	MAX	0.748550	0.771506	0.747506	0.743434	0.728437
	MEAN	0.657963	0.680213	0.680433	**0.689076**	0.684156
	MIN	0.610197	0.574742	0.642839	0.609445	0.659611
	STD	0.059961	0.070427	0.041937	0.049784	0.028045
5	MAX	0.752215	0.786841	0.779099	0.749823	0.777481
	MEAN	0.707533	**0.725946**	0.727725	0.715999	0.724150
	MIN	0.651764	0.662645	0.674935	0.654281	0.667966
	STD	0.048607	0.048407	0.037212	0.038084	0.045016
6	MAX	0.763616	0.801336	0.754420	0.789521	0.792809
	MEAN	0.740601	0.747979	0.738089	**0.760011**	0.729199
	MIN	0.716017	0.703841	0.716292	0.721608	0.665860
	STD	0.019136	0.044840	0.016157	0.031888	0.056669

Note: Bold indicates the best results.

Table 12.2 shows the results of the comparative methods in terms of SSIM for the first image, where the studied threshold values are 2, 3, 4, 5, and 6. SCA method got the best results when the threshold is 2, followed by AOA getting the second-best result, MFO getting the third-best result, AO getting the fourth-best results, and GWO getting the worst outcome. GWO method got the best results when the threshold is 3, followed by AOA, SCA, MFO, and finally, AO got the worst result. SCA method got the best results when the threshold is 4, followed by GWO, AOA, AO, and finally, MFO got the worst outcome. AO method got the best results when the threshold is 5, followed by AOA, GWO, SCA, and finally, MFO got the worst result. SCA method got the best results when the threshold is 6, followed by AO, MFO, AOA, and finally, GWO got the worst outcome. From these results, the comparative method has almost the same ability

TABLE 12.3 The PSNR Results of Test Case 2

		Comparative Methods				
Threshold	**Metric**	**MFO**	**AO**	**AOA**	**SCA**	**GWO**
2	MAX	15.471962	15.295673	15.651540	15.335668	15.131655
	MEAN	14.217528	**14.602174**	14.599283	14.485660	14.255816
	MIN	13.605956	13.433321	13.216679	12.960001	13.670562
	STD	0.802499	0.702546	0.972024	1.089673	0.601336
3	MAX	17.246504	17.546996	17.537882	17.450973	17.590468
	MEAN	15.569204	15.551294	16.003878	**16.314159**	15.990260
	MIN	13.690310	13.472263	13.119474	15.157767	15.326168
	STD	1.609269	1.812119	1.787802	1.015561	0.961340
4	MAX	18.678109	18.248024	19.678195	18.500839	19.043795
	MEAN	17.687410	17.727030	17.653942	16.937269	**17.799022**
	MIN	16.292935	17.314697	16.306597	14.910170	16.837776
	STD	1.071482	0.371702	1.210154	1.350605	0.910718
5	MAX	18.998216	19.684901	18.090098	20.458119	18.495370
	MEAN	17.874212	18.430864	17.184527	**19.161166**	17.514731
	MIN	16.553436	17.356807	16.018679	16.544953	16.613426
	STD	0.975513	1.120105	1.060493	1.545128	0.667684
6	MAX	20.731419	19.962071	21.882511	18.728248	19.692292
	MEAN	18.971349	18.772431	**19.116016**	18.077264	18.551134
	MIN	17.813701	17.066461	17.057802	17.491766	17.709129
	STD	1.302847	1.238189	2.108106	0.560587	0.736740

Note: Bold indicates the best results.

in solving this problem with small different values in which almost each technique got the best results in one case.

Table 12.3 shows the results of the comparative methods in terms of PNSR for the second image, where the studied threshold values are 2, 3, 4, 5, and 6. AO method got the best results when the threshold is 2, followed by AOA getting the second-best result, SCA getting the third-best result, GWO getting the fourth-best results, and MFO getting the worst outcome. SCA method got the best results when the threshold is 3, followed by AOA, GWO, MFO, and finally, AO got the worst result. GWO method got the best results when the threshold is 4, followed by AOA, MFO, AO, and finally, SCA got the worst outcome. SCA method got the best results when the threshold is 5, followed by AO, MFO, GWO, and finally, AOA got the worst result. AOA method got the best results when the threshold is 6, followed by MFO, AO, GWO, and finally, SCA got the worst outcome.

TABLE 12.4 The SSIM Results of Test Case 2

		Comparative Methods				
Threshold	**Metric**	**MFO**	**AO**	**AOA**	**SCA**	**GWO**
2	MAX	0.544748	0.553944	0.573614	0.533009	0.523342
	MEAN	0.508561	0.504913	0.507887	0.499184	**0.509840**
	MIN	0.464967	0.468397	0.448828	0.442427	0.479369
	STD	0.031153	0.031185	0.048142	0.041355	0.017822
3	MAX	0.639242	0.638169	0.643872	0.628374	0.625280
	MEAN	0.542195	0.544572	**0.568489**	0.561394	0.566621
	MIN	0.450369	0.454379	0.538518	0.506380	0.433216
	STD	0.088118	0.080094	0.044088	0.050952	0.081970
4	MAX	0.653305	0.716350	0.712210	0.706915	0.646744
	MEAN	0.608640	0.645870	0.633159	**0.647934**	0.594047
	MIN	0.557592	0.600352	0.551977	0.584550	0.508841
	STD	0.046862	0.042746	0.066259	0.049200	0.052402
5	MAX	0.694442	0.697529	0.635260	0.739098	0.641360
	MEAN	0.629094	0.639029	0.597418	**0.667503**	0.613908
	MIN	0.558075	0.598502	0.550020	0.566200	0.560858
	STD	0.053218	0.041069	0.034756	0.063630	0.032070
6	MAX	0.736841	0.710020	0.737729	0.704790	0.713448
	MEAN	0.655874	**0.672085**	0.670986	0.628487	0.646584
	MIN	0.604704	0.628685	0.586189	0.600177	0.613408
	STD	0.051321	0.030425	0.061193	0.043978	0.040961

Note: Bold indicates the best results.

From these results, the comparative method got almost the same results in which each technique got the best results in one case.

Table 12.4 shows the results of the comparative methods in terms of SSIM for the second image, where the studied threshold values are 2, 3, 4, 5, and 6. GWO method got the best results when the threshold is 2, followed by MFO getting the second-best result, AOA getting the third-best result, AO getting the fourth-best results, and SCA getting the worst outcome. AOA method got the best results when the threshold is 3, followed by GWO, SCA, AO, and finally, MFO got the worst result. SCA method got the best results when the threshold is 4, followed by AO, AOA, MFO and finally, GWO got the worst outcome. SCA method got the best results when the threshold is 5, followed by AO, MFO, GWO, and finally, AOA got the worst result. AO method got the best results when the threshold is 6, followed by AOA, MFO, GWO, and finally, SCA got the worst outcome.

TABLE 12.5 The PSNR Results of Test Case 3

		Comparative Methods				
Threshold	Metric	MFO	AO	AOA	SCA	GWO
2	MAX	13.227682	12.438366	12.884061	13.810486	11.776894
	MEAN	**11.817583**	11.768215	11.332075	11.711131	11.148342
	MIN	10.799602	11.275597	7.743474	8.714551	10.746667
	STD	0.895836	0.475232	2.062259	1.937611	0.427636
3	MAX	14.779363	14.355753	14.514332	16.163099	15.033695
	MEAN	13.841261	13.258394	12.667901	**14.668072**	13.989229
	MIN	13.152600	12.379409	9.110000	13.457329	12.343015
	STD	0.613185	0.870170	2.145122	1.027981	1.115527
4	MAX	16.938915	16.180318	16.090862	16.438659	16.050804
	MEAN	15.153131	14.752003	15.041115	15.197813	**15.292490**
	MIN	13.296259	13.576445	14.041915	13.933162	13.766246
	STD	1.596601	1.024705	0.787327	1.099559	0.940853
5	MAX	15.495683	17.867023	16.270468	17.862500	18.659106
	MEAN	14.627647	**16.905370**	15.274349	16.820489	16.779312
	MIN	14.006040	15.286573	13.858849	15.473721	14.706902
	STD	0.568058	0.969047	1.088886	0.908695	1.876511
6	MAX	19.204815	20.153581	19.696822	19.395118	17.504559
	MEAN	17.886062	18.282895	**18.515897**	18.305542	16.708677
	MIN	16.615842	15.794201	16.763561	16.481046	15.747980
	STD	1.295169	2.244437	1.547865	1.589979	0.889812

Note: Bold indicates the best results.

Table 12.5 shows the results of the comparative methods in terms of PNSR for the third image, where the studied threshold values are 2, 3, 4, 5, and 6. MFO method got the best results when the threshold is 2, followed by AO getting the second-best result, SCA getting the third-best result, AOA getting the fourth-best results, and GWO getting the worst outcome. SCA method got the best results when the threshold is 3, followed by GWO, MFO, AO, and finally, AOA got the worst result. GWO method got the best results when the threshold is 4, followed by SCA, MFO, AOA, and finally, AO got the worst outcome. SCA method got the best results when the threshold is 5, followed by SCA, GWO, AOA, and finally, MFO got the worst result. AOA method got the best results when the threshold is 6, followed by SCA, AO, MFO, and finally, GWO got the worst outcome. From these results, the comparative method got almost the same results. For the

TABLE 12.6 The SSIM Results of Test Case 3

		Comparative Methods				
Threshold	**Metric**	**MFO**	**AO**	**AOA**	**SCA**	**GWO**
2	MAX	0.308588	0.482018	0.307728	0.254234	0.333168
	MEAN	0.251583	**0.269302**	0.233109	0.226810	0.236565
	MIN	0.193802	0.136589	0.113526	0.191899	0.186752
	STD	0.040663	0.135685	0.072620	0.022828	0.057255
3	MAX	0.510643	0.412598	0.439493	0.436661	0.415118
	MEAN	0.338566	0.331385	0.339157	**0.372016**	0.349126
	MIN	0.235450	0.260038	0.268715	0.307266	0.274080
	STD	0.124991	0.056276	0.067463	0.049470	0.060885
4	MAX	0.464586	0.484114	0.479182	0.469044	0.556220
	MEAN	0.434806	0.413367	0.411860	0.423005	**0.446955**
	MIN	0.412696	0.348303	0.333469	0.370638	0.314779
	STD	0.022212	0.061685	0.060029	0.039974	0.094039
5	MAX	0.509901	0.597133	0.508123	0.555407	0.534861
	MEAN	0.439093	**0.528491**	0.430101	0.491413	0.500055
	MIN	0.342481	0.454688	0.382594	0.450659	0.448751
	STD	0.065163	0.055347	0.051012	0.041761	0.032012
6	MAX	0.585753	0.766643	0.576721	0.680670	0.646187
	MEAN	0.562616	0.587097	0.536800	**0.589500**	0.571192
	MIN	0.520235	0.388443	0.466031	0.517927	0.522904
	STD	0.036755	0.189823	0.061456	0.083123	0.065838

Note: Bold indicates the best results.

smaller threshold, MFO got the best results, and for the higher threshold, AOA got the best results.

Table 12.6 shows the results of the comparative methods in terms of SSIM for the third image, where the studied threshold values are 2, 3, 4, 5, and 6. AOA method got the best results when the threshold is 2, followed by MFO getting the second-best result, GWO getting the third-best result, AOA getting the fourth-best results, and SCA getting the worst outcome. SCA method got the best results when the threshold is 3, followed by GWO, AOA, MFO, and finally, AO got the worst result. GWO method got the best results when the threshold is 4, followed by MFO, SCA, AO, and finally, AOA got the worst outcome. AO method got the best results when the threshold is 5, followed by GWO, SCA, MFO, and finally, AOA got the worst result. SCA method got the best results when the threshold is 6, followed by AO, GWO, MFO, and finally, AOA got the worst outcome. From

TABLE 12.7 The PSNR Results of Test Case 4

		Comparative Methods				
Threshold	**Metric**	**MFO**	**AO**	**AOA**	**SCA**	**GWO**
2	MAX	14.362739	14.491257	14.444371	13.337760	13.608691
	MEAN	**13.477875**	12.673522	13.157226	12.077817	11.781985
	MIN	11.523348	10.768210	10.596057	10.907689	10.362312
	STD	1.131371	1.479595	1.533307	0.868865	1.280963
3	MAX	15.816185	14.881273	15.767141	15.569526	15.300949
	MEAN	14.481869	13.735797	**14.591342**	14.476892	14.250221
	MIN	12.572235	11.856343	12.961232	13.091963	12.512715
	STD	1.194472	1.241986	1.054654	1.060070	1.070336
4	MAX	17.394547	16.411155	16.468319	16.340483	17.538119
	MEAN	**16.108989**	15.784820	16.005897	15.790181	14.941529
	MIN	14.374874	14.709080	15.589154	15.227404	12.165059
	STD	1.559009	0.935809	0.441358	0.556644	2.691043
5	MAX	15.714750	17.345913	16.423812	19.151665	19.553117
	MEAN	14.058849	17.054434	15.884887	15.997257	**17.308783**
	MIN	13.205259	16.611989	14.898142	12.097907	15.985702
	STD	1.434281	0.389556	0.855759	3.585398	1.954007
6	MAX	19.474877	18.668386	17.637476	19.366333	17.926761
	MEAN	18.008757	18.192601	16.870358	**18.450832**	16.975858
	MIN	15.517844	17.868571	16.217689	17.581423	16.284999
	STD	2.168451	0.420949	0.716779	0.893347	0.851213

Note: Bold indicates the best results.

these results, the comparative method got almost the same results. For the smaller threshold, AOA got the best results, and for the higher threshold, SCA got the best results.

Table 12.7 shows the results of the comparative methods in terms of PNSR for the fourth image, where the studied threshold values are 2, 3, 4, 5, and 6. MFO method got the best results when the threshold is 2, followed by AOA getting the second-best result, AO getting the third-best result, SCA getting the fourth-best results, and GWO getting the worst outcome. AOA method got the best results when the threshold is 3, followed by MFO, SCA, GWO, and finally, AO got the worst result. MFO method got the best results when the threshold is 4, followed by AOA, SCA, AO, and finally, GWO got the worst outcome. GWO method got the best results when the threshold is 5, followed by AO, SCA, AOA, and finally, MFO got the worst result. SCA method got the best results when

TABLE 12.8 The SSIM Results of Test Case 4

		Comparative Methods				
Threshold	**Metric**	**MFO**	**AO**	**AOA**	**SCA**	**GWO**
2	MAX	0.686323	0.704256	0.702202	0.616001	0.641092
	MEAN	**0.638330**	0.594781	0.620154	0.551220	0.536486
	MIN	0.522396	0.478532	0.449673	0.523203	0.428940
	STD	0.067399	0.088949	0.101332	0.037323	0.080214
3	MAX	0.772559	0.711949	0.826957	0.765816	0.715729
	MEAN	0.691137	0.671897	**0.700271**	0.682044	0.670792
	MIN	0.583049	0.613879	0.598533	0.617409	0.611009
	STD	0.069262	0.041082	0.083147	0.061062	0.041121
4	MAX	0.841120	0.781623	0.804633	0.842562	0.867383
	MEAN	0.776762	0.769559	**0.792258**	0.770380	0.745641
	MIN	0.712128	0.759603	0.780920	0.704761	0.599316
	STD	0.064497	0.011160	0.011891	0.069135	0.135714
5	MAX	0.729153	0.807160	0.830229	0.796472	0.867517
	MEAN	0.671644	0.774535	**0.778356**	0.763405	0.712082
	MIN	0.636691	0.734023	0.748403	0.743556	0.529868
	STD	0.050189	0.037201	0.045103	0.028828	0.170410
6	MAX	0.891931	0.875029	0.783535	0.885937	0.806019
	MEAN	**0.842515**	0.838495	0.757670	0.839925	0.772537
	MIN	0.765339	0.792427	0.723720	0.780201	0.751826
	STD	0.067707	0.042118	0.030716	0.054185	0.029267

Note: Bold indicates the best results.

the threshold is 6, followed by AO, MFO, AOA, and finally, GWO got the worst outcome. From these results, the comparative method got almost the same results in which each technique got the best results in one case. For the smaller threshold, MFO got the best results, and for the higher threshold, SCA got the best results.

Table 12.8 shows the results of the comparative methods in terms of SSIM for the fourth image, where the studied threshold values are 2, 3, 4, 5, and 6. MFO method got the best results when the threshold is 2, followed by AOA getting the second-best result, AO getting the third-best result, SCA getting the fourth-best results, and GWO getting the worst outcome. AOA method got the best results when the threshold is 3, followed by MFO, SCA, AO, and finally, GWO got the worst result. AOA method got the best results when the threshold is 4, followed by MFO,

SCA, AO, and finally, GWO got the worst outcome. AOA method got the best results when the threshold is 5, followed by AO, SCA, GWO, and finally, MFO got the worst result. MFO method got the best results when the threshold is 6, followed by SCA, AO, GWO, and finally, AOA got the worst outcome. For the smaller and higher thresholds, MFO got the best results. The results show that the comparative methods almost have the same ability in solving the given problems with different results according to the test cases.

Figures 12.2–12.6 present the segmented test images by the comparative methods using image 1 and their histogram patterns with the threshold values measured by the comparative methods are shown in Figures 12.7–12.11.

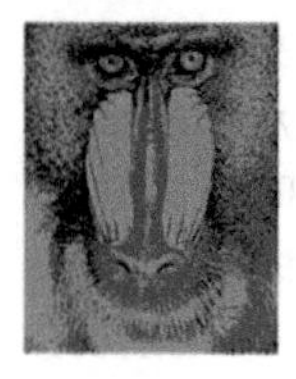
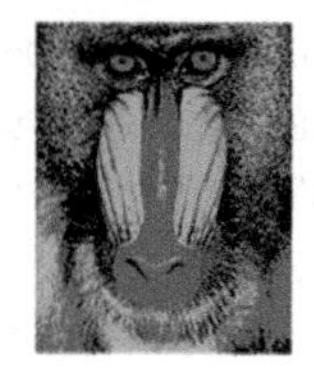

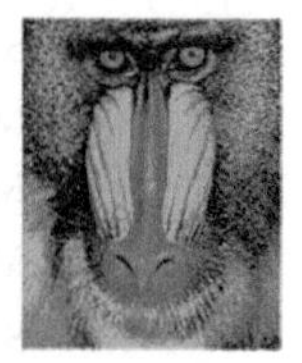

FIGURE 12.2 The segmented images by the MFO with different threshold values from 2 to 6.

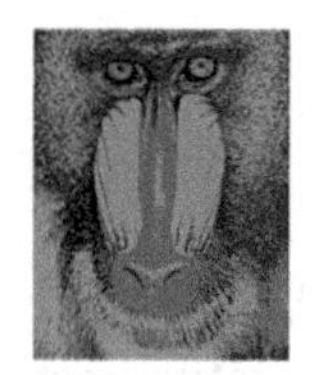
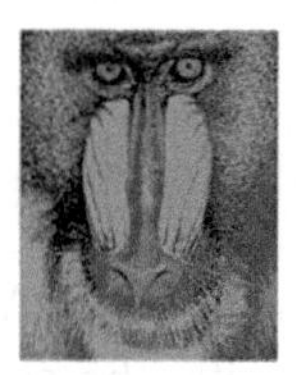

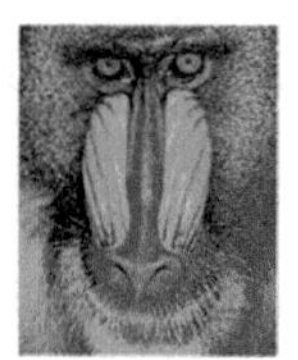

FIGURE 12.3 The segmented images by the AO with different threshold values from 2 to 6.

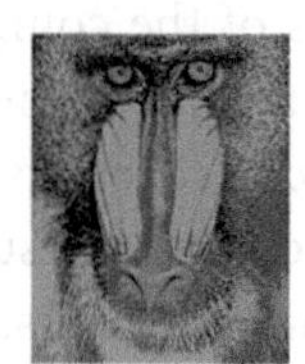

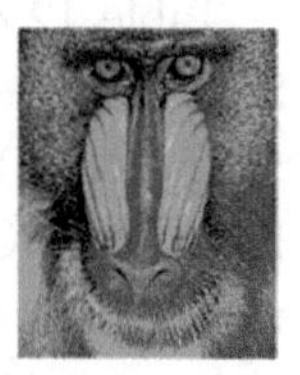

FIGURE 12.4 The segmented images by the AOA with different threshold values from 2 to 6.

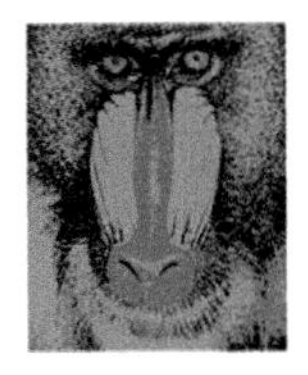   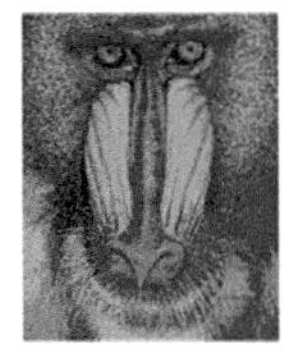 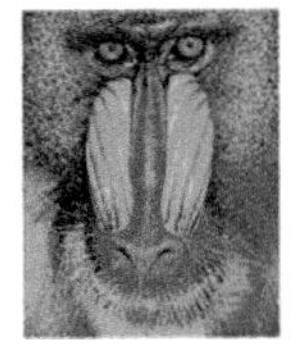

FIGURE 12.5 The segmented images by the SCA with different threshold values from 2 to 6.

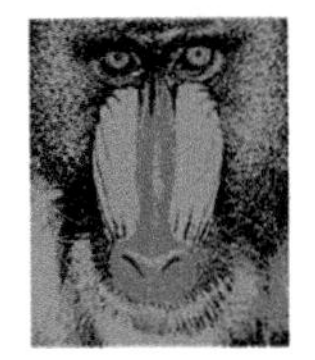  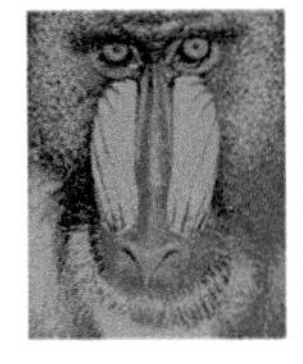 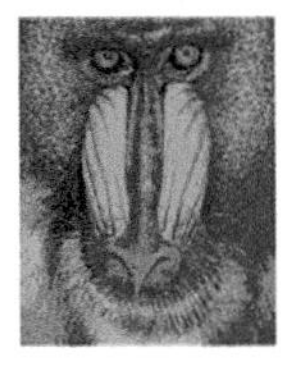 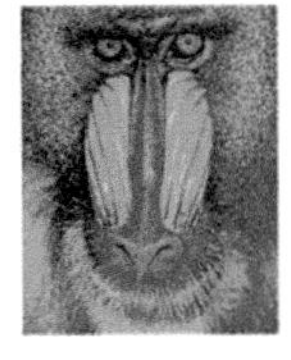

FIGURE 12.6 The segmented images by the GWO with different threshold values from 2 to 6.

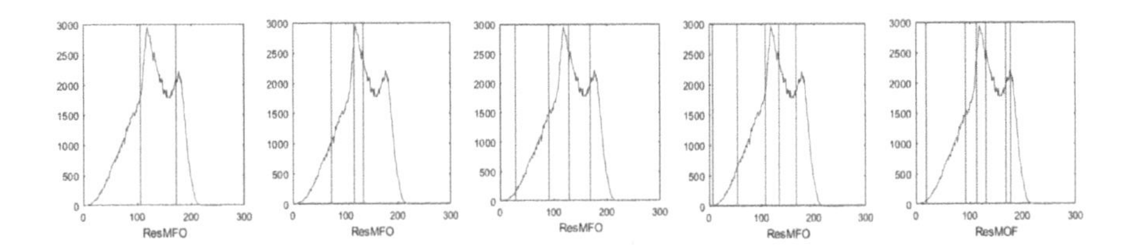

FIGURE 12.7 The histogram pattern with the threshold values (from 2 to 6) measured by the MFO for image 1.

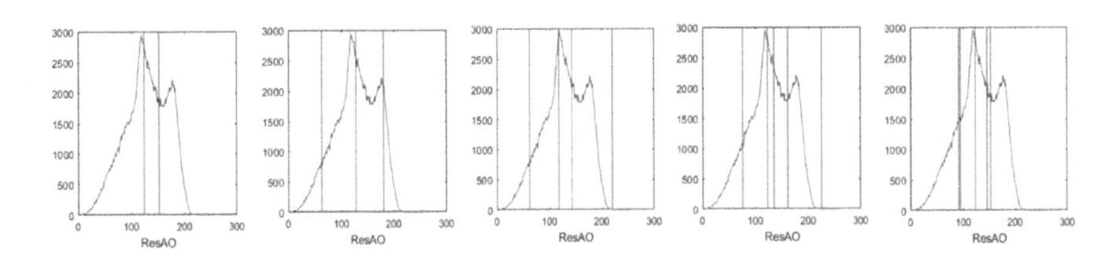

FIGURE 12.8 The histogram pattern with the threshold values (from 2 to 6) measured by the AO for image 1.

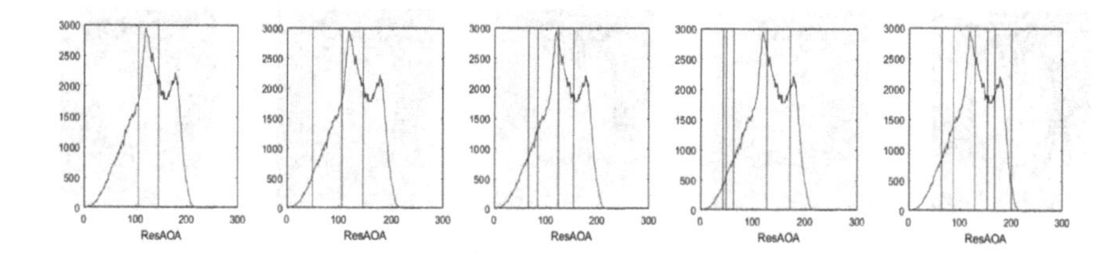

FIGURE 12.9 The histogram pattern with the threshold values (from 2 to 6) measured by the AOA for image 1.

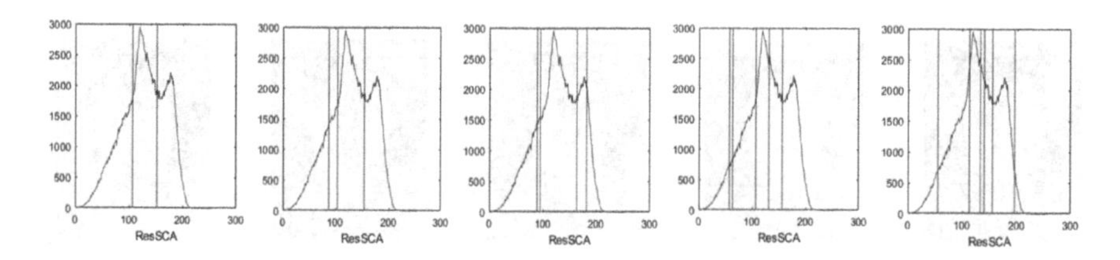

FIGURE 12.10 The histogram pattern with the threshold values (from 2 to 6) measured by SCA for image 1.

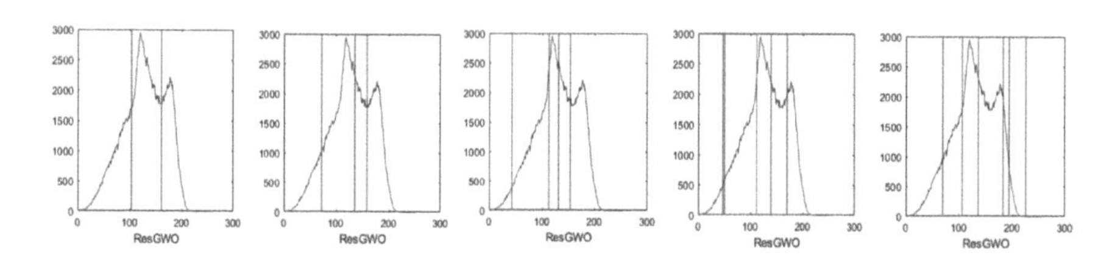

FIGURE 12.11 The histogram pattern with the threshold values (from 2 to 6) measured by GWO for image 1.

12.4 CONCLUSION AND FUTURE WORKS

Image segmentation is a crucial problem used significantly in a number of fields. The threshold method is one of the most important segmentation techniques. A comparative analysis of the results of the most recent and commonly used algorithms for solving multilevel shareholding image segmentation problems is discussed and evaluated in this research. MFO algorithm, AOA, AO, GWO, and SCA are the comparative approaches. The competitors are evaluated using four typical gray and color images to demonstrate their success in solving the segmentation problems. To assess the performance of the comparative approaches, two standard elevation measurements (i.e., PSNR and SSIM) are used to evaluate the obtained segmented images. The results are calculated with a variety of threshold values (i.e., 2, 3, 4, 5, and 6). The findings reveal that the comparable approaches are approximately identical in their ability to solve the given problems, though the results vary depending on the test cases. The comparative methods can be improved to solve the image segmentation problems. These methods can be applied to various optimization problems in the future, including electrocardiogram, text clustering, feature selection, image enhancement, parameter estimation, industrial problems, engineering problems, business problems, and advanced benchmark function problems. Other optimization techniques can also be used to enhance their performance.

REFERENCES

1. Abdelkader Abbassi, Rached Ben Mehrez, Bilel Touaiti, Laith Abualigah, and Ezzeddine Touti. Parameterization of photovoltaic solar cell double-diode model based on improved arithmetic optimization algorithm. *Optik*, 253:168600, 2022.
2. Mohamed Abd Elaziz, Ahmed A Ewees, and Diego Oliva. Hyper-heuristic method for multilevel thresholding image segmentation. *Expert Systems with Applications*, 146:113201, 2020.
3. Mohamed Abd Elaziz, Ahmed A Ewees, Dalia Yousri, Laith Abualigah, and Mohammed AA Al-qaness. Modified marine predators algorithm for feature selection: case study metabolomics. *Knowledge and Information Systems*, 64:1–27, 2022.
4. Mohamed Abd Elaziz and Songfeng Lu. Many-objectives multilevel thresholding image segmentation using knee evolutionary algorithm. *Expert Systems with Applications*, 125:305–316, 2019.
5. Laith Abualigah. Group search optimizer: a nature-inspired meta-heuristic optimization algorithm with its results, variants, and applications. *Neural Computing and Applications*, 1–24, 2020.
6. Laith Abualigah, Mohamed Abd Elaziz, Abdelazim G Hussien, Bisan Alsalibi, Seyed Mohammad Jafar Jalali, and Amir H Gandomi. Lightning search algorithm: a comprehensive survey. *Applied Intelligence*, 51:1–24, 2020.
7. Laith Abualigah, Mohamed Abd Elaziz, Putra Sumari, Zong Woo Geem, and Amir H Gandomi. Reptile search algorithm (RSA): A nature-inspired meta-heuristic optimizer. *Expert Systems with Applications*, 191:116158, 2022.
8. Laith Abualigah and Ali Diabat. A comprehensive survey of the grasshopper optimization algorithm: results, variants, and applications. *Neural Computing and Applications*, 32:1–24, 2020.
9. Laith Abualigah and Ali Diabat. A novel hybrid antlion optimization algorithm for multi-objective task scheduling problems in cloud computing environments. *Cluster Computing*, 24:1–19, 2020.

10. Laith Abualigah and Ali Diabat. Advances in sine cosine algorithm: a comprehensive survey. *Artificial Intelligence Review*, 54:1–42, 2021.
11. Laith Abualigah and Ali Diabat. Improved multi-core arithmetic optimization algorithm-based ensemble mutation for multidisciplinary applications. *Journal of Intelligent Manufacturing*, 1–42, 2022, in press.
12. Laith Abualigah, Ali Diabat, Maryam Altalhi, and Mohamed Abd Elaziz. Improved gradual change-based Harris hawks optimization for real-world engineering design problems. *Engineering with Computers*, 1–41, 2022, in press.
13. Laith Abualigah, Ali Diabat, and Zong Woo Geem. A comprehensive survey of the harmony search algorithm in clustering applications. *Applied Sciences*, 10(11):3827, 2020.
14. Laith Abualigah, Ali Diabat, Seyedali Mirjalili, Mohamed Abd Elaziz, and Amir H Gandomi. The arithmetic optimization algorithm. *Computer Methods in Applied Mechanics and Engineering*, 376:113609, 2021.
15. Laith Abualigah and Akram Jamal Dulaimi. A novel feature selection method for data mining tasks using hybrid sine cosine algorithm and genetic algorithm. *Cluster Computing*, 24:1–16, 2021.
16. Laith Abualigah, Mohamed Abd Elaziz, Ahmad M Khasawneh, Mohammad Alshinwan, Rehab Ali Ibrahim, Mohammed AA Al-qaness, Seyedali Mirjalili, Putra Sumari, and Amir H Gandomi. Meta-heuristic optimization algorithms for solving real-world mechanical engineering design problems: a comprehensive survey, applications, comparative analysis, and results. *Neural Computing and Applications*, 34:1–30, 2022.
17. Laith Abualigah, Amir H Gandomi, Mohamed Abd Elaziz, Husam Al Hamad, Mahmoud Omari, Mohammad Alshinwan, and Ahmad M Khasawneh. Advances in meta-heuristic optimization algorithms in big data text clustering. *Electronics*, 10(2):101, 2021.
18. Laith Abualigah, Amir H Gandomi, Mohamed Abd Elaziz, Abdelazim G Hussien, Ahmad M Khasawneh, Mohammad Alshinwan, and Essam H Houssein. Nature-inspired optimization algorithms for text document clustering: a comprehensive analysis. *Algorithms*, 13(12):345, 2020.
19. Laith Abualigah, Mohammad Shehab, Mohammad Alshinwan, and Hamzeh Alabool. Salp swarm algorithm: a comprehensive survey. *Neural Computing and Applications*, 32:1–21, 2019.
20. Laith Abualigah, Mohammad Shehab, Mohammad Alshinwan, Seyedali Mirjalili, and Mohamed Abd Elaziz. Ant lion optimizer: a comprehensive survey of its variants and applications. *Archives of Computational Methods in Engineering*, 28:1–20, 2020.
21. Laith Abualigah, Dalia Yousri, Mohamed Abd Elaziz, Ahmed A Ewees, Mohammed AA Al-qaness, and Amir H Gandomi. Aquila optimizer: a novel meta-heuristic optimization algorithm. *Computers & Industrial Engineering*, 157:107250, 2021.
22. Laith Abualigah, Raed Abu Zitar, Khaled H Almotairi, Ahmad MohdAziz Hussein, Mohamed Abd Elaziz, Mohammad Reza Nikoo, and Amir H Gandomi. Wind, solar, and photovoltaic renewable energy systems with and without energy storage optimization: A survey of advanced machine learning and deep learning techniques. *Energies*, 15(2):578, 2022.

23. Laith Mohammad Abualigah, Ahamad Tajudin Khader, Essam Said Hanandeh, and Amir H Gandomi. A novel hybridization strategy for krill herd algorithm applied to clustering techniques. *Applied Soft Computing*, 60:423–435, 2017.
24. Laith Mohammad Qasim Abualigah. *Feature Selection and Enhanced Krill Herd Algorithm for Text Document Clustering*. Springer, Berlin, Germany, 2019.
25. Suhaila Farhan Ahmad Abuowaida, Huah Yong Chan, Nawaf Farhan Funkur Alshdaifat, and Laith Abualigah. A novel instance segmentation algorithm based on improved deep learning algorithm for multi-object images. *Jordanian Journal of Computers and Information Technology (JJCIT)*, 7(1):74–88, 2021.
26. Sanjay Agrawal, Rutuparna Panda, Sudipta Bhuyan, and Bijaya K Panigrahi. Tsallis entropy based optimal multilevel thresholding using cuckoo search algorithm. *Swarm and Evolutionary Computation*, 11:16–30, 2013.
27. Jeffrey O Agushaka, Absalom E Ezugwu, and Laith Abualigah. Dwarf mongoose optimization algorithm. *Computer Methods in Applied Mechanics and Engineering*, 391:114570, 2022.
28. Mohammed AA Al-Qaness, Ahmed A Ewees, Hong Fan, Laith Abualigah, and Mohamed Abd Elaziz. Marine predators algorithm for forecasting confirmed cases of covid-19 in Italy, USA, Iran and Korea. *International Journal of Environmental Research and Public Health*, 17(10):3520, 2020.
29. Zaher Ali Al-Sai and Laith Mohammad Abualigah. Big data and e-government: A review. *In 2017 8th International Conference on Information Technology (ICIT)*, 580–587. IEEE, Jordan, 2017.
30. Bisan Alsalibi, Laith Abualigah, and Ahamad Tajudin Khader. A novel bat algorithm with dynamic membrane structure for optimization problems. *Applied Intelligence*, 51:1–26, 2020.
31. Bisan Alsalibi, Seyedali Mirjalili, Laith Abualigah, Amir H Gandomi, et al. A comprehensive survey on the recent variants and applications of membrane-inspired evolutionary algorithms. *Archives of Computational Methods in Engineering*, 1–17, 2022, in press.
32. Mohammad Alshinwan, Laith Abualigah, Mohammad Shehab, Mohamed Abd Elaziz, Ahmad M Khasawneh, Hamzeh Alabool, and Husam Al Hamad. Dragonfly algorithm: a comprehensive survey of its results, variants, and applications. *Multimedia Tools and Applications*, 80:1–38, 2021.
33. Asaju La'aro Bolaji, Mohammed Azmi Al-Betar, Mohammed A Awadallah, Ahamad Tajudin Khader, and Laith Mohammad Abualigah. A comprehensive review: krill herd algorithm (kh) and its applications. *Applied Soft Computing*, 49:437–446, 2016.
34. Ali Danandeh Mehr, Amir Rikhtehgar Ghiasi, Zaher Mundher Yaseen, Ali Unal Sorman, and Laith Abualigah. A novel intelligent deep learning predictive model for meteorological drought forecasting. *Journal of Ambient Intelligence and Humanized Computing*, 1–15, 2022, in press.
35. El-Sayed M El-Kenawy, Abdelhameed Ibrahim, Seyedali Mirjalili, Marwa Metwally Eid, and Sherif E Hussein. Novel feature selection and voting classifier algorithms for covid-19 classification in CT images. *IEEE Access*, 8:179317–179335, 2020.

36. Ali El-Zaart and Ali A Ghosn. Sar images thresholding for oil spill detection. *In 2013 Saudi International Electronics, Communications and Photonics Conference*, 1–5. IEEE, Saudi Arabia, 2013.
37. Ahmed A Ewees, Laith Abualigah, Dalia Yousri, Zakariya Yahya Algamal, Mohammed AA Al-qaness, Rehab Ali Ibrahim, and Mohamed Abd Elaziz. Improved slime mould algorithm based on firefly algorithm for feature selection: a case study on QSAR model. *Engineering with Computers,* 1–15, 2021, in press.
38. Taymaz Rahkar Farshi, John H Drake, and Ender Özcan. A multimodal particle swarm optimization-based approach for image segmentation. *Expert Systems with Applications*, 149:113233, 2020.
39. Amir H. Gandomi, Fang Chen, and Laith Abualigah. Machine learning technologies for big data analytics. *Electronics*, 11(3), 2022.
40. Lifang He and Songwei Huang. An efficient krill herd algorithm for color image multilevel thresholding segmentation problem. *Applied Soft Computing*, 89:106063, 2020.
41. Jagat Narain Kapur, Prasanna K Sahoo, and Andrew KC Wong. A new method for gray-level picture thresholding using the entropy of the histogram. *Computer Vision, Graphics, and Image Processing*, 29(3):273–285, 1985.
42. A Kaveh and Siamak Talatahari. A novel heuristic optimization method: charged system search. *Acta Mechanica*, 213(3):267–289, 2010.
43. Tony Lindeberg and Meng-Xiang Li. Segmentation and classification of edges using minimum description length approximation and complementary junction cues. *Computer Vision and Image Understanding*, 67(1):88–98, 1997.
44. Seyedali Mirjalili. Moth-flame optimization algorithm: a novel nature-inspired heuristic paradigm. *Knowledge-Based Systems*, 89:228–249, 2015.
45. Seyedali Mirjalili. SCA: a sine cosine algorithm for solving optimization problems. *Knowledge-Based Systems*, 96:120–133, 2016.
46. Seyedali Mirjalili, Seyed Mohammad Mirjalili, and Andrew Lewis. Grey wolf optimizer. *Advances in Engineering Software*, 69:46–61, 2014.
47. Husein S Naji Alwerfali, Mohammed AA Al-Qaness, Mohamed Abd Elaziz, Ahmed A Ewees, Diego Oliva, and Songfeng Lu. Multi-level image thresholding based on modified spherical search optimizer and fuzzy entropy. *Entropy*, 22(3):328, 2020.
48. Mohammed Otair, Osama Talab Ibrahim, Laith Abualigah, Maryam Altalhi, and Putra Sumari. An enhanced grey wolf optimizer based particle swarm optimizer for intrusion detection system in wireless sensor networks. *Wireless Networks*, 28:1–24, 2022.
49. Nobuyuki Otsu. A threshold selection method from gray-level histograms. *IEEE Transactions on Systems, Man, and Cybernetics*, 9(1):62–66, 1979.
50. R Venkata Rao, Vimal J Savsani, and DP Vakharia. Teaching–learning-based optimization: a novel method for constrained mechanical design optimization problems. *Computer-Aided Design*, 43(3):303–315, 2011.

51. Hussein Mohammed Ridha, Hashim Hizam, Seyedali Mirjalili, Mohammad Lutfi Othman, Mohammad Effendy Ya'acob, and Laith Abualigah. A novel theoretical and practical methodology for extracting the parameters of the single and double diode photovoltaic models (December 2021). *IEEE Access*, 10:11110–11137, 2022.
52. Mukaram Safaldin, Mohammed Otair, and Laith Abualigah. Improved binary gray wolf optimizer and SVM for intrusion detection system in wireless sensor networks. *Journal of Ambient Intelligence and Humanized Computing*, 12:1–18, 2020.
53. Canan Batur Şahin, Özlem Batur Dinler, and Laith Abualigah. Prediction of software vulnerability based deep symbiotic genetic algorithms: phenotyping of dominant-features. *Applied Intelligence*, 51:1–17, 2021.
54. Mohammad Shehab, Laith Abualigah, Husam Al Hamad, Hamzeh Alabool, Mohammad Alshinwan, and Ahmad M Khasawneh. Moth–flame optimization algorithm: variants and applications. *Neural Computing and Applications*, 32(14):9859–9884, 2020.
55. Mohammad Shehab, Hanadi Alshawabkah, Laith Abualigah, and Al-Madi Nagham. Enhanced a hybrid moth-flame optimization algorithm using new selection schemes. *Engineering with Computers*, 37:1–26, 2020.
56. Mohammad Shehab, Mohammad Sh Daoud, Hani Mahmouad AlMimi, Laith Mohammad Abualigah, and Ahamad Tajudin Khader. Hybridising cuckoo search algorithm for extracting the ODF maxima in spherical harmonic representation. *International Journal of Bio-Inspired Computation*, 14(3):190–199, 2019.
57. Wen-Hsiang Tsai. Moment-preserving thresolding: a new approach. *Computer Vision, Graphics, and Image Processing*, 29(3):377–393, 1985.
58. Dalia Yousri, Mohamed Abd Elaziz, and Seyedali Mirjalili. Fractional-order calculus-based flower pollination algorithm with local search for global optimization and image segmentation. *Knowledge-Based Systems*, 197:105889, 2020.
59. Dalia Yousri, Mohamed Abd Elaziz, Diego Oliva, Laith Abualigah, Mohammed AA Al-qaness, and Ahmed A Ewees. Reliable applied objective for identifying simple and detailed photovoltaic models using modern metaheuristics: Comparative study. *Energy Conversion and Management*, 223:113279, 2020.

CHAPTER 13

Optimal Design of Truss Structures with Continuous Variable Using Moth-Flame Optimization

Nima Khodadadi
Florida International University

Seyed Mohammad Mirjalili
Concordia University

Seyedali Mirjalili
Torrens University Australia
Yonsei University

CONTENTS

13.1 Introduction 266
13.2 Problem Definition 267
13.3 Numerical Examples 268
  13.3.1 The 25-bar Space Truss 268
  13.3.2 The 72-bar Space Truss 272
  13.3.3 The 200-bar Planar Truss 275
References 279

DOI: 10.1201/9781003205326-17

13.1 INTRODUCTION

The trusses are the most used structural systems employed by engineers in designing bridges, arches, domes, etc. However, limitation in resources and high construction costs in such engineering structures require optimal designs, and finding an optimum solution is challenging. In addition, various methods have been developed to deal with this problem, and the utilized optimization techniques can be classified into two main groups of analytical and approximate methods.

Mathematics is the base of analytical approaches. These methods can find the exact solution to the problems and require less computational effort than approximate methods. However, these approaches are sensitive to the initial starting point, and just a correct starting point can lead to a high-quality solution. In addition, when faced with complex optimization problems like the optimal design of truss structures under frequency constraints, these methods are not capable of providing the global minima and only reached local ones.

Most of the approximate techniques are global search methods, which are called Metaheuristics. These methods are developed to address the weaknesses of classical approaches. The metaheuristic algorithms can find near-global or global solutions by employing intelligence of natural phenomena. In the last decades, various metaheuristic algorithms are proposed based on natural processes, collective behavior, or scientific rules such as Genetic Algorithm (GA) [1] and Differential Evolution (DE) [2], Ant Colony Optimization (ACO) [3], Particle Swarm Optimization (PSO) [4], Dynamic Water Strider Algorithm (DWSA) [5], Cuckoo Search (CS) [6], Stochastic Paint Optimizer (SPO) [7], Flow Direction Algorithm (FDA) [8], and Advanced Charged System Search (ACSS) [9]. Many researchers utilized these algorithms to solve structural optimization problems such as frames [10] and the real application of structural engineering [11,12]. These studies demonstrate that metaheuristic algorithms can solve problems with good accuracy in a reasonable time when employed to deal with complex optimization problems. Ease of implementation, simple framework, good accuracy, and reasonable execution time are some advantages of metaheuristic algorithms compared with the analytical techniques.

In the study of structural analysis, the design of truss structures under constraints is a critical subject. These algorithms are designed to reduce the weight of structures while meeting specific design criteria. The goal of new metaheuristic algorithms is to get better answers to issues in a reasonable amount of time.

In recent years, researchers have used several metaheuristic methods to improve truss designs, although with certain constraints. The Big Bang–Big Crunch (BB-BC) optimization of truss structures with continuous and discrete design variables, for example, was developed by Camp [13]. Lamberti [14] proposed the Simulated Annealing (SA) technique for truss structural design optimization. In [15], four benchmark truss problems were considered with discrete design variables and focused on minimization of the structure weight under the required constraints. Jawad et al. [16] obtained a more appropriate design for truss structures using a discrete novel nature-inspired optimization algorithm. Finally, Kaveh et al. [17] developed an effective hybrid optimization technique for optimal truss design based on the invasive weed optimization algorithm and the shuffled frog-leaping algorithm. In this work, the Moth-Flame Optimization (MFO) [18] algorithm is used to optimize the design of several structures.

The rest of this chapter is as follows: The problem definition of structural design optimization is presented in Section 13.2. The results are presented in Section 13.3. This chapter concludes with Section 13.4.

13.2 PROBLEM DEFINITION

This section contains the mathematical formulae for the size optimization used in this study. The optimization of truss structure size entails achieving optimal cross-section (A_i) values that minimize the structural weight of W. This minimum design also has to satisfy the constraints that restrict design variable sizes and structural responses [10]. The optimal design problem is as follows:

$$\begin{aligned} &\text{Minimize } W(\{x\}) = \sum_{i=1}^{nm} \gamma_i \cdot A_i \cdot L_i(x) \\ &\text{Subject } \delta_{\min} \leq \delta_i \leq \delta_{\max} \quad i = 1,2,\ldots,m \\ &\sigma_{\min} \leq \sigma_i \leq \sigma_{\max} \quad i = 1,2,\ldots,n \\ &\sigma_i^b \leq \sigma_i^i \leq 0 \quad i = 1,2,\ldots,ns \\ &A_{\min} \leq A_i \leq A_{\max} \quad i = 1,2,\ldots,ng \end{aligned} \tag{13.1}$$

where $W(\{x\})$ is the structure's weight, n is the number of members forming the structure, m is the number of nodes, ns is the number of compression elements, and ng is the number of design variables or the number of member groups. Here, γ_i is the material density of the member i, L_i is the length of the member i, A_i is the cross-sectional area of the member i chosen between A_{min} and A_{max}, min shows the lower bound and max indicates the upper bound, σ_i and δ_i are the stress and nodal deflection, respectively, and σ_i^b is the allowable buckling stress in the member i when it is in compression. The penalty function is expressed as follows:

$$f_{penalty}(X)=(1+\varepsilon_1\cdot v)^{\varepsilon_2} \quad v=\sum_{i=1}^{n}\max[0,v_i] \tag{13.2}$$

where v is the total amount of the constraints violated, and constants ε_1 and ε_2 are selected considering the exploration and the exploitation rate of the search space. In this case, ε_1 is set to 1, and ε_2 is chosen to minimize penalties and to reduce cross-sections. At the beginning of the search process, ε_2 is set to 1.5 and is subsequently increased to 3 [19].

13.3 NUMERICAL EXAMPLES

This section contains the example's explanations. These well-known benchmark design cases are solved using the MFO algorithm to show that the proposed MFO is more efficient than other earlier metaheuristics for continuous optimization. All runs were done on a MacBook Pro for truss problems with a CPU 2.3 GHz (8-Core an Intel Core i9 computer platform), 16 GB RAM on a Macintosh computer, and the code was written in MATLAB® (macOS Big Sur). Consider the following scenarios to show the effectiveness of the MFO algorithm: a 25-bar space truss, a 72-bar space truss, and a 200-bar planar truss. Each problem's results are compared to solutions developed using a variety of other optimization approaches. The suggested technique has 200 iterations for 25- and 72-bar space trusses and a total of 400 iterations for a 200-bar planar truss, and each example uses a population of 50 particles.

13.3.1 The 25-bar Space Truss

The 25-bar space truss issue has become a common test problem in optimization, and it has been solved using a variety of optimization techniques. Figure 13.1 depicts the 25-bar space truss under consideration. The structural components are organized into eight groups, each with its own material and cross-sectional characteristics. The two load cases that were applied to the truss are shown in Table 13.1.

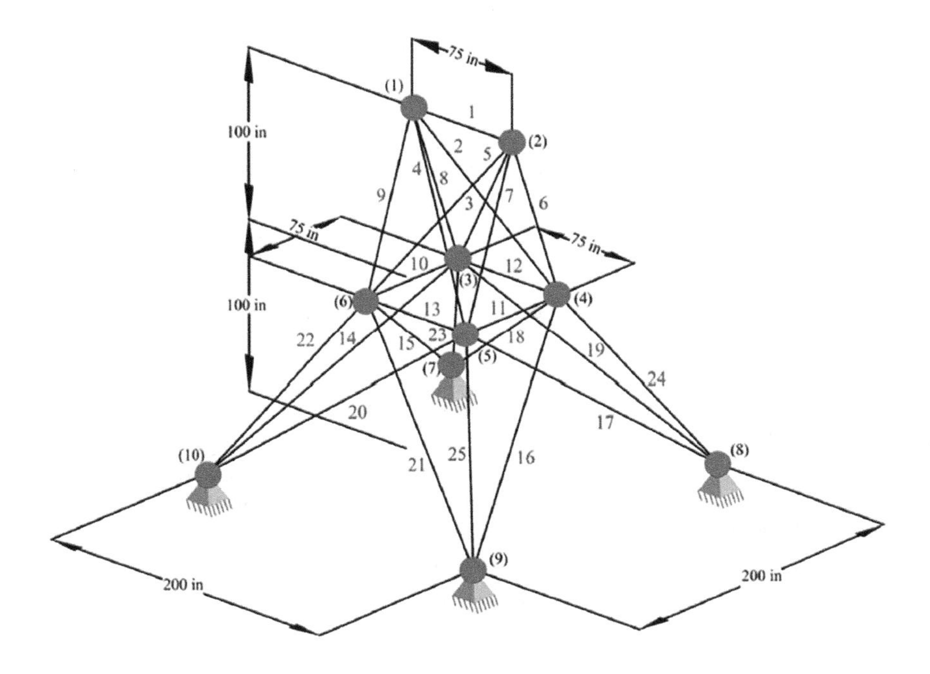

FIGURE 13.1 The 25-bar space truss.

TABLE 13.1 Loading Condition for 25-bar Space Truss

Node Number	Load (kips)					
	Case 1			Case 2		
	P_x	P_y	P_z	P_x	P_y	P_z
1	0	20	−5	1	10	−5
2	0	−20	−5	0	10	−5
3	0	0	0	0.5	0	0
6	0	0	0	0.5	0	0

Table 13.2 shows how the members are divided into groups based on their numbers. As shown in Table 13.3, each node was controlled by a maximum displacement limit of ±0.35 in each direction, as well as axial load constraints for each group. The range of cross-sectional areas is from 0.01 to 3.4 in^2. The material has a modulus of elasticity of 10,000 Ksi and the mass density of 0.1 lb/in^3.

Table 13.4 illustrates that the MFO approach produces significantly better outcomes than the other algorithms. As can be seen, MFO was able to come up with satisfactory solutions after 9,750 analyses. MFO's results are also shown in Table 13.4, where they are compared to those of other

TABLE 13.2 Element Grouping for 25-bar Space Truss

Element Group Number							
1	2	3	4	5	6	7	8
1:(1,2)	2:(1,4)	6:(2,4)	10:(6,3)	12:(3,4)	14:(3,10)	18:(4,7)	22:(10,6)
	3:(2,3)	7:(2,5)	11:(5,4)	13:(6,5)	15:(6,7)	19:(3,8)	23:(3,7)
	4:(1,5)	8:(1,3)			16:(4,9)	20:(5,10)	24:(4,8)
	5:(2,6)	9:(1,6)			17:(5,8)	21:(6,9)	25:(5,9)

TABLE 13.3 Member Stress Limitation for the 25-bar Space Truss

Element Group		Compressive Stress Limitations Ksi (MPa)	Tensile Stress Limitations Ksi (MPa)
1	A1	35.092 (241.96)	40.0 (275.80)
2	A_2–A_5	11.590 (79.0913)	40.0 (275.80)
3	A_6–A_9	17.305 (119.31)	40.0 (275.80)
4	A_{10}–A_{11}	35.092 (241.96)	40.0 (275.80)
5	A_{12}–A_{13}	35.092 (241.96)	40.0 (275.80)
6	A_{14}–A_{17}	6.759 (46.603)	40.0 (275.80)
7	A_{18}–A_{21}	6.959 (47.982)	40.0 (275.80)
8	A_{22}–A_{25}	11.082 (76.410)	40.0 (275.80)

TABLE 13.4 Comparison of MFO-Optimized 25-bar Space Truss with other Algorithms

Member Group	HS [20]	IGWO [21]	WEO [22]	BB-BC [23]	CBO [24]	SFLA-IWO [17]	Present Work MFO
1 (A_1)	0.047	0.012	0.010	0.010	0.010	0.010	0.010
2 (A_2–A_5)	2.022	1.962	1.981	2.092	2.129	1.989	1.989
3 (A_6–A_9)	2.950	3.020	3.002	2.964	2.886	2.990	2.989
4 (A_{10}–A_{11})	0.010	0.026	0.010	0.010	0.010	0.010	0.010
5 (A_{12}–A_{13})	0.014	0.010	0.011	0.010	0.010	0.010	0.010
6 (A_{14}–A_{17})	0.688	0.684	0.682	0.689	0.679	0.680	0.684
7 (A_{18}–A_{21})	1.657	1.686	1.677	1.601	1.607	1.676	1.676
8 (A_{22}–A_{25})	2.663	2.652	2.661	2.686	2.692	2.667	2.667
Best weight (lb)	544.38	545.48	545.16	545.38	544.31	545.15	545.14
Average weight (lb)	N/A	549.67	N/A	545.78	545.25	545.22	545.29
Standard deviation	N/A	2.811	N/A	0.49	0.29	0.07	0.411
Number of analyses	15,000	5,640	19,750	10,000	9,090	10,000	9,750

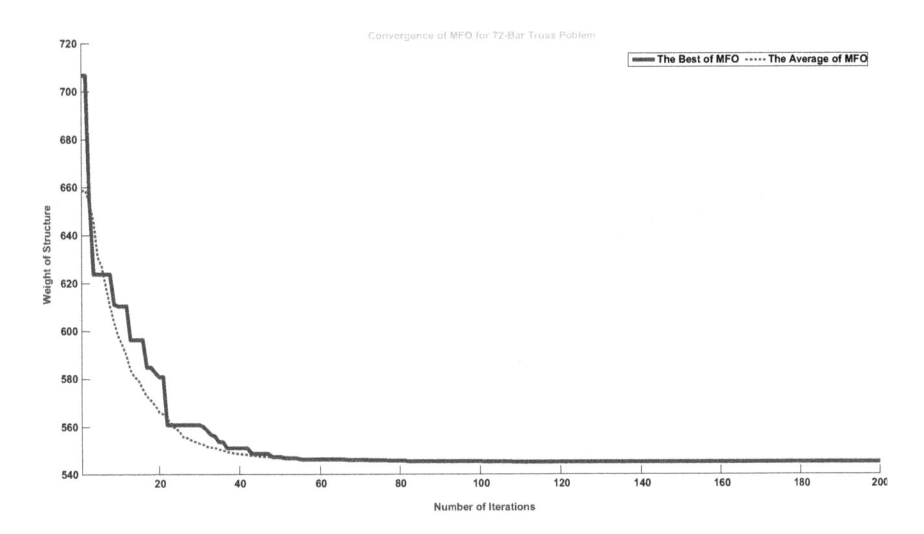

FIGURE 13.2 The best and average convergence curve of MFO for 25-bar space truss.

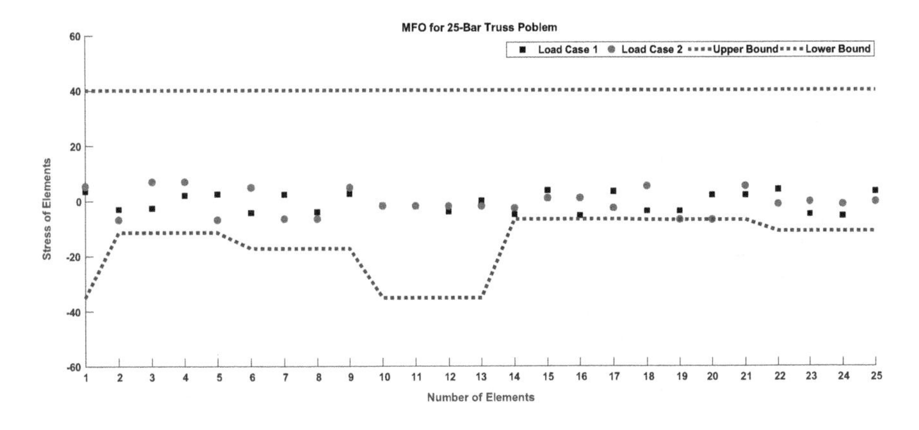

FIGURE 13.3 Best solution stresses of 25-bar space truss against allowable values.

algorithms. Other algorithms, such as HS [20], IGWO[21], WEO [22], BB-BC [23], CBO [24], and SFLA-IWO [17], have also solved this problem. The MFO has the best weight of 545.14 lb. The results achieved by MFO outperform those obtained by the other algorithms. This is an improvement above the proposed algorithm's optimum design. The convergence curves of the best and average MFO runs are shown in Figure 13.2. According to Figures 13.3 and 13.4, stresses and displacements displayed with permissible values for two MFO load instances. Obviously, there are no violations for constraints.

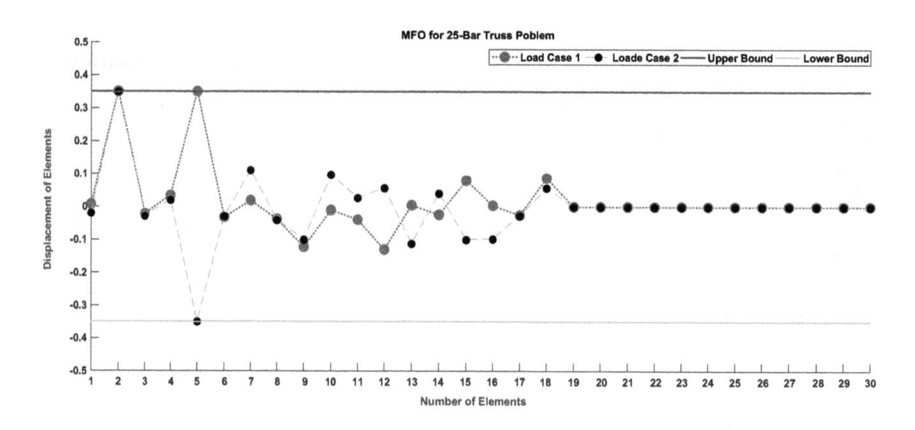

FIGURE 13.4 Best solution displacement of 25-bar space truss against allowable values.

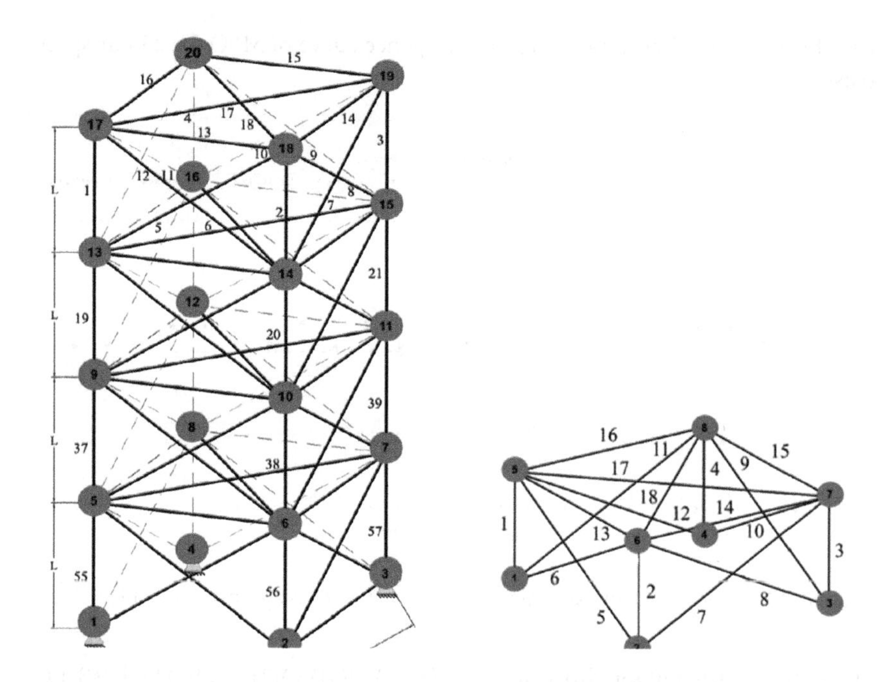

FIGURE 13.5 The 72-bar space truss.

13.3.2 The 72-bar Space Truss

The 72-bar space truss is shown in Figure 13.5 as an example. The material's density is considered to be 0.1 lb/in^3 and its modulus of elasticity is 10,000 Ksi. The elements of this example are divided into 16 groups. All members have a tension and compression stress of 25 Ksi. In both the

TABLE 13.5 Load Condition for the 72-bar Space Truss

Node number	Load (kips)					
	Case 1			Case 2		
	P_x	P_y	P_z	P_x	P_y	P_z
17	5	5	−5	0	0	−5
18	0	0	0	0	0	−5
19	0	0	0	0	0	−5
20	0	0	0	0	0	−5

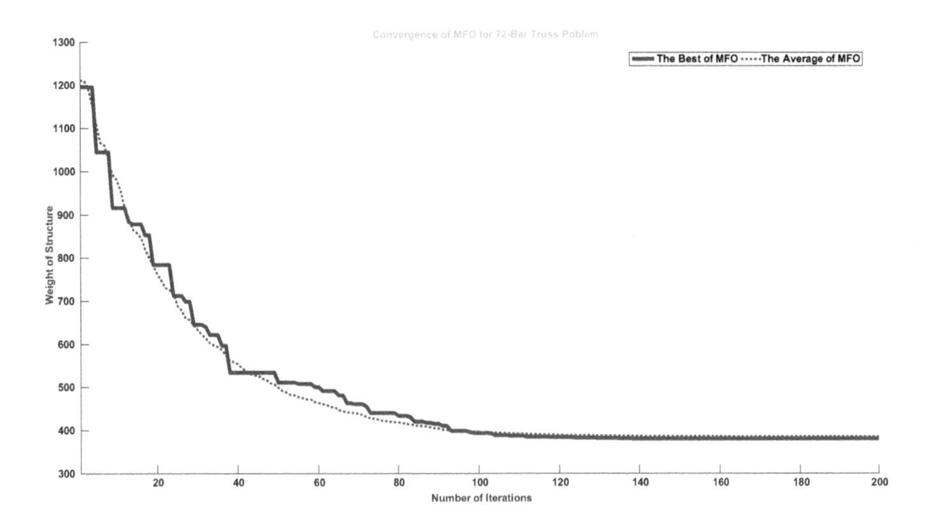

FIGURE 13.6 The best and average convergence curve of MFO for 72-bar space truss.

x and y directions, the top node displacement is <0.25 in. The minimum and maximum permissible cross-section areas of each member are 0.10 and 4.00 in^2, respectively. Table 13.5 shows the numbers and directions for the two different space truss load scenarios.

The best and average convergence curves of the MFO algorithm are shown in Figure 13.6. MFO required 9,850 analyses to converge, which is second to none. Table 13.6 shows the MFO algorithms' optimal solutions. The data show that the current method performs better.

The best outcome of the MFO method is 379.62 lb, while it is 380.24, 379.63, 379.69, 379.85, 379.76, 379.75, 380.45, 379.63, and 379.75 lb for the algorithms such as ACO [25], CS [26], TLBO [27], BB-BC [23], IGWO [21], CBO [24], RO [28], and SFLA-IWO [17], respectively.

TABLE 13.6 Comparison of MFO-Optimized 72-bar Space Truss with other Algorithms

Member group	ACO [25]	CS [26]	TLBO [28]	BB-BC [23]	IGWO [21]	CBO [24]	RO [27]	SFLA-IWO [17]	Present Work MFO
1 (A_1–A_4)	1.948	1.912	1.906	1.857	1.858	1.917	1.836	1.9004	1.8804
2 (A_5–A_{12})	0.508	0.510	0.506	0.505	0.502	0.503	0.502	0.5149	0.5135
3 (A_{13}–A_{16})	0.101	0.100	0.100	0.100	0.100	0.100	0.100	0.1000	0.1000
4 (A_{17}–A_{18})	0.102	0.100	0.100	0.100	0.100	0.100	0.100	0.1000	0.1000
5 (A_{19}–A_{22})	1.303	1.257	1.261	1.247	1.301	1.272	1.252	1.2770	1.2699
6 (A_{23}–A_{30})	0.511	0.512	0.511	0.526	0.515	0.505	0.503	0.5077	0.5092
7 (A_{31}–A_{34})	0.101	0.100	0.100	0.100	0.100	0.100	0.100	0.1000	0.1000
8 (A_{35}–A_{36})	0.100	0.100	0.100	0.101	0.100	0.100	0.100	0.1000	0.1000
9 (A_{37}–A_{40})	0.561	0.522	0.531	0.520	0.531	0.518	0.573	0.5277	0.5179
10 (A_{41}–A_{48})	0.492	0.517	0.515	0.517	0.512	0.536	0.549	0.5139	0.5178
11 (A_{49}–A_{52})	0.100	0.100	0.100	0.100	0.100	0.100	0.100	0.1000	0.1000
12 (A_{53}–A_{54})	0.107	0.100	0.100	0.100	0.103	0.100	0.100	0.1000	0.1000
13 (A_{55}–A_{58})	0.156	0.156	0.156	0.156	0.156	0.156	0.157	0.1565	0.1564
14 (A_{59}–A_{66})	0.550	0.540	0.549	0.550	0.547	0.537	0.522	0.5438	0.5474
15 (A_{67}–A_{70})	0.390	0.415	0.409	0.392	0.420	0.406	0.435	0.4083	0.4158
16 (A_{71}–A_{72})	0.592	0.570	0.569	0.592	0.579	0.574	0.597	0.5738	0.5639
Best weight (lb)	380.24	379.63	379.69	379.85	379.76	379.75	380.45	379.63	379.62
Average weight (lb)	383.16	379.73	380.86	382.08	380.68	380.03	382.55	379.68	384.49
Standard deviation	3.66	N/A	1.85	1.91	0.73	0.37	N/A	0.06	12.54
Number of analyses	18,500	10,600	18,460	19,621	11,960	18,000	19,084	10,000	9,850

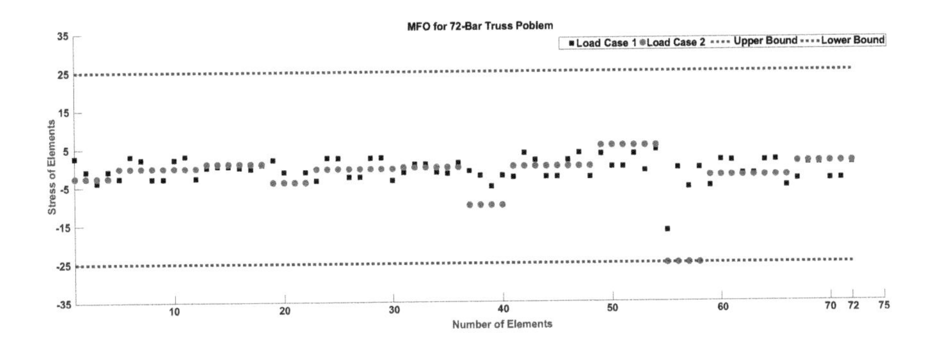

FIGURE 13.7 Best solution stresses of 72-bar space truss against allowable values.

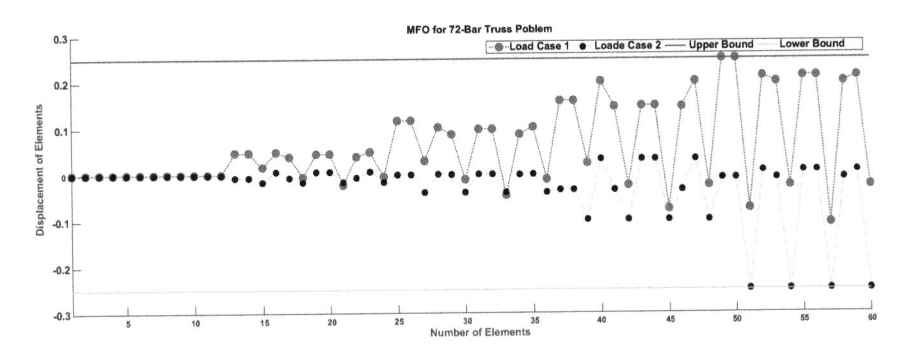

FIGURE 13.8 Best solution displacement ratio of 72-bar space truss.

According to Table 13.6, the MFO algorithm's standard deviation and average weights are 12.54 and 384.49 lb, respectively. To evaluate the constraints violation of MFO, for two load scenarios, Figures 13.7 and 13.8 compare the MFO's permitted and existing stress and displacement constraint values.

13.3.3 The 200-bar Planar Truss

The efficiency of the MFO algorithm is evaluated using a more prominent issue from Kaveh and Mahdavi. Figure 13.9 shows a 200-bar planar truss that has been created to be lightweight. Table 13.7 shows how the components are grouped into 29 groups. Steel is employed in the construction of all elements; the material density and modulus of elasticity are 0.283 lb/in^3 (7933.41 kg/m^3) and 30 Msi (206 GPa), respectively.

Only stress limits of ±10 Ksi (68.95 MPa) are applied to this truss. There are three loading requirements that are independent of each other: (a) 1.0 Kips (4.45 kN) acting in the positive x-direction at nodes 1, 6, 15, 20, 29, 43,

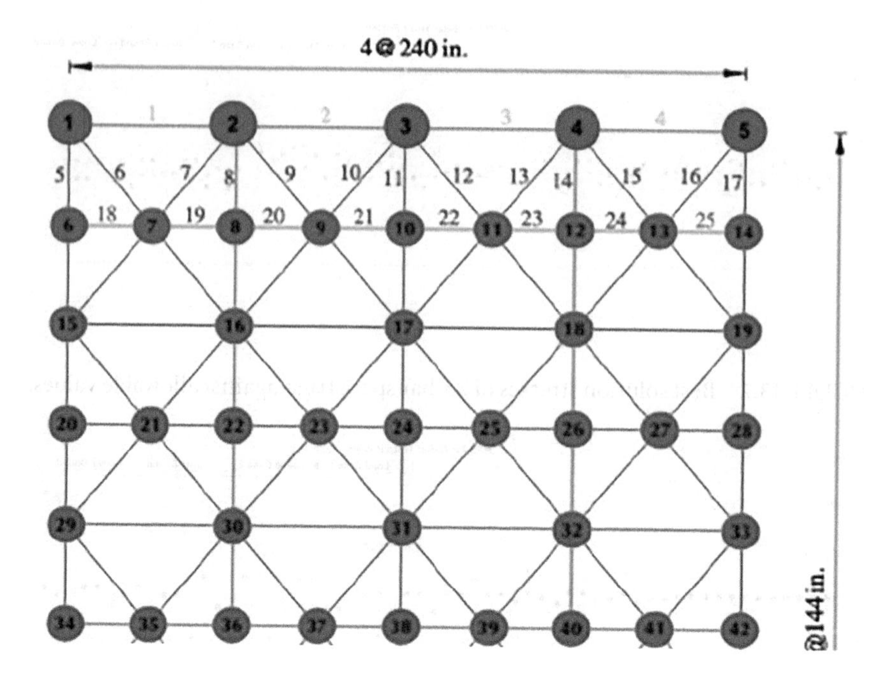

FIGURE 13.9 The 200-bar planar truss.

TABLE 13.7 Element Grouping for 200-bar Planar Truss

Group	Member Number	Group	Member Number
1	1,2,3,4	16	82,83,85,88,89,91,92,103,104,106,107, 109,110,112,113
2	5,8,11,14,17	17	115,116,117,118
3	19,20,21,22,23,24	18	119,122,125,128,131
4	18,25,56,63,94,101,132,139, 170,177	19	133,134,135,136,137,138
5	26,29,32,35,38	20	140,143,146,149,152
6	6,7,9,10,12,13,15,16,27,28,30, 31,33,34,36,37	21	120,121,123,124,126,127,129,130,141, 142,144,145,147,148,150,151
7	39,40,41,42	22	153,154,155,156
8	43,46,49,52,55	23	157,160,163,166,169
9	57,58,59,60,61,62	24	171,172,173,174,175,176
10	64,67,70,73,76	25	178,181,184,187,190
11	44,45,47,48,50,51,53,54,65,66, 68,69,71,72,74,75	26	158,159,161,162,164,165,167,168,179, 180,182,183,185,186,188,189
12	77,78,79,80	27	191,192,193,194
13	81,84,87,90,93	28	195,197,198,200
14	95,96,97,98,99,100	29	196,199
15	102,105,108,111,114		

TABLE 13.8 Comparison of MFO-Optimized 200-bar Planner Truss with other Algorithms

Group Number	GWO	PSO [20] [21]	PSOPC [20]	Present Work MFO
1	0.802	1.336	0.759	0.100
2	2.403	2.753	0.903	0.986
3	4.341	0.592	1.1	0.260
4	5.697	0.526	0.995	0.100
5	3.954	5.028	2.135	2.097
6	0.595	0.495	0.419	0.315
7	5.608	1.751	1.004	0.100
8	9.195	3.373	2.8052	3.126
9	4.513	0.206	1.034	0.100
10	4.601	4.304	3.784	4.124
11	0.555	0.708	0.526	0.417
12	18.751	0.121	0.43	0.100
13	5.994	6.647	5.268	5.332
14	0.100	0.100	0.968	0.901
15	8.156	6.924	6.047	6.335
16	0.271	0.810	0.782	0.807
17	11.152	0.194	0.592	0.100
18	7.126	7.980	8.185	8.193
19	4.465	0.911	1.036	0.100
20	9.164	10.726	9.206	9.198
21	2.762	1.054	1.477	0.808
22	0.554	0.281	1.833	2.265
23	16.164	15.000	10.611	12.267
24	0.497	0.131	0.985	0.100
25	16.225	15.000	12.509	13.288
26	1.004	0.947	1.975	2.180
27	3.610	7.889	4.514	4.053
28	8.368	15.000	9.8	8.329
29	15.562	13.880	14.531	20.000
Best weight (lb)	44081.4	33137.45	28537.8	27388.98
Average weight (lb)	N/A	34561.67	N/A	34810.91
Standard deviation	N/A	991.92	N/A	3732.10
Number of analyses	150,000	13,200	150,000	23,760

48, 57, 62, and 71; (b) 10 Kips (44.5 kN) acting in the negative y-direction at nodes 1, 2, 3, 4, 5, 6, 8, 10, 12, 14, 15, 16, 17, 18, 19, 20, 22, 24,..., 71, 72, 73, 74, and 75; and (c) conditions 1 and 2 acting together. All members have a minimum cross-sectional size of 0.1 in^2 (0.6452 cm^2) and a maximum cross-sectional area of 20 in^2 (129.03 cm^2).

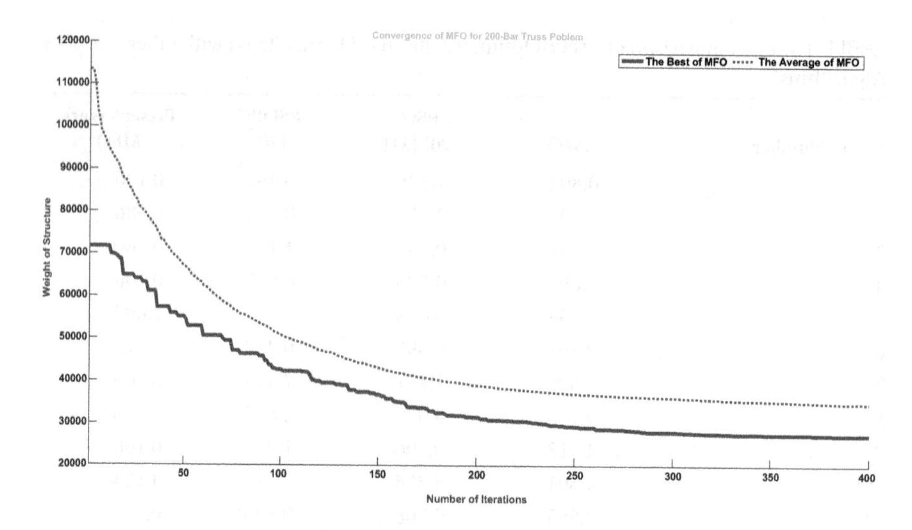

FIGURE 13.10 The best convergence curve of MFO for 200-bar planar truss.

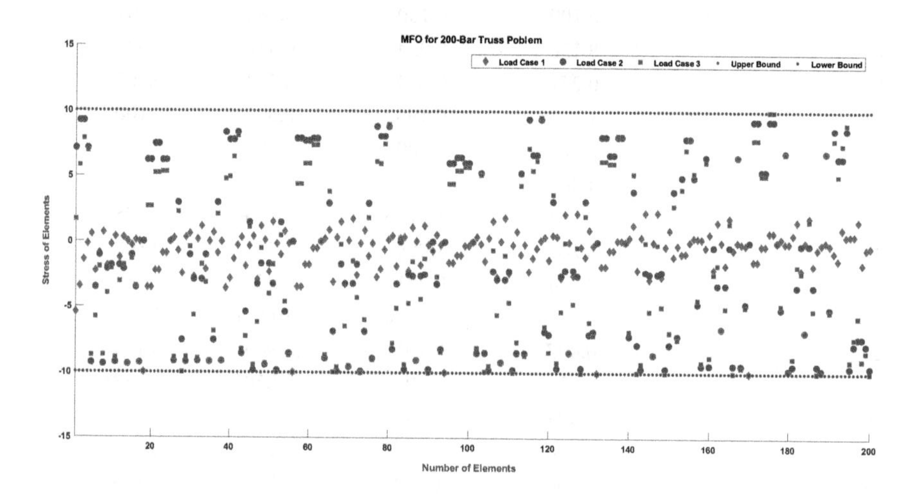

FIGURE 13.11 Best solution stresses of 200-bar planar truss against allowable values.

In comparison with some other research, minimum weight and associated cross-sections of MFO are acquired, and the findings are shown in Table 13.8. The algorithms are PSO [20], GWO [21], and PSOPC [20]. The best convergence curves for these methods are shown in Figure 13.10. It is evident that 23760 analyses are adequate for the new method. In Figure 13.11, it can be shown that the restrictions are met for MFO, with three load situations taken into account.

REFERENCES

1. D. Whitley, "A genetic algorithm tutorial," *Stat. Comput.*, vol. 4, no. 2, pp. 65–85, 1994, doi: 10.1007/BF00175354.
2. K. V Price, "Differential evolution," in *Handbook of Optimization*. Springer, Berlin, Germany, 2013, pp. 187–214.
3. M. Dorigo, M. Birattari, and T. Stutzle, "Ant colony optimization," *IEEE Comput. Intell. Mag.*, vol. 1, no. 4, pp. 28–39, 2006.
4. J. Kennedy and R. Eberhart, "Particle swarm optimization," *in Proceedings of ICNN'95-International Conference on Neural Networks*, 1995, vol. 4, pp. 1942–1948.
5. A. Kaveh, A. D. Eslamlou, and N. Khodadadi, "Dynamic water strider algorithm for optimal design of skeletal structures," *Period. Polytech. Civ. Eng.*, vol. 64, no. 3, pp. 904–916, 2020.
6. A. H. Gandomi, X.-S. Yang, and A. H. Alavi, "Cuckoo search algorithm: A metaheuristic approach to solve structural optimization problems," *Eng. Comput.*, vol. 29, no. 1, pp. 17–35, 2013.
7. A. Kaveh, S. Talatahari, and N. Khodadadi, "Stochastic paint optimizer: Theory and application in civil engineering," *Eng. Comput.*, vol. 13, pp. 1–32, 2020.
8. H. Karami, M. V. Anaraki, S. Farzin, and S. Mirjalili, "Flow Direction Algorithm (FDA): A novel optimization approach for solving optimization problems," *Comput. Ind. Eng.*, vol. 156, p. 107224, 2021.
9. A. Kaveh, N. Khodadadi, B. F. Azar, and S. Talatahari, "Optimal design of large-scale frames with an advanced charged system search algorithm using box-shaped sections," *Eng. Comput.*, vol. 156, pp. 1–21, 2020.
10. A. Kaveh, S. Talatahari, and N. Khodadadi, "The hybrid invasive weed optimization-shuffled frog-leaping algorithm applied to optimal design of frame structures," *Period. Polytech. Civ. Eng.*, vol. 63, no. 3, pp. 882–897, 2019.
11. A. Kaveh, N. Khodadadi, and S. Talatahari, "A comparative study for the optimal design of steel structures using CSS and acss algorithms," *Iran Univ. Sci. Technol.*, vol. 11, no. 1, pp. 31–54, 2021.
12. N. Khodadadi, M. Azizi, S. Talatahari, and P. Sareh, "Multi-Objective Crystal Structure Algorithm (MOCryStAl): Introduction and performance evaluation," *IEEE Access*, vol. 9, pp. 117795–117812, 2021.
13. C. V Camp, "Design of space trusses using Big Bang–Big Crunch optimization," *J. Struct. Eng.*, vol. 133, no. 7, pp. 999–1008, 2007.
14. L. Lamberti, "An efficient simulated annealing algorithm for design optimization of truss structures," *Comput. Struct.*, vol. 86, no. 19–20, pp. 1936–1953, 2008.
15. O. K. Erol and I. Eksin, "A new optimization method: big bang–big crunch," *Adv. Eng. Softw.*, vol. 37, no. 2, pp. 106–111, 2006.
16. F. K. J. Jawad, M. Mahmood, D. Wang, A.-A. Osama, and A.-J. Anas, "Heuristic dragonfly algorithm for optimal design of truss structures with discrete variables," *Structures*, vol. 29, pp. 843–862, 2021.

17. A. Kaveh, S. Talatahari, and N. Khodadadi, "Hybrid invasive weed optimization-shuffled frog-leaping algorithm for optimal design of truss structures," *Iran. J. Sci. Technol. Trans. Civ. Eng.*, vol. 44, pp. 1–16, 2019.
18. S. Mirjalili, "Moth-flame optimization algorithm: A novel nature-inspired heuristic paradigm," *Knowl. Based Syst.*, vol. 89, pp. 228–249, 2015.
19. A. Kaveh, B. F. Azar, and S. Talatahari, "Ant colony optimization for design of space trusses," *Int. J. Sp. Struct.*, vol. 23, no. 3, pp. 167–181, 2008.
20. K. S. Lee and Z. W. Geem, "A new structural optimization method based on the harmony search algorithm," *Comput. Struct.*, vol. 82, no. 9–10, pp. 781–798, 2004.
21. A. Kaveh and P. Zakian, "Improved GWO algorithm for optimal design of truss structures," *Eng. Comput.*, vol. 34, no. 4, pp. 685–707, 2018.
22. A. Kaveh and T. Bakhshpoori, "A new metaheuristic for continuous structural optimization: Water evaporation optimization," *Struct. Multidiscip. Optim.*, vol. 54, no. 1, pp. 23–43, 2016.
23. A. Kaveh and S. Talatahari, "Size optimization of space trusses using Big Bang–Big Crunch algorithm," *Comput. Struct.*, vol. 87, no. 17–18, pp. 1129–1140, 2009.
24. A. Kaveh and V. R. Mahdavi, "Colliding bodies optimization method for optimum design of truss structures with continuous variables," *Adv. Eng. Softw.*, vol. 70, pp. 1–12, 2014.
25. C. V Camp and B. J. Bichon, "Design of space trusses using ant colony optimization," *J. Struct. Eng.*, vol. 130, no. 5, pp. 741–751, 2004.
26. A. H. Gandomi, S. Talatahari, X. Yang, and S. Deb, "Design optimization of truss structures using cuckoo search algorithm," *Struct. Des. Tall Spec. Build.*, vol. 22, no. 17, pp. 1330–1349, 2013.
27. A. Kaveh and M. Khayatazad, "Ray optimization for size and shape optimization of truss structures," *Comput. Struct.*, vol. 117, pp. 82–94, 2013.
28. S. O. Degertekin and M. S. Hayalioglu, "Sizing truss structures using teaching-learning-based optimization," *Comput. Struct.*, vol. 119, pp. 177–188, 2013.

CHAPTER 14

Deep Feature Selection Using Moth-Flame Optimization for Facial Expression Recognition from Thermal Images

Ankan Bhattacharyya

University of Kentucky

Soumyajit Saha

University of Utah

Shibaprasad Sen

University of Engineering and Management

Seyedali Mirjalili

Torrens University Australia

Yonsei University, Seoul,

Ram Sarkar

Jadavpur University

DOI: 10.1201/9781003205326-18

CONTENTS

14.1 Introduction 282
14.2 Literature Survey 285
14.3 Database Used 288
14.4 Methods and Materials 289
14.4.1 CNN Model Used 291
14.4.1.1 Residual Unit 292
14.4.1.2 Transformation Unit 294
14.4.2 Feature Extraction 295
14.4.3 Feature Selection Using MFO 296
14.5 Results and Discussion 302
14.6 Conclusion 307
References 308

14.1 INTRODUCTION

Facial expression can be defined as the facial changes in response to an individual's inner emotional state, intents, or social communication. In most cases, human beings use facial expressions for non-verbal communications among themselves [1]. The research on facial expression has been started long back as early as the 1970s when some guidelines for facial expression recognition (FER) were published by Ekman and Friesen [2] and Ekman et al. [3] as a part of psychological analysis. These guidelines help to identify the facial expression of people depending on the muscle movements and other parameters. However, the guidelines were based on eye estimations, and hence, there was no concrete proof of these guidelines.

With the rapid advancements of Machine Learning (ML) and Deep Learning (DL) approaches, many complex research problems related to human–computer interactions are being solved in recent years. One such research problem considered here is FER, which has various real-life applications like mental diseases diagnosis and human social/physiological interaction detection.

However, for developing and training these stochastic models, a good number of images are necessary. Generally, the aperture and focus of cameras and lens are altered in such a way that a perfect image can be captured in any lighting conditions, so that the image is perceived by our eyes, thus extracting information from the image. These photographs are mostly captured with the help of light in the visible spectrum especially for two major reasons, the first reason is the natural availability of physical light and the second reason is the photosensors of the commercially available cameras generally sense light in the visible spectrum. For FER, lips, chins, and eyebrows of the face need to be prominent in the photograph and should have maximum similarity to the original condition. Various works on FER from human facial images can be found in Refs. [4–8], and the achieved results ensure that the methods are useful for such purposes. However, the lighting condition for such photographs is a major concern. If the required amount of light is not present or captured in an improper illuminated condition, then the photo will either be overexposed or underexposed, thereby leading to information loss. If due to the presence of too much light a photograph is overexposed, then the details of eyebrows and other texture of the face become indistinguishable while detecting the expression. Similarly, due to lesser light (i.e., underexposure), the minute details become indistinguishable.

A solution to such conditions is to use a technology that captures pictures with minute details irrespective of the lighting conditions of the surrounding. One such way is to capture the thermal distribution of the facial muscles by using Infrared Thermal Imaging (IRTI) cameras. As thermal distribution is captured, these images have a widespread applications in different fields. For example, IR images can be used to detect if any human/object is hidden somewhere or not and thus helping in security checks. Thermal images are also used to detect fever or any swollen part of the human body that is resulting in tremendous pain like muscle cramps.

In the present work, thermal images of the face taken from the IR Database have been considered to develop a computer-based automated FER system. A total of eight expressions are present in the said database, namely, Fear, Anger, Contempt, Disgust, Happy, Neutral, Sad, and Surprise.

Researchers from different corners of the world have attempted to recognize facial expressions using ML-based [9–11] and DL-based [12–15] approaches, where they have considered IR facial images as the input to their models. However, it has been noticed that the performance of DL-based approaches is better than ML-based approaches due to the automatic extraction of powerful features from inputs by DL models [12]. Hence, in the current study, we have considered to apply a Convolutional Neural Network (CNN)-based DL model for the extraction of deep features from IR images in order to develop a FER system [16]. It is to be noted that DL-based models are non-linear and such networks come with better flexibility. However, a flip-side of this flexibility is that these models generally learn through a stochastic training method, and due to this, they become very sensitive to the training data. Also, they may find a varied set of weights every time the models get trained, and hence, they generate varied predictions about the input samples.

To this end, in this chapter, we have saved the models at some intervals and extracted features from each of those saved models. Finally, the features from these models are concatenated for the prediction purpose. This concept is known as *snapshot ensemble*. Though by taking the snapshots at different intervals we can incorporate a varied set of weights that help in better prediction of the inputs, the length of the feature vector gets increased resulting in computational overhead. Also, the concatenated feature vector may contain many redundant features as they are extracted from the same input images. Keeping this fact in mind, in the present work, we have introduced an efficient feature reduction method using a nature-inspired metaheuristic called Moth-Flame Optimization (MFO) algorithm. This method tries to eliminate the redundant features generated by the CNN model using the concept of snapshot ensembling. We have chosen the MFO algorithm in the present work because of its successful applications in different domains and some of these are discussed in the next section.

Highlight points of the present work are:

a. Use of a CNN-based framework for FER using IR images.
b. Application of cosine annealing to capture different snapshot models and extraction of the features from the snapshots.
c. Application of the MFO algorithm to reduce the dimension of the feature vector obtained from the ensemble of snapshot models.
d. Experimentation of the proposed model on a publicly available database and comparison of the outcome of the same with state-of-art methodologies.

The rest of the chapter is organized as follows. Section 14.2 describes the recent works related to the current research problem, i.e., FER. Section 14.3 provides the details of the database used in the present experiment. An elaborative discussion of the proposed work has been made in Section 14.4. The observed outcomes along with the analysis of the results have been discussed in Section 14.5. Finally, Section 14.6 concludes the overall work and outlines some possible future directions of this research work.

14.2 LITERATURE SURVEY

There are many works available in the literature for FER. The majority of these works have been on images captured with a camera in the visible light source. This is because the cameras capturing images in visible light are easily available and these cameras are quite affordable. There have been various conventional approaches that focus on the detection of the face region extracting geometric features and appearance features on a target face [17]. In some work, geometric features such as angles and Euclidean distances have been used for the extraction of features from relevant portions of the face [18,19]. The appearance-based features are extracted from the global face region or any other face regions having relevant information [20]. Local Binary Pattern (LBP) [21], Histogram of Oriented Gradients (HOG) [22], Gabor-based filters [23], and many other feature extraction techniques are available in the literature that are fed to different ML-based classifiers for FER. In Ref. [24], the authors have used different classifiers like Multilayer Perceptron (MLP), Decision Tree, and also a CNN model to determine the facial expressions. Chu et al. [25] have proposed an algorithm called multilevel facial Action Unit (AU) for FER which combines spatial and temporal features. In Ref. [26], the authors have introduced a system that uses CNN visualization for understanding the learning of the model using various FER datasets and it helps in enhancing the emotion detection capability. A hybrid CNN-RNN (Recursive Neural Network) framework has been proposed in Ref. [27] for propagating information over a sequence using a continuously valued hidden-layer representation. The authors have also proved that their proposed hybrid CNN-RNN architecture for a facial expression analysis can outperform a previously applied CNN approach using temporal averaging for aggregation. In Ref. [28], a CNN-based DL model has been proposed for the purpose of FER. The model consists of several different structured subnets and each of those is a compact CNN model trained separately. Spatial representations have been extracted using CNN and the temporal dependencies have been modeled by stacking long short-term memories (LSTMs) on top of these representations.

Not only the above-mentioned works, but also most state-of-the-art approaches focus on visible spectrum information for FER. However, this job becomes tedious because of the significant variance of emotions of individuals. Moreover, visible images are susceptible to variation in illumination. Low lighting, pose variation, and disguise have a major impact on the appearance of images and textual information. Though using advanced techniques, some portions can be solved but still the performance is not often satisfactory when compared to the human performance. To overcome these shortcomings, thermal images are preferred to visible images. The reasons behind this include thermal images (a) are less sensitive to different weather conditions, (b) have consistent thermal signatures, and (c) have a temperature distribution formed by the face vein branches.

IR cameras are not widely used as compared to visible light cameras because the former are expensive. However, there are some works that focus on FER using IR or thermal images. The authors in Ref. [29] have proposed a customized CNN called TERNet for the recognition of facial emotions. Their model adopts the features obtained via transfer learning from the VGG-Face CNN model. In Ref. [30], the authors have introduced a high-resolution thermal IR face database for the research community having extensive manual annotations. The authors in Ref. [31] have proposed a pipeline model for thermal face analysis that contains a HOG-SVM-based face detector and different landmark detection methods implemented using feature-based active appearance models, deep alignment networks, and a deep shape regression network. Yoshitomi et al. [32] have proposed a technique for FER which is based on the detection of temperature distribution of the face using IR rays. Jiang et al. [33] have proposed a method for FER using mathematical morphology, drawing, and analysis of the whole geometrical characteristics with some geometrical characteristics of the interesting area of the IRTI. Shen et al. [34] have designed a spontaneous FER method using IR thermal videos. The horizontal and vertical temperature difference sequences of different facial sub-regions have been used for feature extraction which is followed by the selection of a feature subset based on their F-values. Lastly, the Adaboost algorithm with K-Nearest Neighbors (KNN) classifier has been used for the classification purpose.

Analyzing different ML- and DL-based approaches used for FER, it has been noticed that the general trend of the researchers is to produce various high-dimensional feature vectors to achieve better recognition accuracy. For ML-based approaches, different handcrafted features are concatenated for this purpose. Whereas, more future maps are used in DL-based techniques. However, if a high-dimensional feature set is used, the time required to train the model also increases. Moreover, there can be irrelevant and redundant features that may lead to the over fitting of the model. To encounter these problems, many researchers have relied on different feature selection (FS) techniques [35–43] in the domain of ML-based FER approaches, where images are captured through normal cameras. The authors in Ref. [44] have applied Genetic Algorithm (GA) on the features produced by the Gabor filter for FER. The combination of fixed filters and trainable non-linear 2D filters based on the biologic mechanism of shunting inhibition has been proposed in Ref. [45] and FS has been applied using Mutual Information (MI) and class separation scores. In Ref. [46], the authors have initially combined the extracted features from facial expression images using uLBP (Uniform Local Binary Pattern), hvnLBP (Horizontal–Vertical Neighborhood Local Binary Pattern), Gabor filters, HOG, and PHOG (Pyramidal HOG) for FER. Then they have applied a FS method, called LHCMA (late-hill-climbing-based memetic algorithm), on the combined feature sets. In Ref. [47], the authors have proposed a FER system based on hvnLBP features with GA and particle swarm optimization (PSO)-based FS technique. The authors in Ref. [48] have proposed a FS technique for FER, called SFHSA (supervised filter Harmony Search algorithm), which is based on cosine similarity and mRMR (minimal-redundancy maximal-relevance).

MFO algorithm is a metaheuristic algorithm that has proved its effectiveness in solving different complex optimization problems in several domains [49]. The authors in Ref. [50] have claimed to get better performance by MFO algorithm in comparison with PSO and GA on UCI datasets. In Ref. [51], the authors have proposed an enhanced MFO, called EBMFO (enhanced Binary Moth-flame Optimization) with ADASYN (Adaptive synthetic sampling), to predict software faults. They have applied the wrapper-based FS technique and converted the MFO into its binary version. It has been shown that EBMFO is successfully able to augment the overall performance of classifiers. Abu Khurmaa et al. have used the MFO algorithm on 23 medical datasets collected from UCI, Keel, Kaggle data repositories [52] and also reported the superiority of this technique over other FS approaches. MFO has also shown its supremacy in the handwriting recognition domain. In Ref. [53], MFO-based FS approach has been proposed for Arabic handwritten letter recognition which has provided a satisfactory classification accuracy. MFO has also made a significant contribution in the FER domain too. A FER system on images captured through a normal camera has been proposed in Ref. [54] by incorporating a variant of the LFA (Levy-flight firefly algorithm) with MFO. Mehne and Mirjalili have solved the problem of finding finite set of coefficients based on approximating the control input function as a series of given base functions with unknown coefficients [55]. Shehab et al. [56] have made a comprehensive review on MFO algorithm applied on various optimization problems in different fields like image processing, medical applications, and power and energy systems. Muduli et al. [57] have used MFO algorithm for the early detection of breast cancer. Their proposed model uses wavelet transform for the extraction of features from the mammogram images. Then they have reduced the feature dimension with the help of fusion of PCA and LDA methods. Finally, combination of an extreme learning machine and MFO algorithm is considered for the classification purpose. The authors have used MFO for the optimization of hidden node parameters of the extreme learning machine. To the best of our knowledge, till date, MFO has not been used for FS in FER using thermal images. This has been our prime motivation to apply a MFO-based deep FS method in the domain of thermal image-based FER.

14.3 DATABASE USED

For the current work, we have used the IR database published by Kopaczka et al. [58]. The images of this database were recorded using an Infratec HD820 high-resolution thermal IR camera with a 1,024×768 pixel-sized microbolometer sensor having a thermal resolution of 0.03 K at 30°C temperature and equipped with a 30-mm f/1.0 prime lens. Subjects were filmed while sitting at a distance of 0.9 m from the camera, resulting in a spatial resolution of the face of ~0.5 mm per pixel. A thermally neutral backdrop was used for the recordings to minimize background variation. To build the database, video recordings of the subjects acquired with a frame rate of 30 frames/s were manually screened and images were extracted. The database contains a total of 1,782 sample images with eight classes of expressions, namely, Fear, Anger, Contempt, Disgust, Happy, Neutral, Sad, and Surprise. The number of sample images belonging to each class has been shown in Table 14.1. A few sample images from this database have been shown in Figure 14.1.

TABLE 14.1 The Number of Samples for each of the Eight Facial Expressions in the IR Database

Facial Expression	**Number of Samples**
Angry	231
Contempt	189
Disgust	225
Fear	210
Happy	234
Neutral	231
Sad	231
Surprised	231
Total	**1,782**

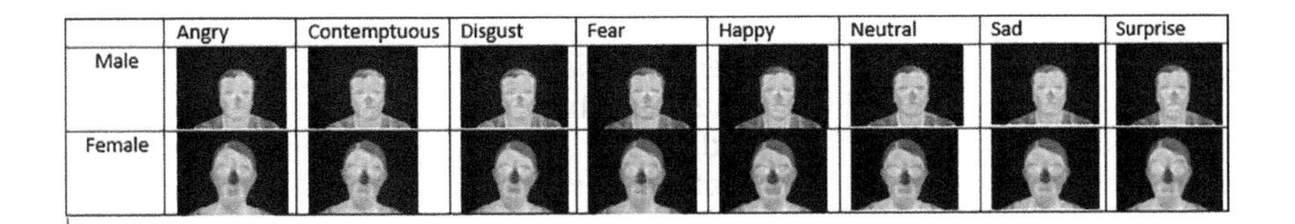

FIGURE 14.1 A few sample thermal images representing different facial expressions for both male and female candidates participated in the data collection of the IR database.

In the proposed work, as a pre-processing step, we have converted the input images into grayscale images and reshaped them into a uniform size of 200×200 pixels.

14.4 METHODS AND MATERIALS

In the present work, the database containing thermal facial images of humans [31] has been used for the development of FER system. As stated earlier that DL-based architectures provide better classification performance than ML-based models, we have applied a CNN model in the initial stage of the proposed approach. Then we have applied MFO-based FS method on the features generated by the CNN model using the concept of snapshot ensembling. The workflow of the proposed model has been shown in Figure 14.2.

In the present work, a CNN model has been trained on the said dataset in such a way that the learning rates are varied according to a sinusoidal function. Over time and epochs, as the learning rates get changed, we have saved the models after every 100th epoch during training. These saved models are known as the snapshot models. For the present experiment, we have generated 10 such snapshots. The features are extracted from the final layer of each snapshot model that produces a feature vector of length 256. The features from these ten snapshot models are combined to form a final feature vector of length 2,560 (10×256). It is understandable that the dimension of the obtained feature vector becomes very high and it results in the consumption of higher time and space complexities. Hence, the combined feature vector is optimized using MFO-based FS method, which selects 1,134 features from 2,560-attributed feature vector by discarding the redundant features. Such features are then fed to the Support Vector Machine (SVM) classifier with a polynomial kernel for the classification of the facial expressions. The detailed architecture of the CNN model and the MFO algorithm have been provided in the following subsections.

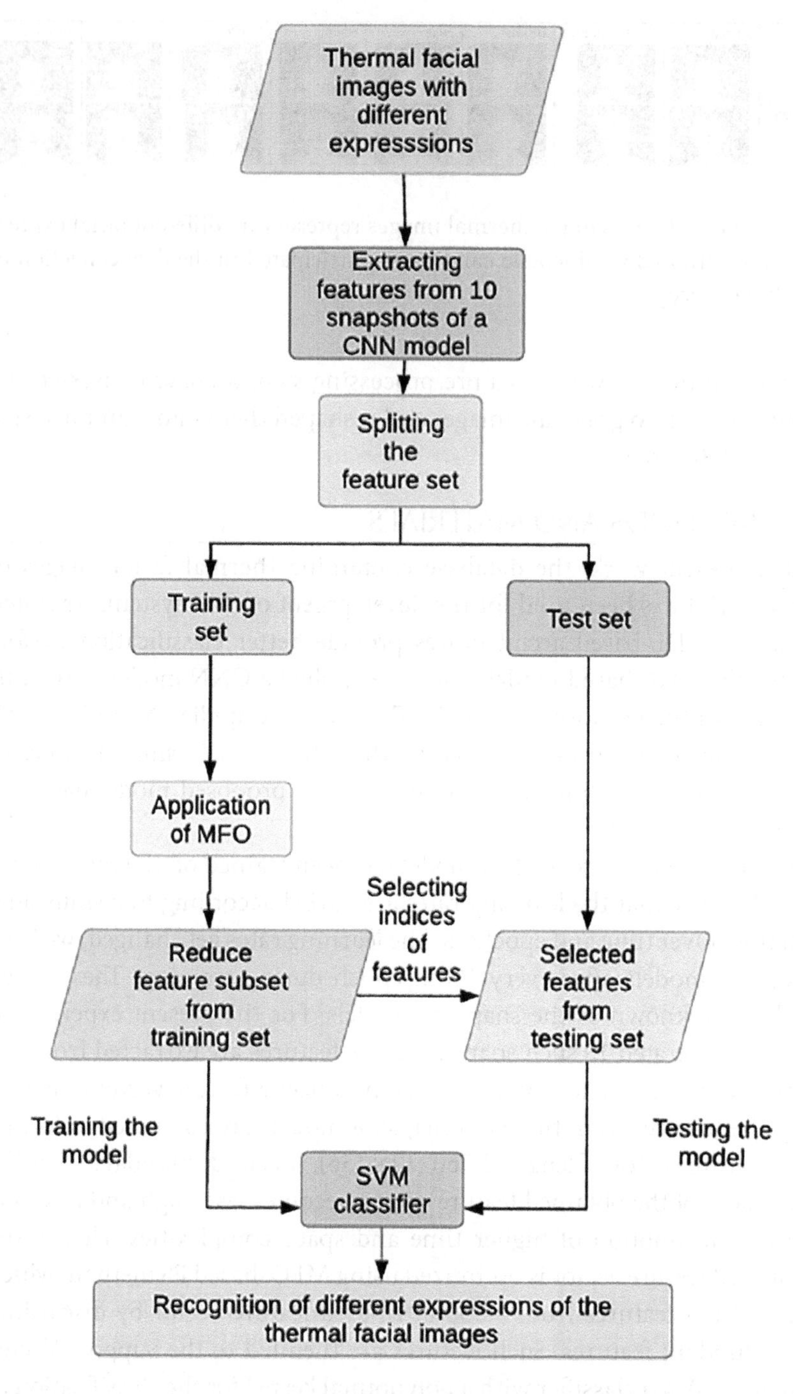

FIGURE 14.2 Schematic diagram of the proposed deep FS method using MFO for the classification of thermal facial expressions.

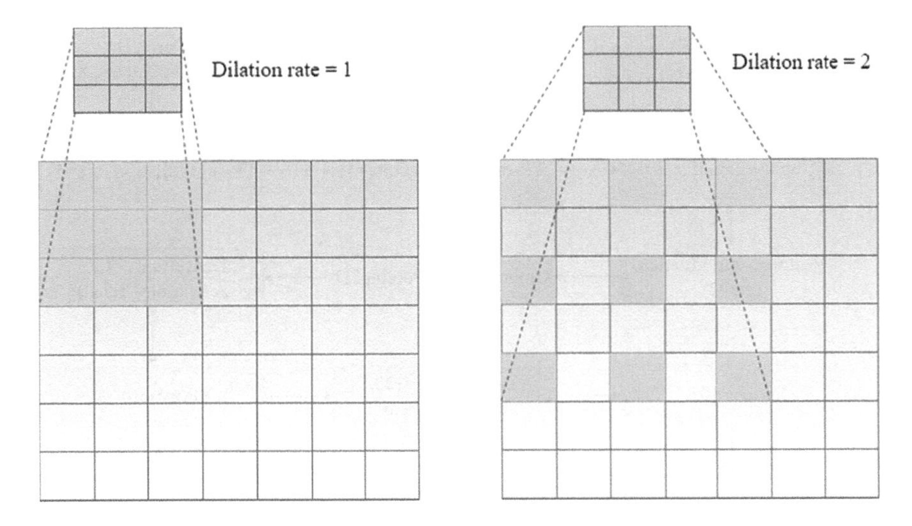

FIGURE 14.3 An illustration of convolution using dilation rates as 1 and 2.

14.4.1 CNN Model Used

The CNN architecture considered here uses depth-wise separable convolution blocks. It is implemented using point-wise convolution operation over depth-wise convolution. A point-wise convolution is a 1×1 kernel, convolving over each point. The kernel has a depth same as the number of channels of the input. Depth-wise convolution is a type of convolution, where a single convolution kernel is applied for each input channel separately. Here, the depth-wise convolution is performed using varying dilation rates. The dilation rates of convolution are like the number of pixels it should skip to convolve with the image. Generally, the dilation rate remains 1. Figure 14.3 shows the process of convolution for dilation rates 1 and 2.

The architecture of the CNN model used here consists of two units, the transformation unit and the residual unit. The architecture of the same has been shown in Figure 14.4.

The residual and transformation units are grouped into a single unit called TR-unit. The TR-units are stacked in the architecture with different dilation rates. The dilation rate of a TR-unit remains the same. Figure 14.5 highlights the diagram of a TR-block. The model consists of three such TR-blocks and a residual block. The output of the residual block is passed through a Global Average Pooling 2D (GAP2D) layer. GAP2D layer outputs a vector of dimension (batch size, 1, 1, d) if the input is of dimension

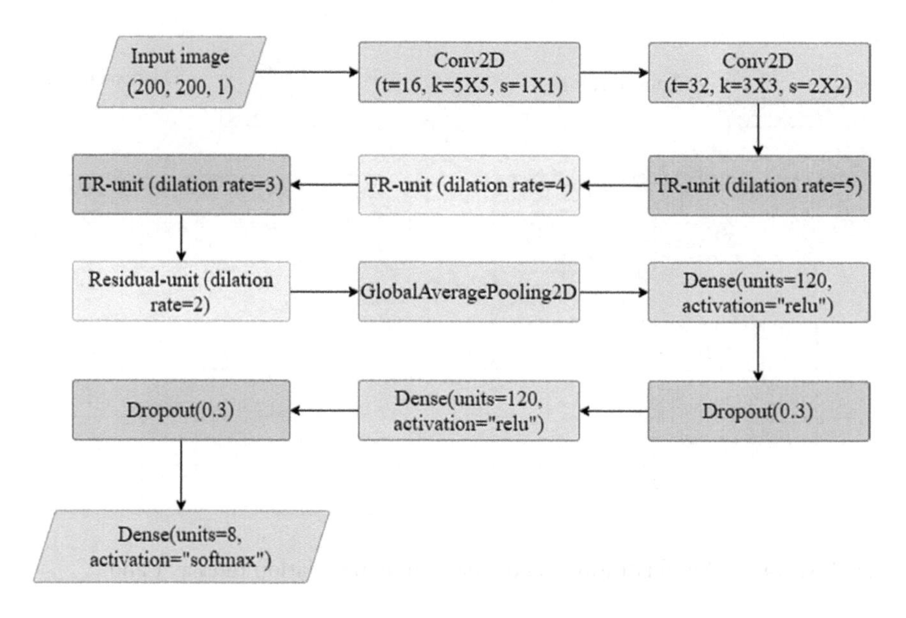

FIGURE 14.4 The block diagram of the CNN architecture used in the present work.

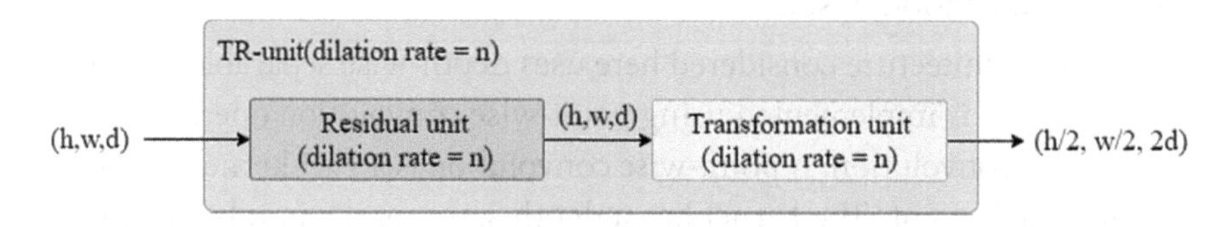

FIGURE 14.5 A block diagram illustrating a TR-unit with dilation rate as "n".

(batch size, *h*, *w*, *d*), where h is the height of the matrix, *w* is the width, and *d* is the depth.

TR-block shown in Figure 14.5 produces a dimension of (*h*/2, *w*/2, 2*d*) when the input is of dimension (*h*, *w*, *d*), where *h* is the height of the matrix, *w* is the width, and *d* represents the depth. The details of the transformation and the residual units have been discussed in the following subsections.

14.4.1.1 Residual Unit

The residual unit uses the concept of residual learning. The purpose of residual learning is to extract features as if the features are produced from a deep network. However, using the residual learning, the apparent depth of the network can be decreased. Figure 14.6 provides the block diagram of

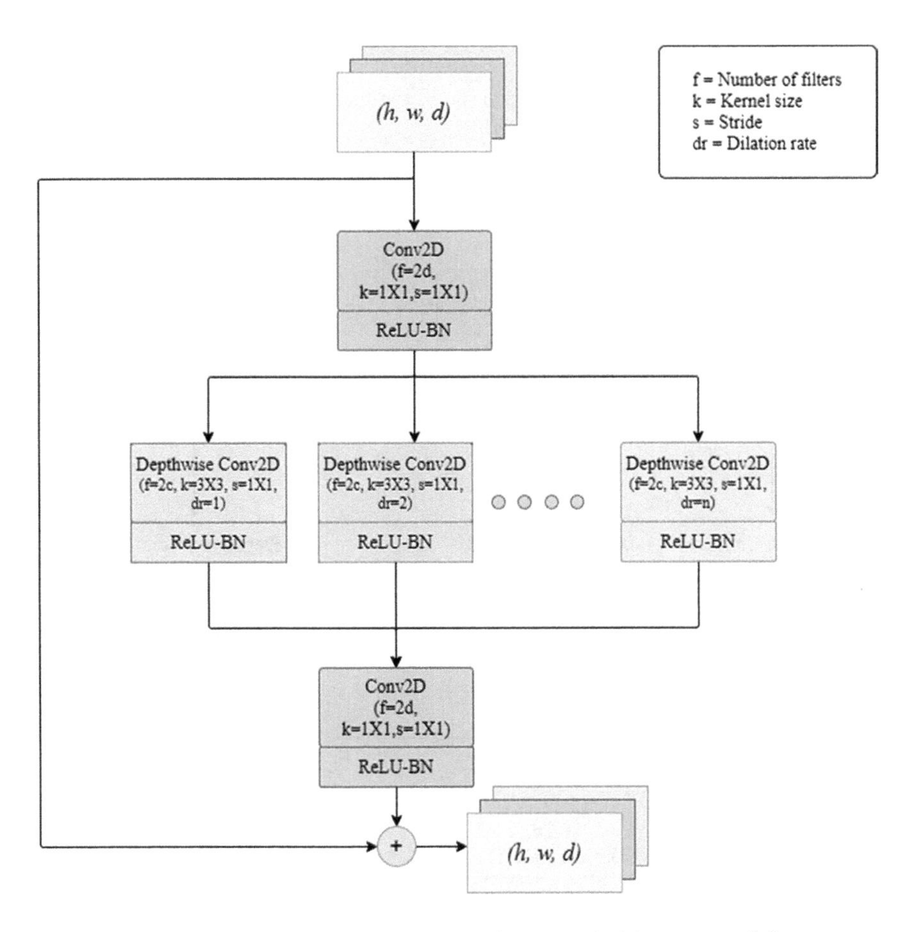

FIGURE 14.6 An illustration of the residual unit with dilation rate "n".

the residual network having a dilation rate of "n". The input to the residual unit is a matrix of dimension ($h \times w \times d$). The input is passed through a point-wise convolution layer termed as Conv2D which has a kernel of size 1×1 and a stride of 1. The filter size of the kernel is 2 times the depth of the input. The output of the point-wise convolution is passed through a Rectified Linear Unit (ReLU) layer and a batch normalization layer. The output passes through the "n" number of depth-wise convolution in parallel. The dilation rates of these "n" depth-wise convolution blocks start from 1 and continue to "n", as shown in Figure 14.6. The outputs of these depth-wise convolution blocks are added and passed to another point-wise convolution block, Conv2D. This Conv2D also has the same configuration as the first one. The output from this Conv2D is added with the input matrix

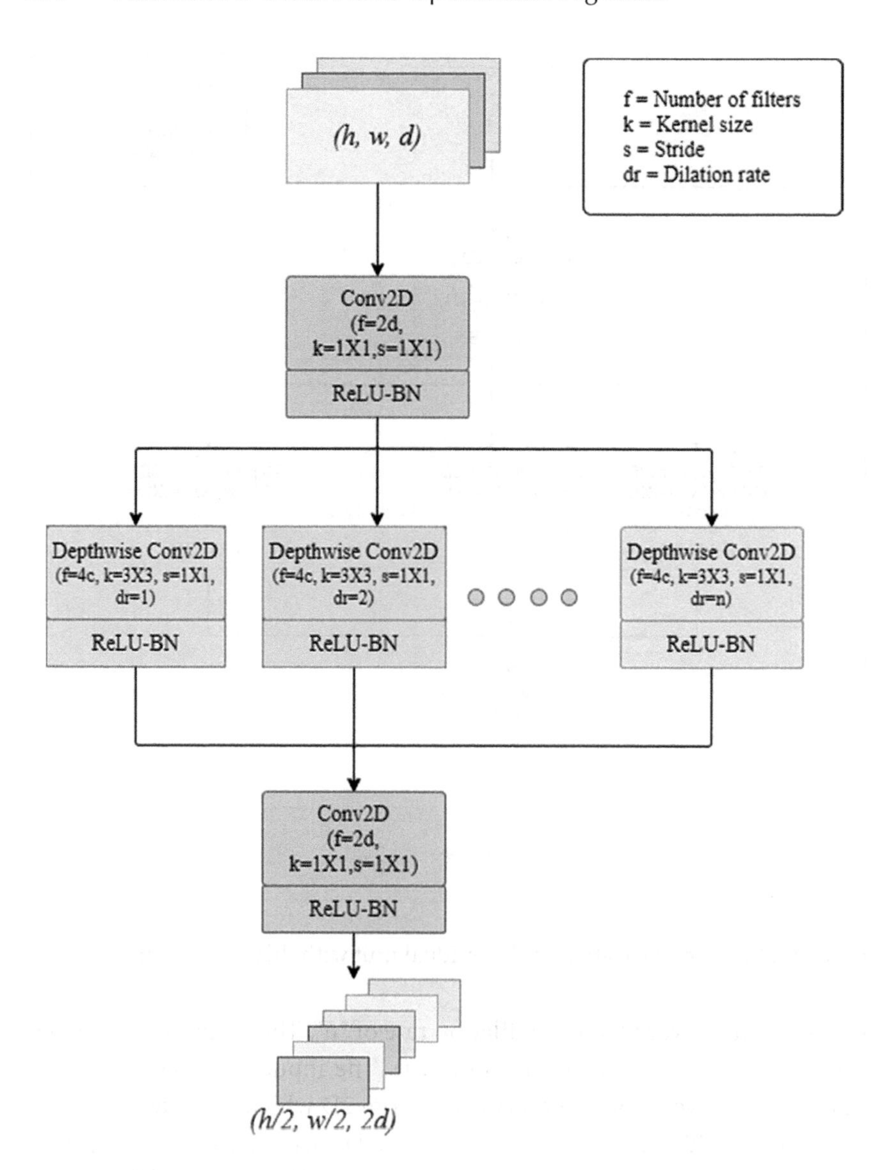

FIGURE 14.7 An illustration of transformation unit with dilation rate "n".

of dimension $(h \times w \times d)$. So, the output dimension remains the same as the input dimension. This allows deeper features using less apparent depth of the network. As a result, fewer parameters are needed.

14.4.1.2 Transformation Unit

The transformation unit is used as a substitution for the pooling layer. Figure 14.7 shows the block diagram of a transformation unit with a

TABLE 14.2 Details about Layer-Wise Feature Extraction

Layer	Input Dimension	Output Dimension
Input	200×200×1	200×200×1
Conv2D (t=16, k=5x5, s=1x1)+BN	200×200×1	200×200×16
Conv2D (t=32, k=3x3, s=2x2)+BN	200×200×16	100×100×32
TR-block	100×100×32	50×50×64
TR-block	50×50×64	25×25×128
TR-block	25×25×128	12×12×256
Residual-block	12×12×256	12×12×256
Global average pooling 2D	12×12×256	1×1×256

dilation rate of "n". The input to a transformation unit is the output coming from a residual unit. A point-wise convolution is executed on the input.

The point-wise convolution has a filter of size $4\times d$, where d is the depth of the input. The kernel and stride are of size 1×1. The output of the point-wise convolution is then passed through a ReLU layer and a batch normalization layer. This output is passed into "n" depth-wise convolution blocks in parallel. The dilation rate varies from 1 to "n", as shown in Figure 14.7. The output from these depth-wise convolution blocks is added and passed to another point-wise convolution. This point-wise convolution has the same properties as that of the first point-wise convolution of the transformation unit. The output of the transformation unit is $h/2\times w/2\times 2d$.

14.4.2 Feature Extraction

Features have been extracted from the GAP2D layer of the said CNN architecture. The details of layer-wise extracted features have been provided in Table 14.2.

In the current experiment, we have considered the image of size 200×200×1 as input. This input image is then passed through a convolution layer with kernel size 5×5, depth 16, and a stride of 1×1. The output of this layer is 200×200×16 as the depth of the kernel is 16 and the stride is 1. The output of the first convolution layer is then passed through another convolution layer. This layer has a depth of 32, stride of 2×2, and kernel size of 3×3. The output of this layer is 100×100×32. This output is passed through three TR-blocks which produce features of a dimension of ($h/2$, $w/2$, $2d$). Hence, after performing three such operations, the dimension of the output from TR-blocks becomes (12×12×256). This output is then passed through the residual block which keeps the dimension unchanged. So, the output from this block has the dimension (12×12×256). Finally, the GAP2D layer takes this as input and produces the dimension of final output as $(1\times1\times256)$.

This model is trained over 1,000 epochs. The learning rates are varied in cosine patterns. Snapshots are captured after every 100 epochs and thereby a total of 10 snapshot models are captured. Cosine annealing confirms that the training of the model reaches the local minima at a lower error rate. This in turn confirms a very high variance, thus preventing the model from overfitting or underfitting. From each of the ten snapshot models, the weights from the GAP2D layer are concatenated and have been used as the final feature vector of length 2,560 (i.e., 256×10).

14.4.3 Feature Selection Using MFO

MFO is a metaheuristic algorithm inspired by the mechanisms used by moths for navigation during the night. The navigation is achieved by a technique called "transverse orientation" [58]. Moths fly by maintaining a fixed angle with respect to the moon. This helps them in traveling long distances in a straight path [59]. The large distance from the moon ensures a straight path flight for moths. However, it can be observed that sometimes they tend to fly in a deadly spiral path around any artificial light generated by humans. This points to the limitation of the transverse orientation mechanism because these lights are more closer than the moon and moths try to maintain the same angle with the artificial light and end up moving spirally and slowly converging toward them [60]. The MFO algorithm uses this mechanism to find an optimal solution for an optimization problem [61].

The main candidates of the algorithm are moths and flames. Flames are considered to be the light sources that moths use to navigate during their flight. The moths' positions in the space serve as solutions to a given problem. According to the work reported in Ref. [49], moths can fly in 1D, 2D, 3D, or hyperdimensional space (of d dimensions) by updating their position vectors. Due to the population-based nature of the algorithm, n number of moths is considered to be in the population and their positions are represented using matrix M as shown in the following equation, where d represents dimension of space in which the moth travels:

$$M=\begin{pmatrix} m_{1,1} & m_{1,2} & \cdots & m_{1,d} \\ m_{2,1} & m_{2,2} & \cdots & m_{2,d} \\ \vdots & \vdots & \ddots & \vdots \\ m_{n,1} & m_{n,2} & \cdots & m_{n,d} \end{pmatrix} \tag{14.1}$$

Another matrix FM is considered for storing the fitness value of the moth present in the population. These fitness values are obtained through the fitness function shown in the following equation:

$$FM = \begin{bmatrix} FM_1 \\ FM_2 \\ \vdots \\ FM_n \end{bmatrix} \tag{14.2}$$

Along with the matrix M, another 2D matrix F is considered for storing the flames. M and F are both 2D matrices and used for storing population vectors of the moths, but the only difference is that F denotes the best population vectors obtained so far during the course of the algorithm and M contains the position of the search agents. F is defined using Equation (14.3). For the flames, their corresponding fitness values are stored in FF as shown in Equation (14.4).

$$F = \begin{pmatrix} f_{1,1} & f_{1,2} & \cdots & f_{1,d} \\ f_{2,1} & f_{2,2} & \cdots & f_{2,d} \\ \vdots & \vdots & \ddots & \vdots \\ f_{n,1} & f_{n,2} & \cdots & f_{n,d} \end{pmatrix} \tag{14.3}$$

$$FF = \begin{bmatrix} FF_1 \\ FF_2 \\ \vdots \\ FF_n \end{bmatrix} \tag{14.4}$$

As the moths follow a spiral path around the flame, it updates its position based on a logarithmic spiral function S as mentioned in Equation (14.5), where D_i denotes the distance of the ith moth for the jth flame, b is a constant for defining the shape of the logarithmic spiral, and t is a random number in the range of $[-1,1]$. The distance D is computed using Equation (14.6), where M_i and F_j indicate the ith moth and the jth flame, respectively.

$$S(M_i, F_j) = D_i \cdot e^{bt} \cdot \cos(2\pi t) + F_j \tag{14.5}$$

$$D_i = |F_j - M_i| \tag{14.6}$$

The update takes place with respect to the flames, and the parameter t determines the distance between the moth and the flame. In the work [29], it has been stated that it creates a hyper-ellipse space around the flame in all directions, and Equation (14.5) ensures that the update in position is within this space. For better exploration and exploitation of the search space, t is assumed to be a random number between $[r,1]$, where r (converging constant) linearly decreases from -2 to -1 as the algorithm iterates further. With the advancement of iteration, the flames get updated and they are sorted based on their fitness values. With the increase of the number of iterations, the number of flames decreases as mentioned in Equation (14.7), where l denotes the current iteration number, N indicates the maximum number of flames, and T denotes the maximum number of iterations. This criterion helps in better convergence to get a global optimal solution and the exploitation of the flames becomes more accurate. Figure 14.8 is a flowchart describing the steps involved in the MFO algorithm.

$$\text{No.of flames} = \text{Round}\left(N - l \cdot \frac{N-1}{T}\right) \tag{14.7}$$

In the present work, we have considered the high-dimensional feature vector (i.e., 2,560 dimensional) obtained from the snapshot ensembling-based CNN model applied on the IR facial images as input to the MFO algorithm. The values of the number of search agents or moths (n), maximum number of flames (N), and the total number of iterations (Max_iter) are set appropriately. We have represented the position of ith moth by $M_i = \{f_1, f_2, f_3, f_4, \ldots, f_d\}$, where $\{f_1, f_2, f_3, f_4, \ldots, f_d\}$ represents a subset of feature attributes and d denotes the dimension of the search space. For the current experiment, we have considered $n(=20)$ random subsets of features having dimensions ranging from 80% to 90% of the original feature set as initial moth and flame population in M and F, respectively. The fitness of the moth has been considered by the classification accuracy obtained by the SVM classifier.

As the algorithm iterates, the number of flames to be taken under consideration gets updated following Equation (14.7) and the fitness of n moths is determined and stored in FM. A condition is applied to

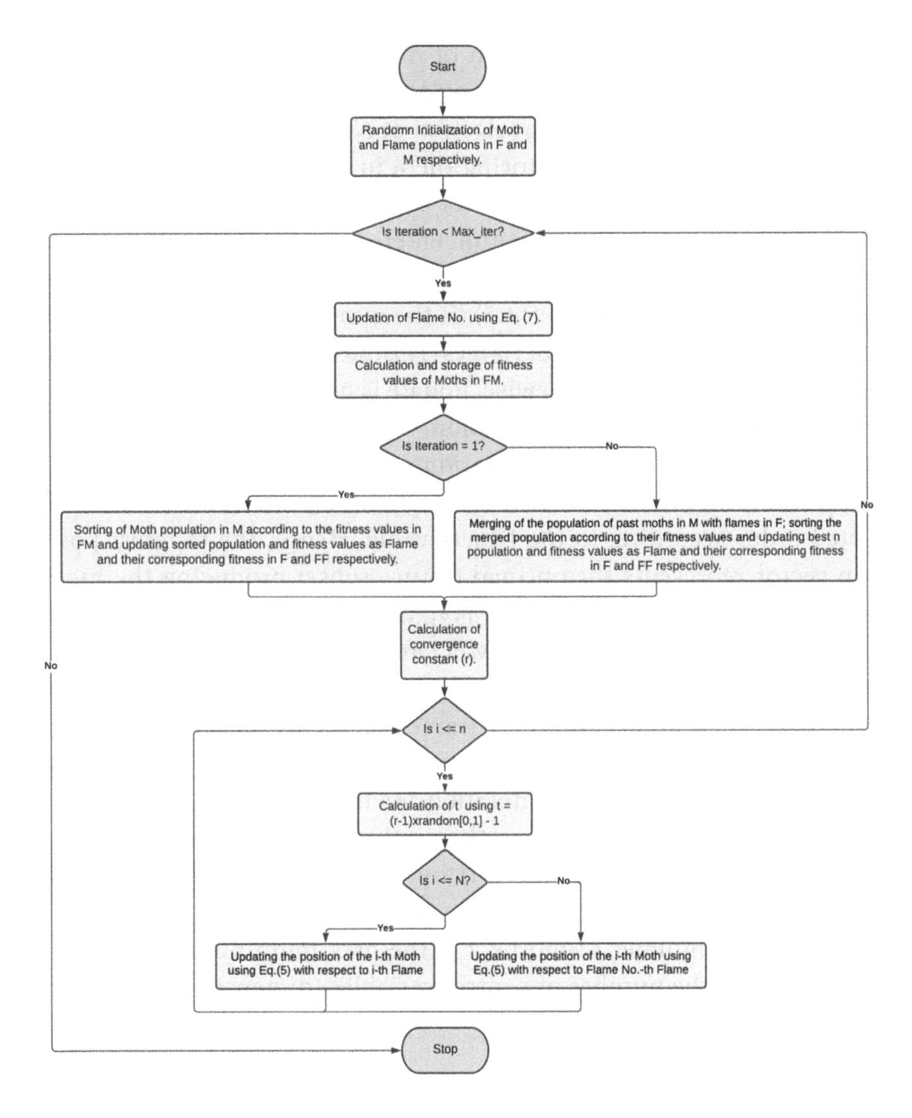

FIGURE 14.8 Flowchart describing the steps of the MFO algorithm.

verify whether the number of iterations is equal to 1. If so, the moth population is sorted with respect to the fitness of each position vector in the population placed in F and their corresponding fitness placed in FF, which denotes the flame population and its corresponding fitness values. Otherwise, the population of moths from the previous iteration is merged with the flame population, sorted according to their fitness values, followed by the selection of $N(=20)$ best positions from the merged population and storing them in F with corresponding fitness values in FF. The calculation of convergence constant (r) is then performed. After this, the algorithm enters into the updating phase. In this phase, the position vectors of n moths are updated with respect to a chosen flame by using Eq. (14.5). If the number of moths (i) is lesser than the maximum number of flames (N), the update is performed with respect to the ith flame; otherwise, update is performed with respect to No. of flames in the flame population. The execution of the algorithm continues until the number of iterations reaches $\text{Max\_iter}(=100)$. After completion, the algorithm returns the position vector of the flames in sorted order according to the fitness value. Hence, the topmost position vector represents the optimal feature subset producing the highest classification accuracy. In Ref. [49], the author has also highlighted the time complexity involved in MFO and has stated that it is the best metric for analyzing the runtime of an algorithm. The time complexity of the algorithm depends on parameters like the number of moths, number of variables, maximum number of iterations, and sorting mechanism of flames in each iteration. As the conventional MFO uses quick sort algorithm for sorting the moth population with respect to its fitness, the sorting has $O(n\log n)$ and $O(n^2)$ in best and worst cases, respectively. Hence, the overall time complexity is given by $O(\text{MFO})$, where n is the number of moths, t is the maximum number of iterations, and d is the number of control variables (dimension of the optimization problem).

$$
\begin{aligned}
O(\text{MFO}) &= O\Big(t\big(O(\text{Quick sort}) + O(\text{Position update})\big)\Big) \\
&= O(t\left(n^2 + n \times d\right) = O\left(tn^2 + tnd\right)
\end{aligned}
$$

Algorithm 14.1 provides the steps involved in the MFO algorithm used in the present experiment.

Algorithm 14.1: Moth-Flame Optimization Algorithm Used in the Current Experiment

INPUT: High-dimensional feature vector, Number of search agents or moths (n), Maximum number of flames (N), and Total number of iterations (Max_iter)
OUTPUT: Optimized feature vector

Randomly initialize M and F with population of n moths and flames having d -dimension
while (No. of iterations Max_iter)
{
Update the No.of flames to be used using Equation (14.7)
Calculate the fitness of n moths and store them in *FM*
if (No. of iterations = 1)
Sort the moths in M and corresponding fitness in *FM* according to their fitness value and place them in F and *FF*, respectively.
else
Merge the population of past moths in M with flames in F. Sort the merged population with respect to their fitness values. Select the best N positions from the sorted merged population and store them in F with their corresponding fitness values in *FF*
end if
Calculate the convergence constant (r)
for each ith moth, where $i \leq n$
{
Calculate t using $t = (r-1) \times r\text{and} - 1$, where r is a random number from uniform distribution in range $[0,1]$
if $(i \leq N)$
Update the position of the ith moth using Equation (14.5) with respect to the ith flame
else
Update the position of the ith moth using Equation (14.5) with respect to No.of flamesth flame
end if
}//**end for**
}//**end while**
Output the best search location and its fitness value

14.5 RESULTS AND DISCUSSION

In the present work, at first, we have designed a CNN model as mentioned earlier and captured ten snapshots of this model during training. The CNN model has been trained on IR database containing eight classes of facial expressions, namely, Angry, Sad, Afraid, Surprised, Happy, Neutral, Contemptuous, and Disgusted. We have extracted 2,560 features from the GAP2D layer of these ten snapshot models of each of the input images. Total images of the dataset have been split into training and test sets in a ratio of 7:3. To test the efficiency of the extracted feature set, the SVM classifier has been used. It is to be noted that we have obtained 97.20% accuracy when SVM uses the default linear kernel and the parameter coef0 value has been set to 0. We have tested the performance of the SVM classifier with different kernels (i.e., linear and polynomial), coef0 values, and the obtained results are shown in Table 14.3. From the entries of the table, it can be observed that the SVM produces the highest classification accuracy as 97.94% when the polynomial kernel has been used with the value of coef0 as 2.

In the present experiment, we have also applied the MFO algorithm on the 2,560-element feature vector (produced after concatenation of features from ten snapshots of the CNN model) to eliminate the redundant features. The MFO algorithm has been run for 100 iterations. At the end of the iteration, MFO algorithm produces a set of 20 solution vectors. These solution vectors are then tested with the SVM classifier keeping the configuration as

TABLE 14.3 Without FS Accuracy of FER after Tuning the Kernel and Coef0 Parameter of the SVM Classifier

Kernel	Coef0	Classification Accuracy (%)
Linear	0	97.20
	1	97.01
	2	97.57
	3	96.64
Polynomial	0	96.82
	1	97.57
	2	**97.94**
	3	96.64

Here a 2,560-element feature vector obtained from snapshot ensembling-based CNN model which is directly fed to the SVM classifier.
Note: Bold indicates best results.

mentioned earlier (i.e., polynomial kernel and coef0=2). The effectiveness of the produced optimized solution sets has been shown in Table 14.4.

After concatenating the features obtained from ten snapshots of the CNN model, 97.94% recognition accuracy has been achieved for 2,560-attributed feature vector. However, after the application of the MFO algorithm-based FS technique, we have considered a pool of 20 subsets. Among these 20 subsets, the subset having 1,134 attributes produces the highest classification accuracy of 99.63%. The number of features of these 20 subsets and corresponding results are reported in Table 14.4. Analyzing the data of table, it has been observed that for the 1,134-attributed feature subset, the SVM classifier lowers the misclassification rate (0.37%) by producing 99.63%

TABLE 14.4 Achieved FER Accuracy When the Optimized Feature Set Obtained through the MFO Algorithm-Based FS Method Is Fed to the SVM Classifier

Reduced Feature Set Produced by MFO	Number of Attributes	Classification Accuracy by SVM (%)
Subset 1	1,112	99.25
Subset 2	1,200	98.50
Subset 3	1,157	98.88
Subset 4	1,195	98.32
Subset 5	1,163	98.69
Subset 6	1,142	97.38
Subset 7	1,200	97.94
Subset 8	1,146	99.25
Subset 9	1,186	98.13
Subset 10	1,163	98.32
Subset 11	1,150	97.20
Subset 12	1,168	97.94
Subset 13	**1,134**	**99.63**
Subset 14	1,156	98.13
Subset 15	1,168	97.57
Subset 16	1,147	96.82
Subset 17	1,176	98.88
Subset 18	1,165	97.76
Subset 19	1,187	99.44
Subset 20	1,146	98.13

Note: Bold indicates best results.

recognition accuracy after reducing 55% feature attributes. Figure 14.9 shows the confusion matrix produced by the SVM classifier.

From the confusion matrix given in Figure 14.9, it can be seen that in 0.01% of cases Anger has been misclassified as Fear, and in 0.02% of cases, Contempt has been misclassified as Disgust. Some samples of confusing facial expressions are shown in Figure 14.10.

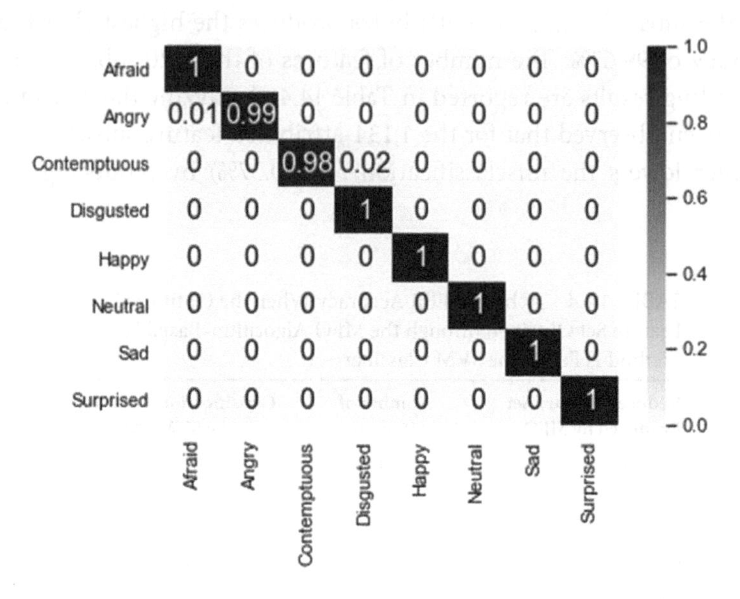

FIGURE 14.9 Confusion matrix produced by the SVM classifier when it is fed with the optimized feature subset produced by the MFO algorithm-based FS method applied over IR database.

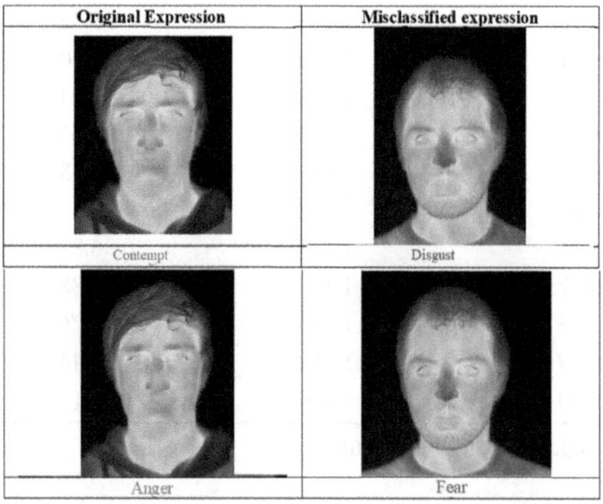

Original Expression	Misclassified expression
Contempt	Disgust
Anger	Fear

FIGURE 14.10 Misclassification occurs due to confused facial expressions.

TABLE 14.5 A Table Representing the AUC-ROC Values for Different Train–Test Splits Using SVM Classifier for Different Expressions

AUC-ROC for	Train–Test Ratio		
	7:3	8:2	9:1
Angry	1	1	1
Sad	1	1	1
Afraid	1	1	1
Surprised	0.99	1	0.99
Happy	1	1	1
Neutral	1	1	1
Contemptuous	1	1	1
Disgust	1	1	1

The maximum accuracy obtained by the SVM classifier is for the polynomial kernel with coef0=2. Here, we consider the Area Under the Receiver Operating Curve (AUC) for the same classifier with three different train–test split ratios of 7:3, 8:2, and 9:1. Table 14.5 represents the AUC values for each expression with respect to the other expressions.

The Receiver Operating Curve (ROC) is a plot of true positive rate (TPR or sensitivity) and false positive rate (FPR or specificity). Figure 14.11 represents the ROC for different expressions with different train–test split ratios. TPR and FPR are calculated using the following equations:

$$\text{TPR} = \frac{\text{TP}}{\text{FN} + \text{TP}} \tag{14.8}$$

$$\text{FPR} = \frac{\text{FP}}{\text{TN} + \text{FP}} \tag{14.9}$$

where

TP = True Positive

TN = True Negative

FP = False Positive

FN = False Negative

TP for an expression determines that the expression detected is originally the expression shown in reality. TN is the original misclassification of expression. However, FP and FN are the misclassified expressions. FP for an expression, Anger, determines that the person is detected as Angry but

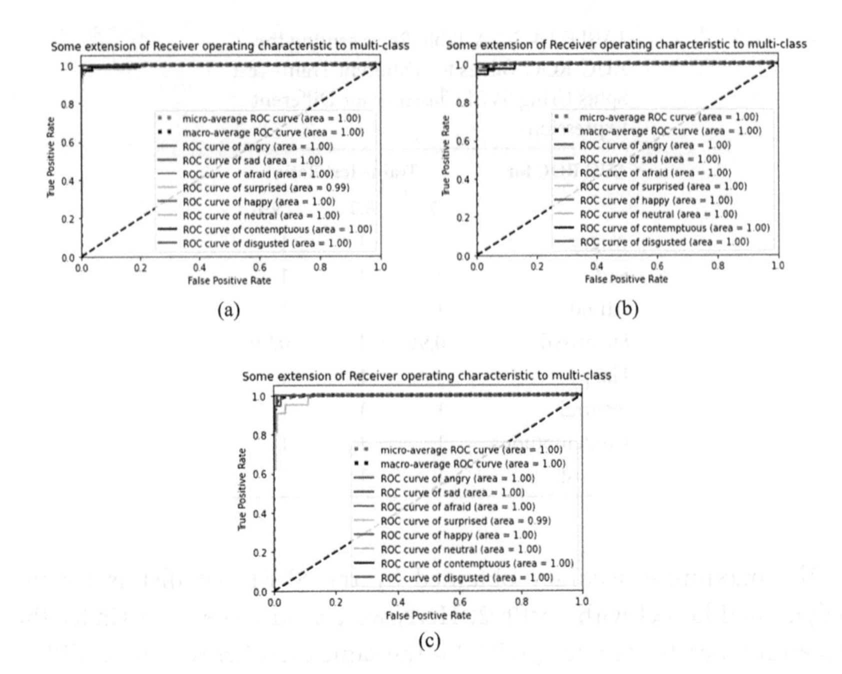

FIGURE 14.11 The ROC graphs for three different train–test splits using the SVM classifier: (a) first train-split, (b) second train-split, and (c) third train-split.

TABLE 14.6 Performance Comparison of Proposed Method with some Past Methods

Work References	**Methods**	**Classification Accuracy (%)**
M. Kopaczka et al. [30]	HOG and SVM	46.70
M. Kopaczka et al. [31]	HOG with Random Forest	65.75
Present work	FS using MFO on features extracted from a CNN model with SVM classifier	**99.63**

Note: Bold indicates best results.

in reality the person may be Disgusted. FN for an expression, Disgust, represents that the person is not disgusted but in reality the person is disgusted.

To determine the efficiency of our proposed method, in Table 14.6, we have compared the performance of the proposed method with some past works found in the literature [30,31], where those authors have also considered the same facial thermal image database. The authors in Ref. [30] have used HOG with SVM classifier to recognize facial expressions. The work mentioned in Ref. [31] has presented a FER system that is based on HOG features and a Random Forest classifier. They have designed a respiratory rate analysis module that calculates the average temperature from an automatically detected region of interest. Analyzing Table 14.6, it can be said that our proposed model has outperformed the previous works by achieving highest classification accuracy which is 99.63%.

Though the proposed model has produced an impressive accuracy, there are a few cases when the model fails to classify the expressions accurately. Thorough observations reveal that confusing facial expressions are reasons for such misclassification. In the future, we will try to overcome this problem by adding some attention mechanism to the CNN model. Some initial pre-processing techniques may help to identify the region of interest on which CNN models can be applied for feature extraction. Another issue which we should address in future is the use of a lighter CNN model. More experimentation can be done to develop a lighter CNN model. As we have trained our customized CNN models for 1,000 epochs to get the various snapshots, a lighter model can reduce the overall training time of the model. Besides, we can work on the FS algorithm also. Though the MFO-based FS algorithm is able to reduce the feature dimension to a significant amount, we can add a local search method to the MFO algorithm to enhance its exploitation capability which might help to get more optimized feature vector.

14.6 CONCLUSION

In the present work, we have developed a model for FER applied on thermal IR images. At first, we have extracted various snapshots of a customized CNN model during the training. Then using the concept of snapshot ensembling, features extracted from different snapshots are concatenated. However, this combined feature vector may have some redundant features as for each snapshot, exactly the same input images are fed to train the CNN model. Hence, in the next stage of this work, we have applied the MFO algorithm-based FS method to eliminate the redundant features. The FS method not only reduces the dimension of the feature vector but also enhances the recognition accuracy of the FER model. Our proposed model produces a classification accuracy of 99.63% which is significantly higher than some previous methods. Though this may be considered as a very good performance, we would like to investigate the reasons for misclassification by minutely observing the wrongly predicted images. We can also apply some data augmentation techniques to increase the training data sample which in turn might help the model to classify similar-type expressions more accurately. Another plan is to apply the model to other FER datasets to establish the robustness of the same.

REFERENCES

1. A. Mehrabian, "Communication without Words," *Psychol. Today*, vol. 2, pp. 53–56 (1968).
2. P. Ekman, W.V. Friesen, *Facial Action Coding System*. Palo Alto, CA: Consulting Psychology Press (1978).
3. P., Ekman, W.V., Friesen, J.C., Hager, *Facial Action Coding System*. Salt Lake City: Research Nexus (2002).
4. H. Harashima, C. S. Choi, T. Takebe, "3-D model-based synthesis of facial expressions and shape deformation", *Human Interface*, vol. 4, pp. 157–166 (1989).
5. K. Mase, "An application of optical flow: Extraction of facial expression", *IAPR Workshop on Machine Vision and Application*, pp. 195–198 (1990).
6. K. Mase, "Recognition of facial expression from optical flow", *Trans. IEICE*, vol. E74, no. 10, pp. 3474–3483 (1991).
7. K. Matsuno, C. Lee, S. Tsuji, "Recognition of facial expressions using potential net and KL expansion", *Trans. IEICE*, vol. J77-D-II, no. 8, pp. 1591–1600 (1994).
8. H. Kobayashi, F. Hara, "Analysis of neural network recognition characteristics of 6 basic facial expressions", *Proceedings of IEEE International Workshop on Robot and Human Communication*, pp. 222–227 (1994).
9. K. Pan, S. Liao, Z. Zhang, S. Z. Li, P. Zhang, "Part-based face recognition using near infrared images, "*2007 IEEE Conference on Computer Vision and Pattern Recognition*, pp. 1–6 (2007). doi: 10.1109/CVPR.2007.383459.
10. S. Z. Li, R. Chu, S. Liao, L. Zhang. "Illumination invariant face recognition using near-infrared images". *IEEE Transactions on Pattern Analysis and Machine Intelligence, 26(Special issue on Biometrics: Progress and Directions)*, vol. 29, pp. 627–638 (2007).
11. S.Z. Li, His Face Team. "AuthenMetricFl: A highly accurate and fast face recognition system". *ICCV2005-Demos* (October 15–21 2005).
12. S. He, S. Wang, W. Lan, H. Fu, Q. Ji, "Facial expression recognition using deep Boltzmann machine from thermal infrared images," *2013 Humaine Association Conference on Affective Computing and Intelligent Interaction*, pp. 239–244 (2013). doi: 10.1109/ACII.2013.46.
13. M. Ranzato, J. Susskind, V. Mnih, G. Hinton, "On deep generative models with applications to recognition," *CVPR 2011* pp. 2857–2864 (2011). doi: 10.1109/CVPR.2011.5995710.
14. M. Ranzato, V. Mnih, J.M. Susskind, G.E. Hinton, "Modeling natural images using gated MRFs," *IEEE Trans. Pattern Anal. Mach. Intell.*, vol. 35, no. 9, pp. 2206–2222 (2013). doi: 10.1109/TPAMI.2013.29.
15. R. Walecki, O. Rudovic, B. Schuller, V. Pavlovic, and M. Pantic, "Deep structured learning for facial expression intensity estimation," *Cvpr* (2017) [Online]. Available: http://openaccess.thecvf.com/content_cvpr_2017/papers/Walecki_Deep_Structured_Learning_CVPR_2017_paper.pdf.
16. S. Zhao, H. Cai, H. Liu, J. Zhang, S. Chen, "Feature selection mechanism in CNNs for facial expression recognition," *British Machine Vision Conference 2018, BMVC 2018*, pp. 1–12 (2019).

17. B.C. Ko, "A brief review of facial emotion recognition based on visual information," *Sensors (Switzerland)*, vol. 18, no. 2 (2018). doi: 10.3390/s18020401.
18. M. Suk, B. Prabhakaran, "Real-time mobile facial expression recognition system-a case study," *IEEE Comput. Soc. Conf. Comput. Vis. Pattern Recognit. Work.*, pp. 132–137 (2014). doi: 10.1109/CVPRW.2014.25.
19. D. Ghimire, J. Lee, "Geometric feature-based facial expression recognition in image sequences using multi-class AdaBoost and support vector machines," *Sensors (Switzerland)*, vol. 13, no. 6, pp. 7714–7734 (2013). doi: 10.3390/s130607714.
20. R.A. Khan, A. Meyer, H. Konik, S. Bouakaz, "Framework for reliable, real-time facial expression recognition for low resolution images," *Pattern Recognit. Lett.*, vol. 34, no. 10, pp. 1159–1168 (2013). doi: 10.1016/j.patrec.2013.03.022.
21. X. Wang, X. Liu, L. Lu, and Z. Shen, "A new facial expression recognition method based on geometric alignment and LBP features," *Proceedings of 17th IEEE International Conference on Computer Science and Engineering CSE 2014, Jointly with 13th IEEE International Conference on Ubiquitous Computing and Communication, IUCC 2014, 13th International Symposium on Pervasive Systems Algorithms, Networks, I-SPAN 2014 8th International Conference on Frontiers of Computer Science and Technology FCST 2014*, pp. 1734–1737 (2015). doi: 10.1109/CSE.2014.318.
22. Y. Tong, L.B. Jiao, X.H. Cao, "A novel HOG descriptor with spatial multiscale feature for FER," *Appl. Mech. Mater.*, vol. 596, pp. 322–327 (2014). doi: 10.4028/www.scientific.net/AMM.596.322.
23. N. Mehta, S. Jadhav, "Facial emotion recognition using Log Gabor filter and PCA."
24. T. A. Rashid, "Convolutional neural networks based method for improving facial expression recognition," *Adv. Intell. Syst. Comput.*, vol. 530, pp. 73–84 (2016). doi: 10.1007/978-3-319-47952-1_6.
25. W. S. Chu, F. De La Torre, J. F. Cohn, "Learning spatial and temporal cues for multi-label facial action unit detection," *Proceedings of 12th IEEE International Conference on Automation of Face Gesture Recognition, FG 2017-1st International Workshop of Adaptive Shot Learning for Gesture Understanding and Production ASL4GUP 2017, Biometrics Wild, Bwild 2017, Heterogeneous Face Recognition, HFR 2017, Jt. Chall. Domin. Complement. Emot. Recognit. Using Micro Emot. Featur. Head-Pose Estim. DCER HPE 2017 3rd Facial Expr. Recognit. Anal. Challenge, FERA 2017*, pp. 25–32 (2017). doi: 10.1109/FG.2017.13.
26. R. Breuer and R. Kimmel, "A deep learning perspective on the origin of facial expressions," pp. 1–16 (2017) [Online]. Available: http://arxiv.org/abs/1705.01842.
27. S. E. Kahou, V. Michalski, K. Konda, R. Memisevic, C. Pal, "Recurrent neural networks for emotion recognition in video," *ICMI 2015- Proceedings of 2015 ACM International Conference on Multimodal Interaction*, pp. 467–474 (2015). doi: 10.1145/2818346.2830596.
28. K. Liu, M. Zhang, Z. Pan, "Facial expression recognition with CNN ensemble," *Proceedings of 2016 International Conference Cyberworlds, CW 2016*, pp. 163–166 (2016). doi: 10.1109/CW.2016.34.

29. K. Shreyas Kamath, R. Rajendran, Q. Wan, K. Panetta, S. S. Agaian, "TERNet: A deep learning approach for thermal face emotion recognition," *Mobile Multimedia/Image Processing, Security, and Applications 2019*, vol. 10993, p. 1099309 (2019).
30. M. Kopaczka, R. Kolk, D. Merhof, "A fully annotated thermal face database and its application for thermal facial expression recognition," *I²MTC, 2018 IEEE International Instrumentation and Measurement Technology Conference: Discovering New Horizons in Instrumentation and Measurement*, pp. 1–6 (2018). doi: 10.1109/I2MTC.2018.8409768.
31. M. Kopaczka, L. Breuer, J. Schock, D. Merhof, "A modular system for detection, tracking and analysis of human faces in thermal infrared recordings," *Sensors (Switzerland)*, vol. 19, no. 19, 2019, doi: 10.3390/s19194135.
32. Y. Yoshitomi, M. Tabuse, T. Asada, "Facial expression recognition using thermal image processing and Neural Network," *Proceedings 6th IEEE International Workshop on Robot and Human Communication*, pp. 57–85, 2012.
33. G. Jiang, X. Song, F. Zheng, P. Wang, A. M. Omer, "Facial expression recognition using thermal image," *Annu. Int. Conf. IEEE Eng. Med. Biol. Proc.*, vol. 7, pp. 631–633 (2005). doi: 10.1109/iembs.2005.1616492.
34. P. Shen, S. Wang, Z. Liu, "Facial expression recognition from infrared thermal videos."
35. V. R. Balasaraswathi, M. Sugumaran, Y. Hamid, "Feature selection techniques for intrusion detection using non-bio-inspired and bio-inspired optimization algorithms," *J. Commun. Inf. Networks*, vol. 2, no. 4, pp. 107–119 (2017). doi: 10.1007/s41650-017-0033-7.
36. L. A. Belanche, F. F. González, "Review and evaluation of feature selection algorithms in synthetic problems," (2011). [Online]. Available: http://arxiv.org/abs/1101.2320.
37. J. Cai, J. Luo, S. Wang, S. Yang, "Feature selection in machine learning: A new perspective," *Neurocomputing*, vol. 300, pp. 70–79 (2018). doi: 10.1016/j.neucom.2017.11.077.
38. C. Deisy, B. Subbulakshmi, S. Baskar, N. Ramaraj, "Efficient dimensionality reduction approaches for feature selection," *in International Conference on Computational Intelligence and Multimedia Applications 2007*, vol. 2, pp. 121–127 (2007). doi: 10.1109/ICCIMA.2007.288.
39. A. Jain, D. Zongker, "Feature selection: Evaluation, application, and small sample performance," *IEEE Trans. Pattern Anal. Mach. Intell.*, vol. 19, no. 2, pp. 153–158 (1997). doi: 10.1109/34.574797.
40. A. K. Jain, B. Chandrasekaran, "Classification pattern recognition and reduction of dimensionality," *Handb. Stat.*, vol. 2, pp. 835–855 (1982). [Online]. Available: http://www.sciencedirect.com/science/article/pii/S0169716182020422.
41. Y. Liu, G. Wang, H. Chen, H. Dong, X. Zhu, S. Wang, "An improved particle swarm optimization for feature selection," *J. Bionic Eng.*, vol. 8, no. 2, pp. 191–200 (2011). doi: 10.1016/S1672-6529(11)60020-6.

42. S. Das, "Filters, wrappers and a boosting-based hybrid for feature selection," Engineering, pp. 74–81 (2001) [Online]. Available: http://citeseerx.ist.psu.edu/viewdoc/download?doi=10.1.1.124.5264&rep=rep1&type=pdf.
43. S. Sen, S. Saha, S. Chatterjee, S. Mirjalili, R. Sarkar, "A bi-stage feature selection approach for COVID-19 prediction using chest CT images," *Appl. Intell.* (2021). doi: 10.1007/s10489-021-02292-8.
44. S. M. Lajevardi, Z. M. Hussain, "Feature selection for facial expression recognition based on optimization algorithm," *Proceedings of 2009 2nd International Workshop on Nonlinear Dynamics and Synchronization, INDS 2009*, pp. 182–185 (2009). doi: 10.1109/inds.2009.5228001.
45. P. Li, S. L. Phung, A. Bouzerdom, F. H. C. Tivive, "Feature selection for facial expression recognition," pp. 35–40 (20100.
46. M. Ghosh, T. Kundu, D. Ghosh, R. Sarkar, "Feature selection for facial emotion recognition using late hill-climbing based memetic algorithm," *Multimed. Tools Appl.*, vol. 78, no. 18, pp. 25753–25779 (2019). doi: 10.1007/s11042-019-07811-x.
47. K. Mistry, L. Zhang, S. C. Neoh, C. P. Lim, and B. Fielding, "A micro-GA embedded PSO feature selection approach to intelligent facial emotion recognition," *IEEE Trans. Cybern.*, vol. 47, no. 6, pp. 1496–1509 (2017). doi: 10.1109/TCYB.2016.2549639.
48. S. Saha et al., "Feature selection for facial emotion recognition using cosine similarity-based harmony search algorithm," *Appl. Sci.*, vol. 10, no. 8, pp. 1–22 (2020). doi: 10.3390/APP10082816.
49. S. Mirjalili, "Moth-flame optimization algorithm: A novel nature-inspired heuristic paradigm," *Knowl. Based Syst.*, vol. 89, pp. 228–249 (2015). doi: 10.1016/j.knosys.2015.07.006.
50. H. M. Zawbaa, E. Emary, B. Parv, M. Sharawi, "Feature selection approach based on moth-flame optimization algorithm," *2016 IEEE Congress on Evolutionary Computation CEC*, pp. 4612–4617 (2016). doi: 10.1109/CEC.2016.7744378.
51. I. Tumar, Y. Hassouneh, H. Turabieh, T. Thaher, "Enhanced binary moth flame optimization as a feature selection algorithm to predict software fault prediction," *IEEE Access*, vol. 8, pp. 8041–8055 (2020). doi: 10.1109/ACCESS.2020.2964321.
52. R. Abu Khurmaa, I. Aljarah, and A. Sharieh, *An Intelligent Feature Selection Approach Based on Moth Flame Optimization for Medical Diagnosis*, vol. 5. Springer: London (2020).
53. A. A. Ewees, A. T. Sahlol, M. A. Amasha, "A bio-inspired moth-flame optimization algorithm for arabic handwritten letter recognition," *Proceedings of 2017 International Conferences on Control, Artificial Intelligence, Robotics & Optimization*, vol. 2018, pp. 154–159 (2017). doi: 10.1109/ICCAIRO.2017.38.
54. L. Zhang, K. Mistry, S. C. Neoh, C. P. Lim, "Intelligent facial emotion recognition using moth-firefly optimization," *Knowl. Based Syst.*, vol. 111, pp. 248–267 (2016). doi: 10.1016/j.knosys.2016.08.018.

55. S. H. H. Mehne, S. Mirjalili, "Moth-flame optimization algorithm: theory, literature review, and application in optimal nonlinear feedback control design," *Nat. Inspired Optim.*, pp.143–166 (2020).
56. M. Shehab, L. Abualigah, H. Al Hamad, H. Alabool, M. Alshinwan, A. M. Khasawneh, "Moth–flame optimization algorithm: Variants and applications," *Neural Comput. Appl.*, vol. 32, no. 14, pp. 9859–9884 (2020).
57. D. Muduli, R. Dash, B. Majhi, "Automated breast cancer detection in digital mammograms: A moth flame optimization based ELM approach," *Biomed. Signal Proc. Control*, vol. 59, pp. 101912 (2020).
58. M. Kopaczka, R. Kolk, D. Merhof, "A fully annotated thermal face database and its application for thermal facial expression recognition," *IEEE International Instrumentation and Measurement Technology Conference (I2MTC)*, 1–6 (2018).
59. K. J. Gaston, J. Bennie, T. W. Davies, J. Hopkins, "The ecological impacts of nighttime light pollution: A mechanistic appraisal," *Biol. Rev.*, vol. 88, no. 4, pp. 912–927 (2013). doi: 10.1111/brv.12036.
60. C. Rich, T. Longcore, "Ecological consequences of artificial night lighting," p. 480 (2006).
61. M. Mafarja, A.A. Heidari, H. Faris, S. Mirjalili, I. Aljarah Dragonfly algorithm: Theory, literature review, and application in feature selection. In: S. Mirjalili, J. Song Dong, A. Lewis (eds) *Nature-Inspired Optimizers: Studies in Computational Intelligence*, vol. 811. Springer, Cham (2020). doi: 10.1007/978-3-030-12127-3_4.

CHAPTER 15

Design Optimization of Photonic Crystal Filter Using Moth-Flame Optimization Algorithm

Seyed Mohammad Mirjalili and Somayeh Davar

Concordia University

Nima Khodadadi

Florida International University

Seyedali Mirjalili

Torrens University Australia
Yonsei University

CONTENTS

15.1 Introduction 314
15.2 PhCs and PhC Filters 316
15.3 Design Optimization of PhC Filter by Using MFO 317
15.4 Results and Discussion 319
References 321

DOI: 10.1201/9781003205326-19

15.1 INTRODUCTION

In the past, the optimization process of optimization problems involved heavy human intervention and supervision. For instance, an engineer would have to build several prototypes for a boat propeller to measure its efficiency and find the best one. It was an expensive, tedious, and error-prone exercise. With the advancements in computer hardware and the popularity of simulators, physical prototyping was minimized and mainly moved to computers. For instance, in the above example, the engineer would now be able to create several 3D models of the propeller and calculate their efficiency to find the best one. This process was much faster than the manual technique. It also reduced human involvement and leveraged the speed and accuracy of CPUs in computers.

However, the process of modeling by tuning the decision variables of an optimization problem was still laborious and error-prone. Another issue was the slow process of optimization. Despite using computers for evaluating different designs and calculating the objective values (merit factors), the designs (3D models) would have to be done by humans, which led to a substantial reduction in the speed of the whole process. This made people think to find ways and tackle this bottleneck.

The idea was to remove the human from the loop and automate the modeling using computer programs. This was a revolutionary idea as, for the first time, computers were able to design and test automatically. Several issues were emerged, including how to use an algorithm to find the best design. This was the beginning of a wide range of computer-aided design processes and the development of optimization algorithms. Optimization problems and algorithms will be around us forever, so this area of research is quite active. Progress in technologies has led to more challenging optimization problems that demand more efficient optimization algorithms.

First-generation optimization algorithms include conventional techniques, which mainly rely on calculating the derivative of a search space. Gradient Descent is a well-known algorithm in this space, which starts with a random solution. It then iteratively updates this solution using the negative of the gradient (in the case of maximization, the gradient). This leads to a convergence toward the best local solutions. Despite the merits of this algorithm and its popularity these days, mainly due to simplicity and fast convergence speed, there are two major drawbacks.

Gradient Descent algorithm always converges toward the best locally optimal solutions, so if the initial solution is not on a slope toward the global optimum, the algorithm will be trapped in locally optimal solutions. There have been mechanisms in the literature to alleviate this drawback, such as running the algorithm multiple times starting from different random solutions and incorporating a stochastic mechanism to help with resolving local optimal stagnation [1]. The second issue is the need to calculate the derivative of the objective function as the gradient. This algorithm becomes impractical if the gradient is not known or difficult to calculate. There are ways to approximate gradient (e.g., Monte Carlo integral) in the literature, but such methods introduce uncertainties in the system.

The drawback of gradient descent and similar optimization algorithms leads to the development of second-generation optimization algorithms and heuristic optimization algorithms [2]. In such algorithms, educated decisions are made during the optimization process to find a reasonably good solution in a reasonable time. They are usually used when there is no exact solution for a given problem. Heuristics algorithm used some sort of heuristic function that assisted the optimization algorithm to rank a set of possible solutions in each step of optimization and choose the most promising one. Since this function tends to have past (existing objective values) and future (heuristic) elements, they tend to outperform exact algorithms on challenging problems.

An excellent example of a heuristic algorithm is A* in path planning. The heuristic function in this algorithm considers the distance traveled so far and the elimination of distance to the destination for each node on the graph. This heuristic function is only suitable for this or very similar graph-related problems. In other words, the heuristic function is problem-specific and should be re-designed when changing the problem. This is considered a drawback of heuristic algorithms, which motivated computer scientists to come up with the third-generation optimization algorithms, metaheuristics [3].

As the third-generation optimization algorithms, metaheuristics are high-level algorithms and problem independent. The candidate solutions for an optimization problem are purely evaluated based on their objective values. They employ stochastic mechanisms to iteratively update a set of candidate solutions to improve their objective values. What makes these algorithms to be considered state-of-the-art is twofold. First, metaheuristics are problem independent. This means that the mechanism in the metaheuristics algorithm provides ways to create and update a set of solutions by evaluating them using an objective function. Second, there is no need to calculate the derivative of the objective function.

The area of metaheuristics has been booming in the last decade with a wide range of recent algorithms. They can be divided into three main classes based on the source of inspiration: evolutionary, swarm intelligence, and physics-based [4]. Evolution algorithms mimic evolutionary processes in nature. For instance, the well-regarded Genetic Algorithm (GA) simulates [5] the survival of the fittest in nature using evolutionary operators such as selection, recombination, and mutation.

In the second class, the sources of inspiration are navigation or foraging of swarms in nature. For instance, Particle Swarm Optimization (PSO) follows [6] the simple principles of cognitive and social intelligence of birds in a swarm when seeking food. Each particle tends to memorize its own best solution obtained so far and is aware of the swarm's best solution. These two "pinpoints" will provide insights for each particle to update its position in each step of the optimization process.

In the third class, physical rules of phenomena that lead to problem-solving or stable/optimal state in nature are mimicked in metaheuristics to solve optimization problems. One of the most popular algorithms in this category is Simulated Annealing [7], which mimics how metals are heated and cooled down slowly to change their physical properties.

Despite many optimization algorithms in the classes mentioned above, the literature has seen many works on variants, improvements, adaptation, tuning, and hybridization of metaheuristics. This is due to such methods' stochastic nature, which makes them unpredictable in solving optimization problems. Also, the No-Free-Lunch theorem logically proves that none of them are a "master key" to solve all optimization problems. Therefore, there is always room for changes in the structure of these algorithms to solve existing or new optimization problems better. In this work, the recently proposed Moth-Flame Optimization (MFO) algorithm is used to optimize the design of photonic crystal (PhC) filters.

The first step of solving an optimization problem is formulated. As discussed in Chapter 1 of this book, all optimization problems have three main components: decision variables, objectives, and constraints. In this chapter, the structure of a PhC filter is first presented and the designing process is formulated to identify these three components. Then, the MFO algorithm is used to solve this problem.

15.2 PhCs AND PhC FILTERS

PhCs are very popular optical nanostructures that can be widely found in nature (e.g., wings of butterflies) or laboratories (e.g., optical waveguides). Such multi-purpose structures are popular since the creation of a defect in the bandgap enables the manipulation of light waves. This allows crafting different devices such as optical filters, sensors, and waveguides. Usually, the whole process of designing a device, changing the parameters, and finding the best design is manual and labor-intensive. Also, the analytical relationship between the parameters of this problem and the output(s) is hard to find. Other challenges are computationally expensive simulations, large-scale designs, and the heavily constrained nature of the objective function [8].

Due to the difficult nature of the problem and the aforementioned challenges, metaheuristics have been employed in the literature to optimize the shape of devices created using PhCs. The following list presents some of the most recent ones:

- The GA is employed for minimization of quality factor (Q) in PhCs in [9–11].
- The GA is applied to PhC waveguides for minimization of bend loss in [12].
- The gray wolf optimizer is used for maximization of extraction ratio and Purcell factor [13].
- The PSO is used to find an optimal shape for notch-filter PhC [14]
- The gray wolf optimizer and multi-verse optimizer are used to enhance the performance of PhC slow light waveguides [15,16].

These works demonstrate the merits of metaheuristics in this problem area, which motivated our attempts to use the MFO algorithm as well.

15.3 DESIGN OPTIMIZATION OF PhC FILTER BY USING MFO

The structure of the PhCs that leads to filter capability is visualized in Figure 15.1. The device consists of a silicon slab with some air holes. It can be seen that light waves enter the device from the left-hand side and exit from the right-hand side. The configuration, size, and shape of holes create super defect regions that lead to filtering light waves. The cavity section of the silicon slab has 11 holes in which the main structural parameters of this problem will be extracted for optimization purposes. The decision variables of this problem are as follows:

$$\text{Parameters:}\ \vec{x} = \left[\frac{R_1}{a}, \frac{R_2}{a}, \frac{R_3}{a}, \frac{R_4}{a}, \frac{R_5}{a}, \frac{R_6}{a}, \frac{R_7}{a}, \frac{R_8}{a}\right]$$

where R_i is the radius of $\#i$ hole in Figure 15.1.

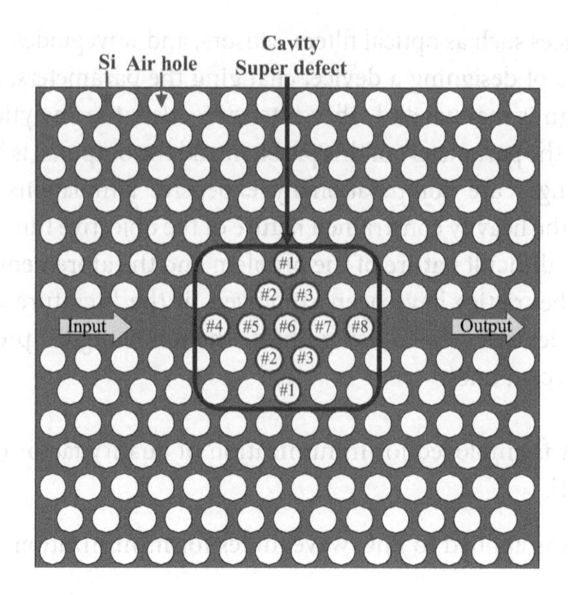

FIGURE 15.1 PhC filter is used as a case study in this work.

The cavity section of the slab is essential, but the crystal lattice is also important to provide a large photonic bandgap. This leads to having a larger wavelength window. A 2D plane Wave Expansion (PWE) is employed in conjunction with slab equivalent index to calculate the bandgap of the lattice [17]. We considered a slab equivalent index of 3.18 for an SOI slab with 400 nm thickness [18].

After identifying the decision variables, we need to discuss the objective function and constraints. To calculate the output of the device (merit factor), the following equation is used [19]:

$$O = \frac{-\text{Amp}_c}{\text{Amp}_s + \text{Deviation}}$$

where 'Amp$_c$' is the maximum amplitude of the output in the main band, 'Amp$_s$' indicates the maximum amplitude of the output in the sidebands, and 'Deviation' represents the deviation of the central wavelength of the output peak (λ_c) to the defined central wavelength of the channel (λ_O) (Deviation $= |\lambda_c - \lambda_O|$). The three elements of the objective function are shown in Figure 15.2.

The constraints of this problem are as follows:

$$\text{Constraints}: C = [C_1, C_2],$$

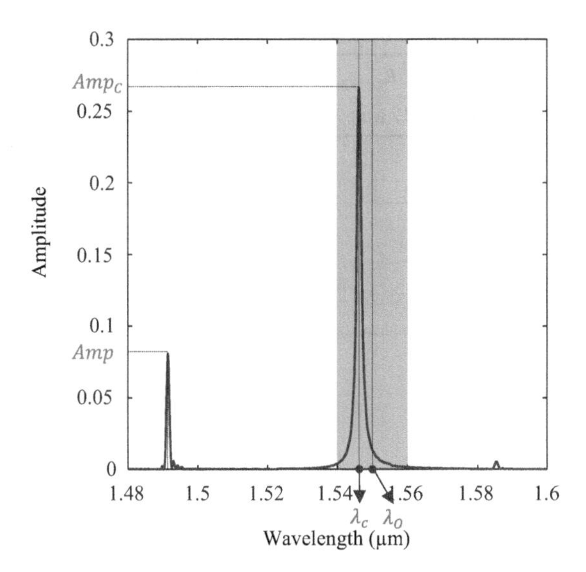

FIGURE 15.2 The output spectral transmission performance of a sample case of a PhC filter.

$$C_1 : 0 \le \frac{R_1}{a}, \ldots, \frac{R_8}{a} \le 0.5,$$

$$C_2 : \ \mathrm{Amp}_s < \mathrm{Amp}_c,$$

15.4 RESULTS AND DISCUSSION

To estimate the global optimum for the optimization problem formulated in the last section, an MFO algorithm with a population of 50 moths and 200 iterations is used. The results are given in Table 15.1 and Figures 15.3 and 15.4.

Table 15.1 shows that the output of the device is –15.42, which is achieved when starting with a design and nearly –1 as the output. This indicates that the MFO algorithm improved a random solution for this problem to almost 15-times better solutions.

The convergence behavior of the MFO algorithm is shown in Figure 15.3. It can be seen that the algorithm shows healthy convergence in the first 50 iterations but suddenly face no improvement for nearly 100 iterations. However, the great exploratory and exploitative mechanisms of MFO finally found a better solution, and the algorithm converged to an excellent solution by the 200th iteration (Figures 15.4 and 15.5).

TABLE 15.1 Obtained Optimal PhC Filter Structure with MFO

R_1	R_2	R_3	R_4	R_5	R_6	R_7	R_8	O
0 nm	102 nm	103 nm	217 nm	169 nm	0 nm	217 nm	142 nm	−15.42

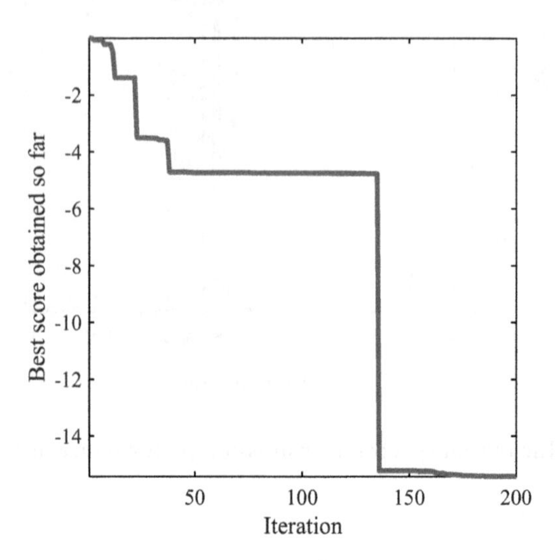

FIGURE 15.3 Convergence curve of MFO for the PhC filter designing.

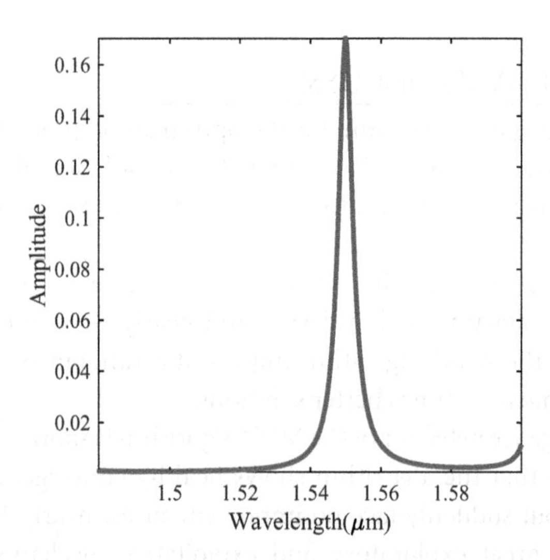

FIGURE 15.4 The output spectral transmission performance of the final optimal PhC filter designs.

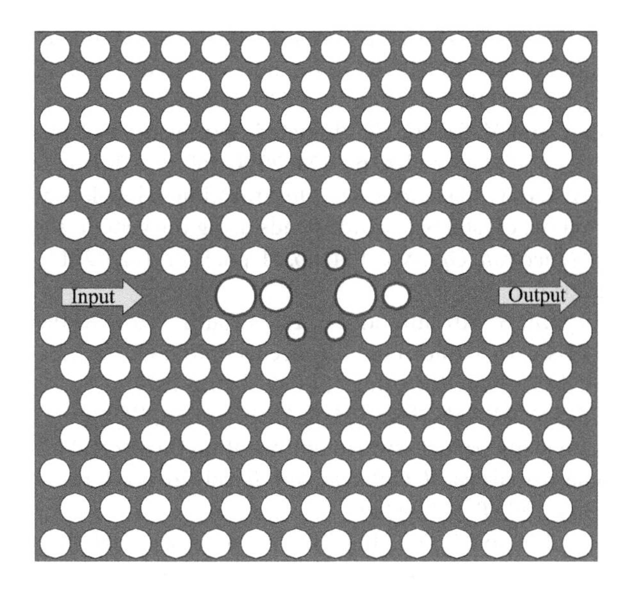

FIGURE 15.5 The physical geometry of the final optimal PhC filter design.

REFERENCES

1. S. Ruder, "An overview of gradient descent optimization algorithms," September 2016, Accessed: January 08, 2022. [Online] Available: http://arxiv.org/abs/1609.04747.
2. E. A. Silver, "An overview of heuristic solution methods," *J. Oper. Res. Soc.*, vol. 55, no. 9, pp. 936–956, 2004, doi: 10.1057/palgrave.jors.2601758.
3. I. H. Osman and J. P. Kelly, "Meta-heuristics: An overview," in *Meta-Heuristics*. Springer Science & Business Media: Berlin, Germany, 1996, pp. 1–21.
4. S. Mirjalili, J. S. Dong, and A. Lewis, *Nature-Inspired Optimizers: Theories, Literature Reviews and Applications*, vol. 811. Springer: Berlin, Germany, 2019.
5. J. H. Holland, "Genetic algorithms," *Sci. Am.*, vol. 267, no. 1, pp. 66–73, 1992.
6. J. Kennedy and R. Eberhart, "Particle swarm optimization," *Proceedings ICNN'95- International Conference on Neural Networks*, vol. 4, pp. 1942–1948, 1995, doi: 10.1109/ICNN.1995.488968.
7. S. Kirkpatrick, C. D. Gelatt, and M. P. Vecchi, "Optimization by simulated annealing," *Science*, vol. 220, no. 4598, pp. 671–680, 1983.
8. S. M. Mirjalili, H. Taleb, M. Z. Kabir, and P. Bianucci, "Design optimization of orbital angular momentum fibers using the gray wolf optimizer," *Appl. Opt.*, vol. 59, no. 20, pp. 6181–6190, 2020, doi: 10.1364/AO.391731.
9. U. P. Dharanipathy, M. Minkov, M. Tonin, V. Savona, and R. Houdré, "High-Q silicon photonic crystal cavity for enhanced optical nonlinearities," *Appl. Phys. Lett.*, vol. 105, no. 10, p. 101101, 2014, doi: 10.1063/1.4894441.

10. N. V. Triviño et al., "Gallium nitride L3 photonic crystal cavities with an average quality factor of 16900 in the near infrared," *Appl. Phys. Lett.*, vol. 105, no. 23, p. 231119, 2014, doi: 10.1063/1.4903861.
11. M. Minkov and V. Savona, "Automated optimization of photonic crystal slab cavities," *Sci. Rep.*, vol. 4, p. 5124, 2014, doi: 10.1038/srep05124.
12. L. Jiang, H. Wu, W. Jia, and X. Li, "Optimization of low-loss and wide-band sharp photonic crystal waveguide bends using the genetic algorithm," *Opt. Int. J. Light Electron. Opt.*, vol. 124, no. 14, pp. 1721–1725, 2013, doi: 10.1016/j.ijleo.2012.06.005.
13. S. M. Mirjalili, S. Mirjalili, and S. Z. Mirjalili, "How to design photonic crystal LEDs with artificial intelligence techniques," *Electron. Lett.*, vol. 51, no. 18, pp. 1437–1439, 2015, doi: 10.1049/el.2015.1679.
14. M. Djavid, S. A. Mirtaheri, and M. S. Abrishamian, "Photonic crystal notch-filter design using particle swarm optimization theory and finite-difference time-domain analysis," *J. Opt. Soc. Am. B*, vol. 26, no. 4, pp. 849–853, 2009, doi: 10.1364/JOSAB.26.000849.
15. S. M. Mirjalili and S. Z. Mirjalili, "Asymmetric oval-shaped-hole photonic crystal waveguide design by artificial intelligence optimizers," *IEEE J. Sel. Top. Quantum Electron.*, vol. 22, no. 2, p. 4900407, 2016, doi: 10.1109/JSTQE.2015.2469760.
16. S. M. Mirjalili, "SoMIR framework for designing high-NDBP photonic crystal waveguides," *Appl. Opt.*, vol. 53, no. 18, pp. 3945–3953, 2014, doi: 10.1364/AO.53.003945.
17. S. Guo and S. Albin, "Simple plane wave implementation for photonic crystal calculations," *Opt. Express*, vol. 11, no. 2, pp. 167–175, 2003, doi: 10.1364/OE.11.000167.
18. A. Säynätjoki, M. Mulot, J. Ahopelto, and H. Lipsanen, "Dispersion engineering of photonic crystal waveguides with ring-shaped holes," *Opt. Express*, vol. 15, no. 13, pp. 8323–8328, 2007, doi: 10.1364/OE.15.008323.
19. M. J. Safdari, S. M. Mirjalili, P. Bianucci, and X. Zhang, "Multi-objective optimization framework for designing photonic crystal sensors," *Appl. Opt.*, vol. 57, no. 8, pp. 1950–1957, 2018, doi: 10.1364/AO.57.001950.

Index

Note: **Bold** page numbers refer to tables and *italic* page numbers refer to figures.

active sonar dataset 60–61, **61**
algorithm
AOMFO (*see* aquila optimizer with moth-flame optimization algorithm (AOMFO))
benchmark 67–69
derivative-based optimization 5–7, *7*
direct 8
evolutionary 80
evolution-based algorithms 9
genetic 55, 56, 179
heuristic 9
hMFOSMA 161, **163–168** (*see also* hMFOSMA algorithm)
mathematics-based 80
metaheuristic 26, 55, 56, 80
nature-inspired 56, 61, 98
non-derivative-based optimization 7–9, *8*
physics-based 9, 80
swarm-based 9, 80
swarm intelligence 13, 244
vanilla MFO 99–102, 104, 108
aquila optimizer (AO)
comparative methods
multilevel thresholding image segmentation problems 248–255, **248–255**, *256*, *257*
critical search technique 180
expanded exploitation 183
expanded exploration 181
narrowed exploitation 183–184
narrowed exploration 181–182
walk and capture prey 183
aquila optimizer with moth-flame optimization algorithm (AOMFO)
aquila optimizer 181–184
benchmark functions 188, **190–191**
comparison approaches 188, 189, **189**, **192–194**, 195, **196–198**
convergence behavior 202, *203*
crossover and mutation 179
flowchart 187, *188*
moth-flame optimization algorithm
exploitation 186
initial population 184–185
updating the positions 185–186
nature's established developmental processes 179
parameter settings 188, **189**
problem-solving approach 178
real-world problems 179
solutions initialization 187
state-of the-art methods 195, **199–201**
structure 187, *188*
various optimization problems 189, **192–194**
Arabic handwritten letter images (CENPARMI) dataset 26
archive 85, 86, 94
arithmetic optimization algorithm (AOA) 216
multilevel thresholding image segmentation problems 248–255, **248–255**, *256*, *258*

arrhythmia classification 131, 133, 135, 148
auto-encoders (AEs) 55

benchmark algorithm 67–69
benchmark function
hMFOSMA algorithm 161, **163–168**
m-MFO 120, **121**
bi-level thresholding 112, 114, 115
binarization concept 40
binary coded modified moth flame optimization algorithm (BMMFOA) 113
Binary Moth-Flame Optimization Algorithm (BMFO) 40–41

cameras capturing images 285
cancer dataset
best cluster centroids 223, **227**
comparative approaches 220, **221**
cardiovascular diseases (CVDs) 130, 131
deep feature selection for 130–132
dataset 134–136, **136**, *136*
experimental results *142–146*, 142–147, **144**, **147**
methodology 136–142, *137–139*, *141*
research topics 132–134
Cauchy distribution function 41, 113
Cauchy mutation 36–37, 49
efficient BMFO algorithm with 41–42
chaotic MFO (CMFO) 26, 28
classic pue mathematical optimization algorithms 179
classifiers 13, 22
cluster analysis 210
clustering based techniques
best cluster centroids
cancer dataset 223, **227**
CMC dataset 223, **227**
glass dataset 223, **227**
Iris dataset 223, **228**
seeds dataset 223, **228**
statlog (heart) dataset 223, **230**
vowel dataset 223, **231**
wine dataset 223, **231**
comparative approaches
cancer dataset 220, **221**
CMC dataset 220, **221**
glass dataset 220, **222**
Iris dataset 220, **222**
seeds dataset 223, **224**
statlog (heart) dataset 223, **224**
vowels dataset 223, **225**
wine dataset 223, **225**
final rank 223, *226*
graphical comparisons 226, *232*
results 223, *229*
CMC dataset
best cluster centroids 223, **227**
comparative approaches 220, **221**
CMFO *see* chaotic MFO (CMFO)
CNN *see* convolutional neural network (CNN)
color image multilevel thresholding 243
comparative approaches
aquila optimizer with moth-flame optimization algorithm 188, 189, **189**, **192–194**, 195, **196–198**
clustering based techniques
cancer dataset 220, **221**
CMC dataset 220, **221**
glass dataset 220, **222**
Iris dataset 220, **222**
seeds dataset 223, **224**
statlog dataset 223, **224**
vowels dataset 223, **225**
wine dataset 223, **225**
hMFOSMA algorithm
performance results 162, **163–168**
schematic view 162, **169**, *169*
moth-flame optimization algorithm-optimized
25-bar space truss 269, **270**
72-bar space truss 273, **274**
200 bar planner truss **277**, 278
multilevel thresholding image segmentation problems
aquila optimizer 248–255, **248–255**, *256*, *257*
arithmetic optimization algorithm 248–255, **248–255**, *256*, *258*
gray wolf optimizer 248–255, **248–255**, *257*, *258*
moth-flame optimization algorithm 248–255, **248–255**, *256*, *257*
sine cosine algorithm 248–255, **248–255**, *257*, *258*

conjugate gradient methods 55
constraints set 4
continuous optimization 5
convergence analysis, m-MFO 123, *126*
convolutional neural network (CNN) 55, 56, *58*, 58–59
architecture, FER system
residual unit 292–294, *293*
transformation unit *294*, 294–295
customized MFO 64, **64**, 65
underwater targets recognition 54
convolutional neural network 55, 56, *58*, 58–59
Deep Neural Network 57, *57*, 58
loss function 65–66, *66*
Moth-Flame Optimization (MFO) 61–63, *62*
searching agents 64, 65
sensitivity analysis 71, **72–73**
simulation results 66–70, **67**, *67–68*, **69–70**
sonar dataset 59–61, *60*, **61**, *61*
time complexity measurement 70–71
CVDs *see* cardiovascular diseases (CVDs)

Darwinian evolutionary theory 244
data clustering (DC) 210
applications 211–213
extracting natural groupings 211
MFO algorithm
exploitation 215
global-optimal combinatorial optimization 214–215
initial population 213–214
MOA function
exploitation phase 217
exploration phase 216–217
supervised learning method 211
unsupervised learning method 211
data mining 12–13
data pre-processing 13
decision variables 3
Deep Convolutional Neural Network (DCNN) 56, 64, 66, *66*, 70–71, 131, 133
deep feature selection
for cardiovascular disease detection 130–132
dataset 134–136, **136**, *136*
experimental results *142–146*, 142–147, **144**, **147**
methodology 136–142, *137–139*, *141*
research topics 132–134
Deep Learning (DL) 55, 130, 131
Deep Neural Networks (DNNs) 57, *57*, 58, 130, 131
derivative-based optimization algorithms 5–7, *7*
deterministic classification 54
direct algorithms 8
discrete or combinatorial optimization 5
discriminative neural network 57
DL *see* Deep Learning (DL)

efficient BMFO algorithm (EC-BMFO)
with Cauchy mutation 41–42
graph coloring solution representation 42–43
objective function 43–44
simulation results 44, **45–48**, 49
electrocardiogram (ECG) 130, 131, 135
engineering design functions, MMFO 89, **90**, *90*, 91, *91*, **92–93**, 94
engineering problems
hMFOSMA algorithm
gas production facilities design 172–173, **173**, **174**
gas transmission compressor design 171–172, **172**
evaluation, feature selection 15
evolutionary algorithms 80
evolutionary fine-tuning method 56
evolution-based algorithms 9

face detector 286
facial emotions 286
facial expression recognition (FER) system
computer-based automated 283
convolutional neural network architecture
residual unit 292–294, *293*
transformation unit *294*, 294–295
feature selection 296–301, *299*
human–computer interactions 282

facial expression recognition (FER) system (*cont.*)
images and textual information 285
IR database *288*, 288–289, *289*
K-nearest neighbors classifier 286
layer-wise extracted features **295**, 295–296
metaheuristic algorithm 287
misclassification 304, *304*
ML-based approaches 286
real-life applications 282
receiver operating curve 305
sinusoidal function 283
state-of-the-art approaches 285
support vector machine classifier 289
survey 285–288
train–test splits 305, **305**, *306*
transverse orientation 296
visible spectrum 283
Feature Maps (FMs) 58
feature selection 13, *14*, 29
definition of 14
description of 13, 14–16, *15*
FER system 296–301, *299*
MFO algorithm 16, 22, **22**
inspiration 16–17, *17*, *18*
mathematical model 17, 18–21
studies 22, 23–26, **27**, 28
fitness evaluation 103
fitness function 64, 139
flowchart, AOMFO 187, *188*
FMs *see* Feature Maps (FMs)
Friedman rank test 170, **170**, *171*, 180
clustering based techniques 220

GA *see* genetic algorithm (GA)
Gabor filter 286, 287
GCP *see* Graph Coloring Problem (GCP)
generational distance, MMFO 88
generative neural network 57
genetic algorithm (GA) 55, 56, 179
glass dataset
best cluster centroids 223, **227**
comparative approaches 220, **222**
global face region 285
global-optimal combinatorial optimization 185
global optimization problem
heuristic optimization techniques 156
exploration and exploitation 157
optimization challenges 157
mathematical optimization approaches 156
GPUs *see* Graphics Processing Units (GPUs)
gradient descent algorithm 6, 7, *7*
gradient descent-based training algorithms 55
Graph Coloring Problem (GCP) 36, 40
formulation of 37
graph coloring solution representation, EC-BMFO 42–43
Graphics Processing Units (GPUs) 55
gray wolf optimizer (GWO)
multilevel thresholding image segmentation problems 248–255, **248–255**, *257*, *258*

Harmony Search (HS) algorithm 56
Hessian-free optimization 55
heuristic algorithm 9
heuristic optimization techniques
exploration and exploitation 157
optimization challenges 157
hMFOSMA algorithm
benchmark functions 161, **163–168**
comparative approaches
performance results 162, **163–168**
schematic view 162, **169**, *169*
complicated challenges 174
convergence analysis 162, *169*
engineering problems
gas production facilities design 172–173, **173**, **174**
gas transmission compressor design 171–172, **172**
framework 159–161
Friedman rank tests 170, **170**, *171*
future studies 175
motivation 158–159
multi-modal function 161–162
real-life optimization problems 174
real-world engineering challenges 174
statistical analysis **170**, 170–171, *171*
unimodal function 161
Wilcoxon Rank Sum Test 170, **170**
hybrid feature extraction 133

hybrid intelligent approach 24–25
hyper-parameter configuration 56

IGD *see* inverted generational distance (IGD)
ImageNet dataset 136
image segmentation, m-MFO 112, 120, 123, *123*, *124*, **125**
image thresholding segmentation
  bi-level threshold 243
  Kapoor's method 243
  multilevel threshold 243
  Otsu method 243
  Tsallis entropy method 243
IMH *see* Intelligent Machine Hearing (IMH)
implementation techniques 98
Independent Component Analysis (ICA) 145
infrared thermal imaging (IRTI) cameras 283
initialization, feature selection 15
Intel(R) Core (TM) i9-9900K CPU 104, 108
Intelligent Machine Hearing (IMH) 55
inverted generational distance (IGD) 89
IR cameras 286
Iris dataset
  best cluster centroids 223, **228**
  comparative approaches 220, **222**

Kapur's entropy 112, 115–116
kernel extreme learning machine (KELM) 26
K-Nearest Neighbor (KNN) classifier 23–26, 28
Knowledge Discovery in the Database (KDD) 12–13
Krylov subspace descent algorithm 55
layer-wise extracted features, FER system **295**, 295–296
leader selection feature 85, 86, 94
logarithmic spiral function 39
logarithmic spiral, MFO 21
long short-term memories (LSTMs) 285
loss function 65–66, *66*

machine learning techniques 13
Marine Predators Algorithm (MPA) 133
Massachusetts Institute of Technology-Beth Israel Hospital (MIT-BIH) Arrhythmia Database 133–135, 142, 147
mathematics-based algorithms 80
math optimizer accelerated (MOA) function
  exploitation phase 217
  exploration phase 216–217
maximum spread, MMFO 88–89
Mean Square Error (MSE) 65, 116
metaheuristic 9–10, 266
metaheuristic algorithms 26, 55, 56, 80
  FER system 287
metaheuristic optimization algorithms 98
  physics-based approach 244
  swarm intelligence algorithms 244
MFO algorithm *see* moth-flame optimization (MFO) algorithm
MFOAOA *see* moth-flame optimization with arithmetic optimization algorithm (MFOAOA)
minimization problem, MOO 83
MIT-BIH database *see* Massachusetts Institute of Technology-Beth Israel Hospital (MIT-BIH) Arrhythmia database
m-MFO *see* modified moth-flame binary-coded optimization algorithm (m-MFO)
MMFO *see* Multi-objective Moth-Flame Optimization (MMFO)
modified moth-flame binary-coded optimization algorithm (m-MFO) 113, 119–120, *120*, 125
  benchmark function 120, **121**
  convergence analysis 123, 126
  image segmentation 120, 123, *123*, *124*, **125**
  Moth-Flame Optimization 116–118
MOMFA *see* Multi-objective Moth-Flame Optimization Algorithm (MOMFA)
MOO *see* multi-objective optimization (MOO)

moth-flame optimization (MFO) 61–63, *62*, 72, 86, 113, 158
for cardiovascular diseases 137, *139*, 139–142, *141*
experiments 104, **105**, *106*, 106–108, *107*
exploratory capacity 159
modified moth-flame binary-coded optimization algorithm 116–118
and vanilla implementation 99–102
vectorization 102–104, *103*
moth-flame optimization (MFO) algorithm 9–10, 13–14, 37–40
comparative approaches
multilevel thresholding image segmentation problems 248–255, **248–255**, *256*, *257*
25-bar space truss 269, **270**
72-bar space truss 273, **274**
200-bar planner truss **277**, 278
exploitation 186, 215
feature selection 16, 22, **22**
inspiration 16–17, *17*, *18*
mathematical model 17, 18–21
studies 22, 23–26, **27**, 28
global-optimal combinatorial optimization 214–215
initial population 184–185, 213–214
updating positions 185–186
moth-flame optimization with arithmetic optimization algorithm (MFOAOA)
cancer dataset
best cluster centroids 223, **227**
comparative approaches 220, **221**
CMC dataset
best cluster centroids 223, **227**
comparative approaches 220, **221**
experiments settings 219
final rank 223, *226*
flowchart *218*, 219
glass dataset
best cluster centroids 223, **227**
comparative approaches 220, **222**
graphical comparisons 226, *232*
Iris dataset
best cluster centroids 223, **228**
comparative approaches 220, **222**
results 223, *229*
seeds dataset
best cluster centroids 223, **228**
comparative approaches 223, **224**
solutions initialization 217–218
statlog (heart) dataset
best cluster centroids 223, **230**
comparative approaches 223, **224**
structure 218–219
UCI benchmark datasets 219, **219**
vowels dataset
best cluster centroids 223, **231**
comparative approaches 223, **225**
wine dataset
best cluster centroids 223, **231**
comparative approaches 223, **225**
moth's spiral trip 186
MPA *see* Marine Predators Algorithm (MPA)
MSE *see* Mean Square Error (MSE)
multi-agent metaheuristic optimization 98, 99, *99*
multilevel thresholding 112, 114, 116
problem formulation 114–115
Kapur's entropy method 115–116
segmentation performance measure 116
multilevel thresholding image segmentation problem
aquila optimizer
peak signal-to-noise ratio results 248, **248**, 250, **250**, 252, **252**, 254, **254**
structural similarity index measure results 249, **249**, 251, **251**, 253, **253**, 255, **255**
arithmetic optimization algorithm
peak signal-to-noise ratio results 248, **248**, 250, **250**, 252, **252**, 254, **254**
structural similarity index measure results 249, **249**, 251, **251**, 253, **253**, 255, **255**
definitions 245–246
experiments 247, *247*
gray wolf optimizer

peak signal-to-noise ratio results 248, **248**, 250, **250**, 252, **252**, 254, **254**
structural similarity index measure results 249, **249**, 251, **251**, 253, **253**, 255, **255**
moth-flame optimization algorithm
peak signal-to-noise ratio results 248, **248**, 250, **250**, 252, **252**, 254, **254**
structural similarity index measure results 249, **249**, 251, **251**, 253, **253**, 255, **255**
sine cosine algorithm
peak signal-to-noise ratio results 248, **248**, 250, **250**, 252, **252**, 254, **254**
structural similarity index measure results 249, **249**, 251, **251**, 253, **253**, 255, **255**
Multi-objective Ant Lion Optimizer (MOALO) 89
multi-objective formulation 15, 25
Multi-objective Moth-Flame Optimization (MMFO) 81, 82, 85–87, *87*, 89, 94
engineering design functions 89, **90**, *90*, 91, *91*, **92–93**, 94
performance metrics 87
generational distance 88
inverted generational distance 89
maximum spread 88–89
spacing 88
Multi-objective Moth-Flame Optimization Algorithm (MOMFA) 82
multi-objective optimization (MOO) 81, 83–85, *84*, *85*
multi-objective version of MFO based on quantum behaved moth (MOQMFO) 82

nature-inspired algorithms 56, 61, 98
navigation strategy 37
Neural Network (NN) 57
No Free Lunch (NFL) theorem 56, 81, 157, 316
non-derivative-based optimization algorithms 7–9, *8*
numpy library 102, 103, 108

objective function 3, 4
EC-BMFO 43–44
objective space 4
objectives set 3, 4
1D histogram 243
opposition-based learning 25
optical nanostructures 316
optimization 3–5
derivative-based algorithms 5–7, *7*
metaheuristics 9–10
non-derivative-based algorithms 7–9, *8*
optimization problems (OPs) 3, 37, 80, 81, 178
candidate solutions for 315
classification based on differentiability of search space 5, *6*
formulation of 4, 5
MFO algorithm 139
in nature-inspired approaches 112
optimization process of 314
search landscape of 5, *6*
for segmentation 245
structural 266

Pareto dominance 83–84, *84*
Pareto optimal front 85, *85*
Pareto optimality 84, 86
Pareto optimal set 84, *85*
Particle Swarm Optimization (PSO) algorithms 56, 133
passive sonar dataset 59, *60*
pattern detection systems 210
photonic crystal (PhC) filters
computer-aided design processes 314
convergence behavior 319, *320*
design optimization 317–319, *318*, *319*
drawback 315
evolutionary operators 316
first-generation optimization algorithms 314
gradient descent algorithm 314–315
human intervention and supervision 314

photonic crystal (*cont.*)
metaheuristics in 316–317
particle swarm optimization 316
physical geometry 319, *321*
second-generation optimization algorithms 315
third-generation optimization algorithms 315
physics-based algorithms 9, 80
PhysioNet Challenge dataset 131
pick signals to noise ratio (PSNR) 116
pooling model 58–59
pre-processing 135–136, **136**, *136*
primary moth's navigation 158
PSNR *see* pick signals to noise ratio (PSNR)
PSO algorithms *see* Particle Swarm Optimization (PSO) algorithms

Random Forests (RFs) classifier 26
R-domination (R-IMOMFO) 82
roulette-wheel method 86

searching agents 64, 65
search landscape 5
search space 3, 5
seeds dataset
best cluster centroids 223, **228**
comparative approaches 223, **224**
segmentation performance measure, multilevel thresholding 116
Sejnowski & Gorman's Underwater Sonar Dataset 59
sensitivity analysis, customized MFO 71, **72–73**
72-bar space truss
best and average convergence 273, *273*
best solution displacement ratio 275, *275*
best solution stresses 275, *275*
numbers and directions 273, **273**
sigmoid transfer function 23, 24
signal characteristics 54
sine cosine algorithm (SCA)
multilevel thresholding image segmentation problems 248–255, **248–255**, *257*, *258*
single-objective optimization 81
single-objective problems 83
single-solution stochastic algorithm 8
slime mold algorithm (SMA) 158
comparison results 162, *169*
wrap food mechanism 161
SmellIndex 160
snapshot ensemble 284
snapshots 284
sonar dataset 59–61, *60*, **61**, *61*
spacing, MMFO 88
Speedup Ratio (SR) 107
statistical analysis, hMFOSMA algorithm **170**, 170–171, *171*
statlog (heart) dataset
best cluster centroids 223, **230**
comparative approaches 223, **224**
stochastic classification 54, 55
stopping criteria, feature selection 16
subset discovery, feature selection 15
supervised learning approach 130, 131
support vector machine (SVM) 132, 133
classifier 13, 23, 24
FER system 289, **303**
swarm-based algorithms 9, 80
swarm intelligence algorithms 13, 244
swarm intelligence optimization techniques 134

tanh transfer function **22**, 24, 58
thermal facial images 289
3D histogram 243
thresholding 112
time complexity measurement 70–71
TL *see* transfer learning (TL)
transfer functions 22, **22**, 23, 29
transfer learning (TL) 131
transverse orientation 16–17, *17*, 20, 37, 119, 137
truss structural design optimization
analytical approaches 266
approximate techniques 266
hybrid optimization technique 267
metaheuristic algorithms 266
problem definition 267–268
simulated annealing (SA) technique 267
25-bar space truss

best and average convergence 271, *271*
best solution displacement 271, *272*
best solution stresses 271, *271*
element group number 269, **270**
limitation 269, **270**
72-bar space truss
best and average convergence 273, *273*
best solution displacement ratio 275, *275*
best solution stresses 275, *275*
numbers and directions 273, **273**
200-bar planar truss
best convergence curve 278, *278*
best solution stresses 278, *278*
element grouping 275, **276**
25-bar space truss
best and average convergence 271, *271*
best solution displacement 271, *272*
best solution stresses 271, *271*
element group number 269, **270**
limitation 269, **270**
2D filters 286
2D plane Wave Expansion (PWE) 318
200-bar planar truss
best convergence curve 278, *278*
best solution stresses 278, *278*
element grouping 275, **276**

underwater acoustic datasets 56–57
underwater target classification 54

validation process, feature selection 16
vanilla MFO algorithm 99–102, 104, 108
vectorized Moth-Flame Optimization (vMFO) 99–102, 104, 106, *107*, 108
VGG16 131, 132, 136, 137, *138*, 142, 143, 148
visible light cameras 286
vowels dataset
best cluster centroids 223, **231**
comparative approaches 223, **225**
V-shaped transfer function 40

walk and capture prey 183
wavelet-transformation 133
Wilcoxon rank sum test 67, 170, **170**
clustering based techniques 220
wine dataset
best cluster centroids 223, **231**
comparative approaches 223, **225**
wrapper feature selection technique 23, 28

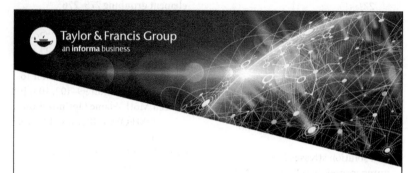